# SUSTAINABLE AGRICULTURE FOR FOOD SECURITY

*A Global Perspective*

# SUSTAINABLE AGRICULTURE FOR FOOD SECURITY

*A Global Perspective*

*Edited by*

**Acharya Balkrishna**

AAP | APPLE ACADEMIC PRESS

First edition published 2022

**Apple Academic Press Inc.**
1265 Goldenrod Circle, NE,
Palm Bay, FL 32905 USA

4164 Lakeshore Road, Burlington,
ON, L7L 1A4 Canada

**CRC Press**
6000 Broken Sound Parkway NW,
Suite 300, Boca Raton, FL 33487-2742 USA

2 Park Square, Milton Park,
Abingdon, Oxon, OX14 4RN UK



**Library and Archives Canada Cataloguing in Publication**

CIP data on file with Canada Library and Archives

**Library of Congress Cataloging-in-Publication Data**

CIP data on file with US Library of Congress

ISBN: 978-1-77463-756-2 (hbk)
ISBN: 978-1-77463-757-9 (pbk)
ISBN: 978-1-00324-254-3 (ebk)

# About the Editor

**Acharya Balkrishna, PhD**

*Vice-Chancellor, Patanjali University, India*

Acharya Balkrishna is a Vice-Chancellor at Patanjali University, India. He is a highly ascetic entrepreneur with a versatile personality who holds expert knowledge of Yoga, Ayurveda, Sanskrit language, Indian holy scriptures, and the Vedas. After dedicating his life to ancient healing and lifestyle traditions, Acharya Balkrishna Ji has become a great source of inspiration for traditional medicinal practitioners and worldwide-known personalities. His maverick leadership as Chairman and MD of Patanjali Ayurveda and his relentless efforts in Yoga and Ayurvedic scientific research have earned him much-deserved awards such as Ayurveda Expert, Manav Ratan, Bharat Gaurav, Ten Versatile and Dynamic Young Men of India, and "UNSDG 10 Most Influential People in Healthcare Award," to name a few.

With a passion for plants from an early age, he has become a most renowned and respected herbal expert. He explored four rare and extinct *Aṣṭavarga* plants called Sanjeevani, Soma, Swarnakshiri, and Swarnadraka, and has taken many initiatives for biodiversity conservation. Presently, he is working for the establishment of an herbal garden and herbarium, working on *World Herbal Encyclopedia, Wealth of Food Crops,* and identifying rare herbs and vegetation survey of India.

He has published more than 130 research articles in various journals, reserved approximately 41 patent rights, as well as authored and edited around 150 books on Yoga and Ayurveda. His immense faith and knowledge in natural healing methods have effectively cured more than 1.5 million patients with a number of stubborn, chronic, and noncommunicable diseases. With the vision of philanthropic quality education, he has established the Ayurvedic College and University to help children realize their dreams and serve the nation.

For his love of plants and Earth's preservation, Dr. Acharya Balkrishna has set up the Patanjali Organic Research Institute to propagate the idea of organic farming. To disseminate his method, he conceptualized the idea of writing a world-level book, i.e., *Sustainable Agriculture for Food Security: A Global Perspective.* Through its publication, Acharya Ji intends to bring another agricultural revolution.

Apart from this, Acharya Ji is actively participating in driving agricultural transformation through organic practices, helping the nation of India in emergency needs like post-disaster needs assessment. With his humanitarian attitude, Acharya Balkrishna continues his phenomenal journey of making world records, uplifting mankind through medicine and lifestyle improvement, and preserving nature's gifts in the form of literature and a sustainable living approach.

# Contents

# Contributors

**Acharya Balkrishna**
Board of Directors, PYP Yog Foundation, Inc., NFP, 323 W Alkire Lake Dr., Sugar Land, Texas, USA; Trustee, Patanjali Yogpeeth (UK) Trust, 40, Lambhill Street, Kinning Park, Glasgow, G41 1AU, Scotland, UK; Vice-Chancellor, University of Patanjali, Haridwar, Uttarakhand, India; Chief Secretary, Patanjali Research Foundation Trust, Haridwar, Uttarakhand, India; Mobile: +91-9897405550, E-mail: pyp@divyayoga.com

**Arun Kumar Kushwaha**
Scientist-B, PHRD Division, Patanjali Research Institute, Haridwar, Uttarakhand, India, E-mail: arunkumar.kushwaha@prft.co.in

**Ashish Dhyani**
Scientist-B, PHRD Division, Patanjali Research Institute, Haridwar, Uttarakhand, India, E-mail: ashish.dhyani@prft.co.in

**Ashok Kumar Mehta**
Ex-ADG at ICAR, New Delhi, India; Ex- Senior Extension Specialist and Ex-Associate Director, Punjab Agricultural University, Ludhiana, Punjab, India

**Bhoomika Sharma**
Ex-Research and Development Scientist, Toronto; Ex-Consultant, Vancouver; Ex-Clinical Researcher, Mississauga; Ex-Medical Writer and Ex-Medical Analyst, Brampton, Canada

**Devika Sharma**
Assistant Scientist, PHRD Division, Patanjali Research Institute, Haridwar, Uttarakhand, India, E-mail: devika.sharma@prft.co.in

**Divya Jyoti**
Publishing Associate, PHRD Division, Patanjali Research Institute, Haridwar, Uttarakhand, India

**Gunjan Sharma**
Scientist-B, PHRD Division, Patanjali Research Institute, Haridwar, Uttarakhand, India, E-mail: gunjan.sharma@prft.co.in

**Himani Malhotra**
Research Scholar, Department of Bio-Informatics, Lovely Professional University, Phagwara, Punjab, India

**Jawahar Lal Dwivedi**
Ex-Senior Consultant-R&D, JK Agri Genetics Ltd., Hyderabad, Telangana, India; Ex-Professor and Head, Genetics and Plant Breeding and Officer-In-Charge, N.D. University of Agriculture and Technology, Faizabad, Uttar Pradesh, India

**Kailash Choudhary**
Union Minister of State for Panchayati Raj, Agriculture and Farmers Welfare, Room No. 199Q, Krishi Bhavan, New Delhi, India

**Megh R. Goyal**
Retired Professor in Agricultural and Biological Engineering, University of Puerto Rico, USA

**Nidhi Sharma**
Scientist-B, PHRD Division, Patanjali Research Institute, Haridwar, Uttarakhand, India

**Pallavi Thakur**
Scientist-B, PHRD Division, Patanjali Research Institute, Haridwar, Uttarakhand, India

**Parshottam Rupala**
Union Minister of State for Panchayati Raj, Agriculture and Farmers Welfare, Room No. 322, A-Wing, Third Floor, Krishi Bhavan, New Delhi, India

**Pawan Kumar**
Ex-Chief Executive of the Uttarakhand Livelihood Project for the Himalayas Supported by IFAD and the Government of Uttarakhand, Uttarakhand, India

**Priyanka Chaudhary**
Ex-Scientist-B, PHRD Division, Patanjali Research Institute, Haridwar, Uttarakhand, India

**Rashmi Mittal**
Scientist-B, PHRD Division, Patanjali Research Institute, Haridwar, Uttarakhand, India,
E-mail: rashmi.mittal@prft.co.in

**Razia Parveen**
Assistant Scientist, PHRD Division, Patanjali Research Institute, Haridwar, Uttarakhand, India

**Rishi Verma**
Ex-AVP, LN Bangar Group, Kota, Rajasthan, India; Ex-Sr. Project Manager (JFM), Louise Berger Group, (Rajasthan Forest Department), Jaipur, Rajasthan, India; Ex-Assistant General Manager (Agroforestry), DCM Shriram Consolidated Ltd., Kota, Rajasthan, India

**Ritika Joshi**
Scientist-B, PHRD Division, Patanjali Research Institute, Haridwar, Uttarakhand, India,
E-mail: ritika.joshi@prft.co.in

**Sherry Bhalla**
Postdoctoral Research Fellow at Mount Sinai Health System, New York, USA

**Shivam Singh**
Scientist-B, PHRD Division, Patanjali Research Institute, Haridwar, Uttarakhand, India

**Swami Narsingh Dev**
Scientist-B, PHRD Division, Patanjali Research Institute, Haridwar, Uttarakhand, India

**Tanuja Gahlot**
University Lecturer, IPGGPG College, Haldwani Kumaun University, Nainital, Uttarakhand, India

**Vedpriya Arya**
Scientist E, PHRD Division, Patanjali Research Institute, Haridwar, Uttarakhand–249405, India,
Mobile: 7060472471, E-mail: vedpriya.arya@prft.co.in

**Vemuri Ravindra Babu**
Ex-Director and Principal Scientist, ICAR-IIRR, Hyderabad, Telangana, India; Ex-Assistant Professor at Agricultural College, Bapatla, Andhra Pradesh, India; Ex-Assistant Research Officer, RRS, Bapatla, Andhra Pradesh, India

# Abbreviations

| | |
|---|---|
| ADCS | amino deoxy chorismate synthase |
| AgMRC | Agricultural Marketing Resource Center |
| AHER | animal health events repository |
| AI | artificial intelligence |
| AMF | arbuscular mycorrhizal fungi |
| APHIS | Animal and Plant Health Inspection Service |
| APMC | Agricultural Produce Marketing Committee |
| AQSIQ | Administration of Quality Supervision, Inspection, and Quarantine |
| AROS | Asia Regional Organic Standard |
| As | arsenic |
| ATP | Agricultural Trade Promotion |
| BAP | best aquaculture practices |
| BFA | biological farmers of Australia |
| BRC | British Retail Consortium |
| CAC | Codex Alimentarius Commission |
| CAGR | compound annual growth rate |
| Cd | cadmium |
| CFDA | China Food and Drug Administration |
| CGIAR | Consultative Group on International Agriculture Research |
| CIN | canal irrigation network |
| CSA | community supported agriculture |
| CSA | Climate Smart Agriculture |
| CTS | cattle tracing system |
| EAOPS | East African Organic Product Standards |
| EFSA | European Food Safety Authority |
| ELISA | Enzyme-Linked Immunosorbent Assay |
| EMA | economically motivated adulteration |
| EMP | Emerging Markets Program |
| E-NAM | electronic-national agricultural markets |
| ET | evapotranspiration |
| EU | European Union |
| FAO | Food and Agricultural Organization |

| | |
|---|---|
| FAS | Foreign Agricultural Service |
| FBO | food business operator |
| FC | field capacity |
| FCI | Food Corporation of India |
| FDA | Food and Drug Administration |
| FEPCA | Federal Environmental Pesticide Control Act |
| FLW | food loss and wastage |
| FMD | foreign market development |
| FMPP | Farmers Market Promotion Program |
| FPO | fruit products order |
| FSA | Food Standards Agency |
| FSANZ | Food Standards Australia New Zealand |
| FSEP | Food Safety Enhancement Program |
| FSIS | Food Safety and Inspection Service |
| FSMA | Food Safety Modernization Act |
| FSMS | food safety management system |
| FSSA | Fitness and Sports Sciences Association |
| FSSAI | Food Safety and Standards Authority of India |
| FSSC | Food Safety System Certification |
| FSSR | Food Safety and Standards Regulations |
| FTIR | Fourier transform infrared spectroscopy |
| GC | gas chromatography |
| GCDPP | global collaborative for development of pesticide for public health |
| GFSI | global food safety initiative |
| GHG | greenhouse gas |
| GI | geographical indication |
| GIS | geographic information system |
| GLN | global location numbers |
| GMP | good manufacturing practices |
| GNSS | global navigation satellite system |
| GPG | good practice guidelines |
| GS1 | Global Standards 1 |
| GTIN | Global Trade Item Numbers |
| GTPCHI | GTP-cyclohydrolase I |
| HACCP | hazard analysis and critical control point |
| Hg | mercury |
| ICP | integrated crop management |

| | |
|---|---|
| ICT | information and communication technology |
| IE | irrigation efficiency |
| IFOAM | International Federation of Organic Agriculture Movements |
| IFS | International Featured Standard |
| $I_G$ | gross irrigation |
| INCSW | Indian National Committee on Surface Water |
| INIA | National Institute for Agrarian Innovations |
| IoT | internet of things |
| ISO | International Organization for Standardization |
| IT | information technology |
| ITC | India tobacco company |
| K | potash |
| LEPA | low energy pressure application |
| LODs | limits of detection |
| MAD | maximum allowable depletion |
| MAP | Market Access Program |
| MAPA | Ministry of Agriculture, Livestock, and Food Supply |
| MRIN | market research and information network |
| MRL | maximum residue limit/level |
| MS | mass spectroscopy |
| MSDS | material safety data sheet |
| N | nitrogen |
| NABARD | National Bank for Agriculture and Rural Development |
| NASAA | National Association for Sustainable Agriculture, Australia |
| NGN | Nigerian naira |
| NMR | Nuclear Magnetic Resonance |
| NPOF | National Project on Organic Farming |
| O&M | operation and maintenance |
| OCIA | Organic Crop Improvement Association |
| OIECC | Organic Industry Export Consultative Committee |
| P | phosphorus |
| P4P | purchase for progress |
| PA | precision agriculture |
| PAN | Pesticide Action Network |
| PAW | plant available water |
| PDP | Pesticide Data Program |

| | |
|---|---|
| PHL | post-harvest losses |
| PHM | post-harvest management |
| PIN | pipe (pressurized) irrigation network |
| PPP | public-private partnership |
| PRK | pesticide record-keeping |
| PSY | phytoene synthase |
| PWP | permanent wilting point |
| PWR | plant water requirement |
| QMS | Quality Management System |
| QR | quick response |
| QSP | Quality Samples Program |
| RASFF | rapid alert system for food and feed |
| RAW | readily available water |
| RFA | rain forest alliance |
| RFID | radio frequency identification |
| RMP | recommended management practices |
| RUP | restricted use pesticide |
| RWHM | rainwater harvesting and management |
| SDI | subsurface drip irrigation |
| SFAC | Small Farmers' Agribusiness Consortium |
| SMS | short message service |
| SPME-GC-MS | solid-phase micro-extraction-gas chromatography-mass spectroscopy |
| SPS | sanitary and phytosanitary |
| SQF | safety quality food |
| SSA | Sub-Saharan Africa |
| TASC | Technical Assistance for Specialty Crops |
| TBT | technical barriers to trade |
| UES | unified export strategy |
| UNIDO | United Nations Industrial Development Organization |
| USA | United States of America |
| USDA | United States Department of Agriculture |
| VFM | virtual farmer's market |
| VRT | variable rate technology |
| WFP | World Food Program |
| WHO | World Health Organization |
| WHOPES | WHO Pesticide Evaluation Scheme |
| WTO | World Trade Organization |

# Foreword 1

Agriculture has endured through time and has ensured the survival of humans and animals on planet earth. But beyond the assumption, from the past few decades, it took a decimate turn, and now currently, it is grievously distressed worldwide. On realizing the situation, it can be clearly depicted that we are heading towards the path of global food crisis. Agriculture practices getting followed globally are negatively affecting both the physical and biological parameters of our planet. From different agricultural surveys, I observed that traditional agricultural practices are no more in use and rather have been replaced by new technological interventions without giving a second thought about their impact on our environment. But in the case of India, I experienced that our roots are still intact, and we are following our tradition and culture in the agriculture field also; and along with this, we have adapted new technological advancements to meet the pace of growing food demand. Sustainable development with ecological balance has always remained our first priority, and for this reason, several acts and policies have been initiated by the current Government in India to support the farmer to practice organic farming and to initiate several other ventures to attain our objectives.

I would like to thank Acharya Balkrishna Ji for his tremendous contributions to human welfare. His efforts for boosting the social, economic, and financial status of farmers are highly applaudable. His digitalized agriculture approach and the motto of *'Jaivik Kheti, Samridh Kisan'* and *'Swastha Bharat'* has the potential to benefit farmers and has already taken a step to improve the agriculture system. Apart from his practical approach, I would also like to congratulate him for drafting this phenomenal piece of work titled *Sustainable Agriculture for Food Security: A Global Perspective.* As I believe that, a documented piece plays a pivotal role in spreading awareness in every corner of the world. This work will surely help the farmers by introducing new advancements in this field and would help the global audience to understand the legitimate gaps acting as barriers in attaining the goal. I convey my wishes to Acharya Ji for the publication and hope that farmers and other countrymen will benefit from his endeavors in the future also.

**—Narendra Singh Tomar**
Honorable Minister of Agriculture & Farmers' Welfare,
Govt. of India

# Foreword 2

The agriculture sector ensures the livelihood of the global population. Need of the hour has emerged to take appropriate measures to ensure food security. Censorious analysis revealed that primarily: reduction in cultivation cost, value-added organic farming produce, and group farming should be more prioritized. The digitalized revolutionary approach will be of great help to farmers. New policies and strategies need to be adopted by the governments of respective countries to improve the farmer's condition and overall agriculture system.

Pujya Acharya Balkrishna Ji has initiated new agricultural ventures over the past few years for increasing organic farming produce for the betterment of farmers. Training and awareness programs organized by him has benefited several farmers and provided them all the required information for initiating organic agriculture system in different parts of the world. When I read his recent work on agriculture, a document title *Sustainable Agriculture for Food Security: A Global Perspective,* I felt that it not only updated my knowledge on new technological interventions in the field but also helped me to gain an insight into the agricultural economic status of different countries as well.

I would like to congratulate Acharya Balkrishna Ji for his marvelous contribution to the global farming community.

—**Sanjay Agarwal**
Secretary,
Department of Agriculture, cooperation & farmers welfare,
Govt of India, New Delhi

# Foreword 3

Present-day agricultural operations and location-specific modern farming practices differ greatly from traditional agricultural systems, which remained in practice till a few decades ago. The introduction of modern technologies has eased off several manual operations. However, various astounding facts of the modern farming system, which remained unattended, reflect a contrasting scenario. Multiple crops are being practiced in the field with the help of agrochemicals, hybrid seeds, and other advanced techniques, and these are getting adapted to sustain the needs of growing populations. Still, a major section of the human global population is suffering from micro-nutrient deficiency and other related issues. Technological advancements have helped in improving the production and quality of food to support the healthy livelihood of individuals in the country. In comparison to other developed/ developing countries, the condition is slightly better in India. However, it requires major transformation globally to ensure a healthy, nutrient-enriched, and adequate diet for one and all. In order to improve the condition of our present agricultural system, certain prominent measures need to be adopted. The use of modern technologies and tools would improve the efficiency of the agricultural system and raise the status of the farming community.

I compliment Acharya Balkrishna Ji for his significant contributions in the field of agriculture. *'Patanjali Farmer Samridhi Program'* initiated by him has trained a large number of farmers in India and apprised them of new farming techniques. Initiatives taken up in the agriculture supply chain management are benefiting farmers as mediators between the producers and consumers have been removed. His book titled *Sustainable Agriculture for Food Security: A Global Perspective* will provide important information about the agricultural systems. I hope the book will be useful for farmers, researchers, academicians, and policymakers both in India and abroad.

I wish him all the success and appreciate his efforts for the betterment of society, especially in the domain of agriculture.

—**T. Mohapatra**
Director General, ICAR, New Delhi, India

# Preface

With the growth and development of human civilization, the need for sustenance has also taken evolutionary shape. But, despite technological advancements and scientific experiments in the food industry, we humankind could not reach a unanimous solution to the problems of decline in agriculture career, mono-cropping, resources limitation, food safety, and insecurity. Currently, the world's food systems are facing the challenge of feeding the enormous population of the world. More than 800 million world's population is not getting enough they need to live an active, healthy life. In fact, one in every nine people goes to bed hungry each night. The statistical data about the sustainability of food are devastating. The world is in utmost requisite of changes in food practices and production. I get deeply disheartened looking at the poor socio-economic status of farmers who are diligently working to make our livelihood complacent without seeking any limelight or recognition. It makes me feel even more sadness to know that the 'Annadata' is being pushed to end their precious life because of such circumstances.

*Sustainable Agriculture for Food Security: A Global Perspective* is such an analytical book that speaks volumes on issues related to current agriculture practices. It widely covers global geographical data and key statistical reports to help readers compare, understand, and make decisions in the agricultural field. The book is divided into eleven chapters, each dealing with a specific problem and their solutions, keeping in mind the context of geographical variations. As the book proceeds with Part I, readers will find information on the historical overview and socio-economic importance of agriculture around the world, along with a discussion on threats and opportunities in the agricultural sector. Then, rain-fed agriculture practices for different continents are explored and highlighted with water harvesting technique. The book throws light on various sustainable irrigation practices in a global context and application of irrigation water management. The guidance on the usage of agrochemicals and solutions to their detrimental effects from non-standardized consumption is also jotted down in this book. And then, it elaborates on organic farming methods in the current scenario, certification standards for their product validation, and key restrictions in the concept implication.

Furthermore, in Part II, postharvest management practices prevailing in different continents are discussed with analysis on the role of technology and main obstacles in the implementation of the plan. The need for commercialization of agriculture through food processing, branding, and retailing is also put forth with key suggestions on available marketing platforms and promotional aspects. It goes on to outline the situational analysis of food adulteration as well as norms and regulations of food security at the international level. Also, the concept of biofortification is elucidated with methods and strategies of biofortified food production and its universal acceptance. The impact of biofortification on the socio-economic development of the country is also analyzed. Lastly, the concept of Global food of traceability and virtual farmer's market is also discussed in detail.

I am much obliged to have the blessings and enlightenment from Param Pujya Yogrishi Swami Ramdev Ji, whose motivational and financial assistance has led us to put efforts on this spiritual path. I would like to appreciate the thoughts and objectives of Mr. Ashish Kumar, Director, Apple Academic Press (AAP), Inc., for publishing with ethics and standards to serve mankind. Also, I am particularly grateful to Dr. Megh Raj Goyal, Senior Editor-in-Chief at AAP, Inc., for his immense support and guidance behind the conceptualization of this book. His constant motivation and direction to our team of researchers and writers have kept their spirit high and thus drafting this precious handbook addressing the need of the hour. Work is ostensibly supported by the Ministry of Agriculture, senior journalists, and Indian farmers. I would like to praise the commendable efforts of my team for rationalizing my ideas into an appealing frame.

Today, our world is in dire need of another agricultural revolution that can be invoked through digital mediums. We, as a world of responsible torchbearers to our future generations, need to change our pernicious agricultural practices and shift towards the ways that boost sustainability as soon as possible.

—Acharya Balkrishna

# PART I

# Global Agriculture: Fears and Facts

# CHAPTER 1

# Global Perspective of Agriculture Systems: From Ancient Times to the Modern Era

ACHARYA BALKRISHNA, GUNJAN SHARMA, NIDHI SHARMA, PAWAN KUMAR, RASHMI MITTAL, and RAZIA PARVEEN

## ABSTRACT

The ancient agriculture system has experienced several historical events and has ensured the survival of humans and animals on planet earth till today. Food production by plant cultivation, animal husbandry, food processing has contributed tremendously to human civilization from the past 10,000 years. Agriculture independently first initiated in three places of the world: Mesoamerica, North China, and Africa, and later its roots also started spreading in other regions of the world. On the basis of domestication, agriculture is believed to be initiated around 10,500 and 4,500 years ago, which has led to the emergence of the economy globally. But in the modern era, the definition of agriculture is to produce more and more to justify the needs of the growing population without ensuring its impact on the environment. The aim of the present chapter is to provide a glimpse of agriculture history and its transformation up to modern times, its socio-economic importance for the development of human civilization, and the threats and opportunities associated with it.

## 1.1 INTRODUCTION

Agriculture is considered as the art and science of cultivating plants and animals for satisfying the prime necessity of human life. Food is one of the prime necessities of life, and the invention of agriculture is a wise output of the human brain. Commencement of agriculture was considered as a significant evolutionary development of *Homo sapiens* in the Neolithic period (10,000 BC) after the discovery of fire in the Paleolithic era about 1 million BC. Agriculture remained in practice independently in different geographical regions of the globe, through which the early human groups switch to their nomadic behavior to the well-settled societies. This shift from hunting to cultivation is also known as the "First Neolithic Revolution," which was observed to be based on the domestication of wild plants and animals [22, 57, 62]. The present chapter has been framed to review the global agriculture scenario from the dawn of civilization to the current era. Overall, the agriculture scenario has been discussed in detail, which is happening around the globe. In the early days, man used agriculture as a tool to fulfill the basic necessity of life, and most of agriculture has remained limited from farm to mouth. Gradually, the commencement of agriculture trade came into light, due to the interlinking of geographical boundaries, which has led to the economic transformation of the agricultural scenario, which has greatly uplifted its status. Subsequently, the industrial revolution takes place during the modern period, which leads to tremendous development in agriculture technology (Figure 1.1).

Previously agriculture status remained confined to the crops which is grown, harvested, or processed for food production. Agriculture value is defined as the value which is sold in markets worldwide for the same product. This is so because of the worldwide importance of agriculture and is found to be directly related to food production. Those who are engaged in producing foods for retail purposes are referred to as retailers and distributors. The worldwide importance of this sector is also referred to as the retail level, which is related to the food produced and the monetary benefits availed from it. Almost all the countries around the world have started to increase their food production by increasing their produce, but few exceptions also do exist in it. FAO listed about 44 countries which are unable to meet the required food supply due to several environmental and political reasons, including an exceptional shortfall in overall food production/supplies [16].

Globally, agriculture has become a major player for the determination of any country's prosperity. Statistical analysis indicated that globally, farm products account for nearly two-thirds of the value of world trade. In fact, all food is brought to us from another location of the food chain. Food processing has also emerged as a new dimension of the agriculture sector,

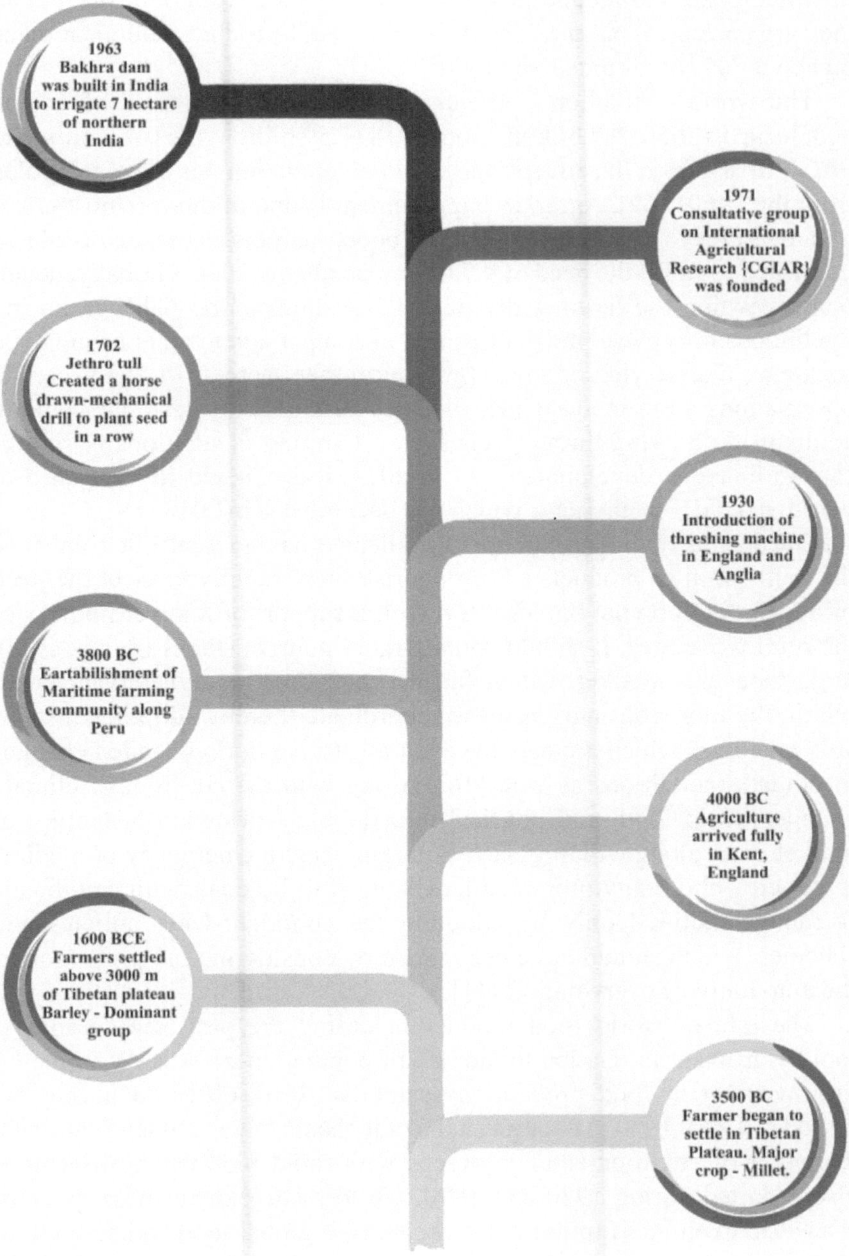


**FIGURE 1.1** Historical events in the timeline of agriculture.

in which plant and animal products enter the processing industry before they are packaged and then sent to stores. Dairy products and meat make up about 50% of all processed foods.

The world population is projected to rise by more than 2.5 billion people, up to 2050 (9.7 billion) from 2018 (7.59 billion) [25]. According to FAO (2018), from the advent, the world's population has almost doubled from the 1960s [69]. Agrarian improvement is one of the foremost effective measures to end extreme poverty, boost shared success, and could be nourished to fulfill the need of 9.7 billion people by 2050. Globally, development within the farming division is 2–4 times more viable in raising livelihoods in between the poor people in comparison to other economical segments like service, manufacture, and other sectors. In 2016, certain observations were made which revealed that 65% of destitute working adults made a living through this sector. Farming is additionally significant to financial development. As in 2014, it accounted for one-third of worldwide GDP with 3.6%, which has become 4% in 2018 [59].

During this century, a number of challenges has emerged out to enhance the cultivation of products and to nourish them rapidly to meet the need of developing population. That too with a support of a lower number of the rural workforce. It should contain more nourish stocks for a possibly large bioenergy market, which further contribute towards advancement within the numerous agribusiness subordinate thereby adopting sustainable strategies which are capable even to survive during sudden changes in climatic conditions as well [16]. Along with the confront of climate change, world food, and political uncertainty, worldwide instability of market costs along with regularly emerging health emergency or medical issue, etc., and many more challenges need to be dealt with cautiously. It can be rectified only by adopting the continent-wide policies and schemes, which should have been made by considering the protection of the autonomy of every nation [41].

The rate of world food production during the past few decades is rapidly growing in relation to the rate of population growth. An immense improvement in food production sector has been observed during the period of 1970–1980. Although on the other hand, water and land shortage, depletion of environmental resources with rapid biodiversity loss were also reported during 1980 and 1990. Up to some extent, these types of issues led to an interruption in the progress of global food and agriculture sector, which was also one of the main reasons of food insecurity [35].

Thus, we can understand a strong relationship between food security and agricultural growth. Agriculture stands at a leading position in the holistic development of the economy of underdeveloped, developing, and developed countries. However, agricultural development is not found similar in every nation of a particular region because of food insecurity, hunger, and poverty. It was also projected that; global population will be raised by 9–10 billion by 2050, and the majority of the population will be from developing nations with the growing food demand. Therefore, it is also expected that, at that time, the demand of agriculture produce will get increase by 70% till 2050 [60, 61, 68].

Development of agriculture is directly related to the availability of environmental resource, and total agriculture production depends upon the raw matter, including the availability of farm produce and fisheries. Therefore, it is necessary to conserve and ensure sustainable utilization of natural resources, and through this, we would be able to produce sufficient food for the present as well as for future generations globally. Not only the adequate use of environmental resources is directly related to agriculture output, but it is also kept in mind to save the agriculture land and other resources by utilization of other sectors or economic activities like rapidly growing urbanization with the higher demand of land and water. Environmental resources are also vulnerable to climate change. It was assumed that, by 2050, the average global production of major crops such as wheat, rice, and maize would be reduced by 6–10% than the demand, due to climate change. Similarly, increasing water crises may affect the production of crops by 5–7% by 2050, especially in the north-western part of India, China Northeast region, and the South-western part of the USA.

The agricultural system is also influenced by technological interventions involved in production, which can also raise some negative impact on soil, water, genetic resource, and the ecosystem. About 12% of global greenhouse gas (GHG) emission emerges from the agriculture practices. Agrochemicals needs a huge amount of water to perform well, but also is found to be majorly responsible for environmental detrition due to irrational use of fertilizers and agrochemicals. Extensive use of agrochemical has posed a major threat to biodiversity of water and ecosystem. The present chapter is an attempt to review the importance of agriculture in both ancient and modern era and create an understanding of related challenges, threats, and opportunities of this sector globally.

## 1.2 HISTORICAL OVERVIEW OF AGRICULTURE

### *1.2.1 ANCIENT PERIOD (10000 BCE TO 650 BCE)*

The first fossil evidence of *Homo sapiens* (anatomically modern humans) was dated to be around 1,96,000 years old, also known as a prehistorical period. Human ensures their survival by gathering plants and hunting animals from their surrounding regions. As the evolutionary process proceeded, Paleolithic people came into existence, and after a long span of time, the origin of agriculture got connected with Epipaleolithic peoples, who were observed to be responsible for the first transitions from wandering and hunting nature to collect and initiate agriculture during the period between 14,500 and 12,000 years ago, particularly evidenced from Southwest Asian region. It is also known as the Ice age or the last Pleistocene glacial period, which is expected to be about 11,700 years ago. From the late Epipaleolithic archaeological period 'Natufians culture,' existed in the Levant region of Asia, it had been observed that the people initiated the use of stone apparatus like sickles. They also gathered wild plants such as barley (*Hordeum spontaneum*) and earlier, they hunted *Gazella* species and gradually raise wild goats and sheep. Some scientists called this region as 'Fertile Crescent' due to its enriched soil and environmental conditions. This region has been documented for possible first domestication of plants by the time period of 12,000–11,000 before the present. Around 10,000 years ago, animal domestication began. In the region of East Asia millet (*Setaria* and *Panicum* sps.) and rice (*Oryza* sp.) were also being domesticated during 8,500–8,000 years ago. Likewise, by 9,500–7,500 BCE, barley was reported as the dominant crop in South Asia (Indian subcontinent) and was apparently supplemented with some species of wheat. Similarly, at this time approximately in 10,000–9000 BCE, squash (probably *Cucurbita pepo* and *C. moschata*) was observed in Southern Mexico and Northern Peru. Around 5000–3000 years ago, evidences of agriculture were procured from eastern North America [15], and the Paleolithic period shifted to the Neolithic period, which is known in history as the 'Neolithic revolution.' Numerous regions experienced settled agriculture system. Along with this, livestock domestication also used to occur. Thus, agriculture is considered as the first technological invention by human in the form of farming tools. The shift towards agriculture was observed to have occurred independently worldwide, including Asia, Europe, America, Africa, and even in New Guinea [14, 15, 28, 56].

The first domesticated plant was probably the millet from the Middle East; rice from China and India or corn from the Americas. With the help of plant breeding methods, farmers of South Asia and Egypt developed a new variety of wheat, which was stronger than previously available cereal grains. Likewise, domesticated animals were reported to be dogs, which were used for hunting, and later on, sheep, goat, cattle, and pigs were probably domesticated from various parts of the globe. Perhaps, once all these animals were hunted for various purpose including meat. But after domestication, most of them were used as a source of milk and butter. Eventually, people used domestic animals such as oxen for plowing, pulling, and transportation [15, 19, 21, 40].

Gradually, extensive cultivation has led to the production of surplus food, which could be used during food scarcity due to crop failure and for the trade purpose. The surplus food production gives rise to the necessity of storage methods. Farmers stored their produce in earthen pots, jars of pits for shorter period. This situation has led to the need of innovation of other non-farming works, which could probably support a well-developed civilization. Now by the passing of time, these civilizations interlinked via trade and other economic or cultural activities. Evidence of one of the foremost civilizations were found near the Tigris and Euphrates Rivers. People of this region used an irrigation system by making small channels from river or streams by the time period of 5,500 years before present. Later on, in Egypt and China, farmers also used different local irrigation practice [5, 13, 15, 34].

It was also documented that; fire is used by native Americans to control berry production. Farmers plowed their fields by the use of animal power and using stone tools like axes, digging sticks, etc. These farming tools were observed to be made of bone, stone, and metal, including bronze and iron to increase their output.

Eventually by the course of growing civilization across the world, some landmark events took place. Along with people's migration, culture, and technology also got exchanged. Approximately 3,000 years ago, trading of non-perishable items was found to begin by the water route, for example, from rivers and sea coasts in Nile river (Egypt), Tigris, and Euphrates rivers (Middle East), and Yellow River (China). At the same time trade of spices also began between Asia and the Middle East, which was evidenced from the Neolithic carvings of an ancient port in India. Furthermore, around 3,500 years ago, land route commerce of goods was also reported. One of the biggest examples of this trading is known as Silk Road in the world history. This route was majorly responsible for the exchange of not only goods but also of culture between China, India, and Eastern Mediterranean region, now known as Iran, Syria, and Turkey. This route had been used till 15th century AD [2].

All the presented events connecting to the origin of agriculture were clearly documented by various historians, and here the origin of agriculture is depicted in the accompanying table as agriculture timeline (Table 1.1).

**TABLE 1.1** Timeline of Beginning of Agriculture Practices Globally

| S. No. | Early Civilization Regions | | Time Period (BCE) |
|---|---|---|---|
| 1 | Fertile Crescent/Near Eastern Levant/Egypt and Mesopotamia | | 11,000–10,000 |
| 2 | Asia | South China/Yangtze basin | 10,000–7,000 |
| | | North China | |
| | | Early Harrappan/Indus valley | |
| | | India Ganges | |
| 3 | New Guinea | | 8,000–6,000 |
| 4 | West Africa/Sub Saharan Africa | | 5,000 |
| 5 | Europe | | 6,000–5,000 |
| 6 | The Americas | | 4,000–3,000 |

Therefore, the ancient civilizations have adopted different culture of the various regions, which were responsible for the holistic development of human societies. This gives us the fullest realization regarding the origin of agriculture. One of the earliest recorded agricultural tools was the basic structure and the idea of farming existed during the early civilization. Subsequently, people used various primary tools, and they managed to create different kinds of vegetables and fruits at the local level. At that time, they were using only a few primitive tools like plows, tiller, wheelbarrow, pits, ground stones, and many others. Ancient agriculture can also separately be viewed under some major centers of early agriculture including subheads covering regions of Asia, America, and Europe.

### *1.2.1.1 WEST ASIA*

The first village farming system was found to be existing in some of the West Asian region, which had been widespread across all the part of Southwest Asia and initiated a settled farming culture by 10,000–9,000 years before the present. People of that era are called 'Upper Paleolithic people,' who lived in shelters and accumulated several grass seeds and other food items from its surroundings. In Israel (Netiv Hagdud site) by 11,500 years ago, wild barley was considered to be the most common plant food found among the grass, legume, nut, and other plant remains. In Syria, the Abū Hureyra site is the largest known site from the era when plants and animals were initially domesticated. People harvested wild einkorn (the progenitor of domesticated wheat), rye (*Secale* species), and gazelle, lentils (*Lens* species), and vetch (*Vicia* species) by the period of 12,000 before the present or maybe domesticated by 10,500 years before present. Gradually, they cultivated a wide variety of plants, for example, barley, rye, legumes, and also domesticated two early forms of wheat known as emmer (*Triticum turgidum dicoccun*) and einkorn (*T. monococcum*). There was some evidence about the use of crop rotation system in this region. Usually, they followed and hunted wild animals, including sheep, goats, and harvested grasses or food plants. Further they also hunted gazelles along with goat and sheep for their primary animal food requirement. After all these events, a developed agricultural system was supported in making of a complex form of political organization of Mesopotamia, which is one of the early developed civilization having predominantly grain-based economies in the western Fertile Crescent. People used hoes or digging sticks for field formation and seed broadcast technique were used by the help of livestock animals, known as 'treading in' which ultimately resulted to have surplus food production. So, they also invented food storage techniques by the use of pit silos and granaries. Crop irrigation was also practiced in drier areas of the region.

In Mesopotamia, agriculture flourished in some early civilizations, for example, in Sumer (5,000 years before) along with barley as the main crop, wheat, flax (*Linum* sp.), dates (*Phoenix* sp.), plums (*Prunus* sp.), and grapes (Vitaceae species) were also cultivated. People breaded sheep and goats to procure animal products like milk, butter, cheese, and meat. Similarly, in the Neil valley civilization of ancient Egypt (7,000 years before the present), two main villages are found in existence, i.e., Al-Fayyūm and Al-Badarī. Due to the drier climatic conditions at Al-Fayyūm village, people used to store their farm produce in village silos. The use of agriculture tools was quite obvious in it. At the village Al-Badarī in Upper Egypt, animals were beard and treated as a part of their family.

Subsequently, Egypt experienced the agricultural scenario as a sophisticated around 4,525 years ago. A bureaucratic system came into existence, which dealt with all the matters related to the economy and of course, agriculture also lies under the same. A well-developed hierarchy system had been followed in agriculture, i.e., ruler > grand vizier > group of people responsible for the administration of agriculture > chief of the field who looked after the farming output and animal. Irrigation using Nile river water had been reported by making a large masonry dam dated about 4,875 years. Not only this, water was also diverted through a channel into Lake Moeris, for its utilization according to the requirements. Animal power such as oxes or asses were used to plow and to support other agricultural practices; seed was sown by a funnel on the plow or; crops were cut with sickles; grain was threshed by asses or cattle and winnowed by tossing in the wind. The grain was then stored in great silos. Along with barley and emmer, wheat was the main crop, and along with this flax, beans, lentils, and onions were also cultivated. Moreover, animals like cattle, sheep, goats, pigs, ducks, and geese were also used to obtain animal products. Animal breeding was also practiced, for example, two cattle breeds were generally preferred for milk and meat. A hunting dog breed was another example of animal breeding [14, 15].

#### *1.2.1.2 EAST ASIA*

China is considered as one of the ancient regions involved in the origin of agriculture in East Asia. It is because of the oldest cultivation centers which can be located in the region. As per some archaeological data, agriculture in China had been originated date back between 12,000 and 9,000 years ago. It was evident from the traces obtained from the Inner Mongolian region and the Huangtu Gaoyuan (Loess Plateau), near the Yellow River, and another from northeastern China near the Liao river. Thus, agriculture communities rise approximately 8,000 years ago from the present day in China. At that time Chinese agriculture sector was divided into two regions of the Qin Mountains; one is the Northern part which was dominated by wheat and millet cultivation, and the second is the Southern part having rice as a crop. In the north region, foxtail and broomcorn millet were domesticated. The South region is located near the Yangtze basin, a wetland area. Over there, rice was domesticated by the people of Pengtoushan culture approximately 9,500–8,100 years ago. People used to burn land for getting more area for rice cultivation near Hangzhou Bay. The south region was favorable for rice domestication, so rice was reported to be first cultivated from this part of East Asia. Along with rice, corns are also reported as a popular food. Animal use was also reported, for example, dogs were widely used for hunting. Eventually, rice and millet agriculture was also adopted by the people of Korea and Japan. About 3,000 years ago, bean crops, for example, adzuki, soybean, and bean were domesticated and reported as the first crop in Korea. At the same time, Wild buckwheat (*Fagopyrum* sp.), millet *(Echinochloa esculenta* or *E. crus-galli utilis*) were domesticated in Japan. Chinese people also practiced agriculture by using farming tools like wooden plow and other methods like row crop farming. Rice variety of Vietnam had been adopted by almost all parts of China. By the time, people also cultivated soybeans and fruit crops (persimmon and peach), beefsteak plant (*Perilla frutescens*), water lily (*Euryale ferox*), water chestnut (*Trapa* sp.), lotus (*Nelumbo nucifera*), tea (*Camellia sinensis*), as well as silk (through sericulture). Domestication of animals had been followed, for example, water buffalo (*Bubalus bubalis*), swine, chickens, pigs, and cattle [15, 39, 52].

### *1.2.1.3 SOUTH ASIA*

This is one of the most fascinating areas because the biggest theories about the origins of agriculture in India first emerged out in the Indus valley civilization. The civilization was flourishing on the sub-continent and has spread to the Middle East. The Indus valley civilization was the first civilization in the history of South Asia. It was also noted to be the first farming cultures in South Asia, which was initially originated at the hilly place now in Baluchistan (Pakistan) to the western part of the Indus river. The well-known region related to this culture is known as Mehrgarh, during the time of 6,000–6,500 years ago. People domesticated wheat along with several animals, including cattle. The evidences of pottery were reported around 5,500 years before from present times. Indus valley culture emerged from this culture by the time of 4,000 years ago also known as pre-Harappan culture. This civilization extensively spreads into alluvial plains of Indus river to Ghaggar and Hakra river. Indus valley civilization is also named as Harappan civilization, as the first city excavated from the Harappa region is now available in Pakistan. By the time of 3,000 years ago, this civilization was bestowed with the development of several larger cities, for example, Harappa, Mohenjo-Daro, and others. Indus valley civilization developed parallelly to Mesopotamia, Sumerian civilization in Egypt and other regions of the Middle East, in the year 3,000 BC.

The crop rotation and multi-cropping methods were first developed by the people of this civilization, earlier from its parallel civilization such as Ancient Egypt and China. It may be noted that this region received double rain in both the growing season, i.e., summer and winter. Thus, it was possible to grow crops in two growing seasons. Millets, rice, and beans like beans urad and horsegram in summer and barley, wheat, and pulses like peas were first grown in winter between 2,800 and 2,100 years ago. Recently, some scholars revealed through their studies. They also suggested that, Indus area received two rains in a year, one in summer and another in winter. Along with major crops, domestication of various crops was also reported, for example, peas, sesame seeds, dates, and cotton. Similarly, several species of animals were also domesticated at that time, like the water buffalo.

Overall, the Indus valley civilization was highly developed, and this has allowed them to cover many areas and the civilization was also consisted of a high population of laborers and farmers. This enabled them to grow cotton, silk, fruits, and other foodstuffs in the different terrains of Indian society. All the incidences happened during the Indus valley era. Indus valley era is considered to be the most ancient civilization in the history of India. There are many more reasons that helped this civilization to emerge out as the ancient source of agriculture. The earliest among the evidences of agriculture in India can be found in the Harappan era. It is because this era is the most famous period in the history of this country. A number of cities can be identified in this ancient civilization. Many of them were built with basic tools, and some of them were totally covered by walls. These were the factors that helped them to become the first civilization in the history of India. People invented a scriptwriting system, which is supposed to be the root of Hinduism. They began to establish trade networks with other regions as well [3, 6, 42, 54, 55].

Another well-versed region is South Asia near the Neolithic Ganga basin. It is characterized by sedentism, domestication of plants (rice and barley), and animals (cattle) as early as 6,500 before the current era. By the time period of 3,000 years before, agriculture in this region was greatly influenced by the adjacent cultures and arrivals. The invasion of the Indo-Aryan people was also supposed to happen at this time. Gradually, a shift had also been observed from Indus to the Ganges, possibly known as the disintegration of Harappan civilization to the center of India during the early 2,000 years before. As per some archaeological evidence from the period of 1,900–1,300, a spread of settlement was observed from the Indus basin to the Ganga-Yamuna basin. However, the Mesolithic culture had been discovered in Ganges basis as early to 10,000 years before the present.

Plants and animals were observed to be essential to ensure the survival of people of this region. People practicing a developed farming systems and practices such as mixed farming and their practices were thoroughly narrated in ancient manuscripts of India. In the Vedic texts (*Rigveda*) *c.* 3,000–2,500 years ago, there are perennial references of agricultural technology and practices, together with iron implements; the cultivation of a large vary of cereals, and fruits; utilization of meat and milk products; and animal husbandry. Additionally, meteorological observation was also taken in consideration in relation to crop prospects. "Agriculture in the *Vedic* period was thus a religio-social activity with all its ancillary aspects from soil to weather forecasts." Various workers supported and described the agriculture technology which was later on followed by Vedic people of India [13, 15, 26, 31, 37, 50, 52, 53, 63].

Overall, a wide range of plant species were utilized by the people of this region between 3,000 years before to the 2nd century. Agriculture land is highly arable with the cultivation of rice (*Oryza sativa*), and sorghum. Later on in the Iron age, some other oil and fiber crops were also included in farming such as *Sesamum indicum*, *Gossypium* sp., *Cannabis sativa*, *Brassica* sp., and many more. Along with this horticulture crops and medicinal crops were also evidenced during the time of 1,000 years ago, indicating the footprints of well-developed and prosperous civilization [15, 30].

The glory of this region had been mentioned by several indigenous as well as foreign authors. Megasthenes (ca. 350–290 BC) was also the author of the book Indica. In his book, he mentioned that:

> "*India has many huge mountains which abound in fruit trees of every kind, and many vast plains of great fertility. The greater part of the soil, moreover, is under irrigation, and consequently bears two crops in the course of the year. In addition to cereals, they grow throughout India much millet, and many pulses of different sorts, and rice also, and what is called bosporum [Indian millet].*" And "*Since there is a double rainfall (i.e., the two monsoons) in the course of each year...the inhabitants of India almost always gather in two harvests annually.*"

There are numerous evidences of the influence of agriculture in ancient civilizations of the Indus and Ganges basin. At the same time, they can be considered as the oldest civilization in the history of India. So, in order to learn the history of agriculture in India, it is highly needed to determine the ancient times [15, 20, 26, 29].

### *1.2.1.4 THE AMERICAS*

America is the biggest continent having a range of environmental conditions, from Canada to southern South American regions. This allows native American people to develop a variety of agriculture systems. Agriculture had been developed independently in Mexico, Central America, and South America. There was various evidence about the origin of agriculture in America. The traces of plant domestication and farming were reported from different time periods at various places. An early evidence of 10,000 years ago of crop production in central America was reported from the Guilá Naquitz site in Southern. Along with the probability of the domestication and use of acorn (*Quercus* sp.), mesquite seeds (*Prosopis* species), prickly pear (*Opuntia* sp.), pine nut (*Pinus edulis*), wild bean, grass seeds, and squashes were also reported.

Between the period of 9,000–8,000, the first crops appear from Mexico as well as from some other parts of South America. Eventually, corn has appeared as a staple crop of American people. However, a proper civilization or villages were found to be appeared after the maize domestication at the time of 3,800 years ago. Gradually, development of big Empire culture or territorial units such as Maya, Toltec, and Aztec were observed about 2,000 years ago. Mesoamerican people extensively depend on corn and other plants such as cacao, chili peppers, avocados, cotton, beans, manioc, tomatoes, and quinoa. They also domesticated animals, for example, dog, duck, and turkey. The use of farming tools was reported for land clearing and seed sowing. Irrigation was also practiced along with agriculture system like terracing and artificial islands (chinampas) which were adapted in low rainfall region. Storage of food crops in pits and granaries were also reported from this region.

Moreover, in South America, potatoes were domesticated in a site of south-central Chile by 5,000 years ago, and the animal domestication (cavy, or guinea pig) was predicted to be around 7,000 years before. However, wild camelids were also hunted by the early people of this region approximately before 10,000 years ago. Later on, by the time of 7,500–6,000 years before, llama, and alpaca were reported to be domesticated. By the time of 7,500 and 6,000 years ago, quinoa and cotton were found to be harvested in northern Peru. Although, evidences of squash between 10,400 and 10,000 years before present and peanuts about 8,500 years before were also reported from this region. Agriculture eventually emerged out as a major component in the development of the Inca empire and other South American cultures about 8,000–7,000 years ago. People cultivated a wide range of food crops like corn, avocados, bottle gourd, cotton, squash, cacao, manioc, papayas, chili peppers, sweet potatoes, and tobacco in the low and high land region of the Amazon basin.

In North America, corn and squash were first reported from New Mexico about 3,200 years ago, and beans were reported around 1,500 years ago. Agriculture was basically developed within the three parts of North America. Two of them were upper and lower Sonoran complex, and the third one appeared from the eastern region of North America. People of Archaic culture majorly used some plants such as squash, corn, cotton, bottle gourd, and beans. Early people of this culture did not practice agriculture, but they developed an agroecosystem, in which they heavily planted plants of their own use in their adjacent area to get more food plants near to their settlements. Now, people are actively engaged in various agricultural activities. They practiced irrigation through dams and contour terrace farming. Irrigation was also performed with the help of a canal system. On the other hand, in the third agriculture region of North America, the first settlement was reported about 5,900 years before. People first domesticated squash, sunflower, amaranth, chenopod, barley, etc. Walnuts, acorns, hickory nuts, deer, shellfish, and fish were also consumed by them [12, 15].

### 1.2.1.5 EUROPE

Agriculture in Europe first appeared near the coastal region of Mediterranean island, which was developed probably by means of invasion during the period of 6,000–5,000 years ago. People from the adjacent culture may be responsible for the development of agriculture in this region. However, it was reported in some resources, that early people of Europe grow peas, lentils, and wild emmer during the period of 9,000–8,000 years. They also used animals such as sheep, goats, pigs, and cattle. Likewise, food crops and animals were first introduced in the Iberian Peninsula. During the time period of 5,000 years before, a village agriculture appeared at Macedonia. However, wheat (four types), peas, oats, barley, and millet were found to be the earliest evidence of agriculture in the northwest region of the Black sea by 7,500 years ago, Gradually, Linear Bandkeramik or LBK culture, has emerged out as an earliest culture of central European region during 7,300–6,900 years before present. People of this culture practiced agriculture methods like land shifting and slashing and burning techniques. Moreover, in British Isles, crop production was found around 5,000 years ago. By the time of the beginning of century period, Romans expanded their empire, and adapted the best farming techniques, from whom they conquered. They also documented various agriculture methods and techniques for the existing civilization (Asia and Africa) at that time in the form of written scripts or manuals [10, 15, 27, 32, 49].

## 1.2.2 CLASSICAL PERIOD (500 TO 1500 BCE)

During this era, agriculture development was based on the prior practices followed by the early people. In the Middle East and North Africa, agriculture became a part of experiment and enjoy its golden time by the time of the 1st century. They learned crop rotation and many more agricultural techniques. In China, agriculture was at its peak by the 4th century. Farmer shifted their cereal (wheat and millet) field into rice fields and received surplus food from their fields. Farmers practiced various techniques related to cultivation and irrigation to get an enormous quantity of food. At the time period of 8th century, China exported the food grains by way of lower Yangtze into the northwest part. Similarly, during this period, agriculture flourished in America also. In the South Asian region, agriculture was further developed in terms of sustainable agriculture practices and a wealthy civilization.

There are many promising dynasty, kingdoms, and republics were found between 600 and 1500 B.C. During this time, agriculture was immensely flourished in the Indian subcontinent, including the region of Indus and Ganges valley civilizations. In Maurya period, an all-round development of agriculture, industry, and trade took place. Various crops were cultivated, such as rice of different varieties, wheat, sesame, pepper, pulses, linseed, mustard, vegetables, fruits like plantain, pumpkins, gourds, sugar-canes, grapes, etc. The importance of cattle-breeding was recognized as a means of agricultural development. Agriculture solely remained dependent on rainfall. The State derived a great part of its revenues from the variety of levies on agriculture and agricultural produce. Further, India had developed an advanced system of agriculture, industry, and trade during the period of the Gupta dynasty. In spite of the richness of the soil and availability of abundant water, Indian agriculture was dependent mainly on rainfall. An ancient Indian literature, i.e., "Varahamihira's Brihat-Samhita" consisting of information regarding, careful forecasts of excessive, sufficient, and scanty rainfall were made in light of astronomical and meteorological data. Irrigation was the concern of the State and it was an important source of revenue, and subsequently, water – rates were charged for the supply of water for irrigation. Indian textile industry dealt with several countries like Syria and Egypt in addition to several others in the West, even with parts of East Africa. In Greek, crop and fallow methods were followed, while Romans experienced the crop rotation method. They introduced coffee, tea, and indigo from Asia [15, 17, 21, 26].

### 1.2.3 MODERN PERIOD (1500 TO 1945 BCE)

This period brings an immense development of agriculture in the 17th century, particularly for Great Britain, Belgium, Luxembourg, and the Netherlands. An immense increase in food production had been observed in Europe and its colonies due to several new agricultural inventions. The earliest technique was invented in England for the purpose of drilling of seed by the power of the horse. This was widely adopted by almost all the part of Europe till the end of the 18th century. Likewise, in the USA, various machines were invented for the development of agriculture, such as cotton gin in 1794, mechanical reaper, horsepower thresher in 1830, steel plow in 1837. Selective animal breeding was also applied to achieve more productive livestock in Europe and America as early in 18th century.

Plant breeding was also practiced on selective basis during this. One of the milestone research had been published by Gregor Mendel in 1866 in the form of heredity studies in pea. This study opens the world of immense opportunities in the field of crop improvement through plant genetics.

During this time period, a new crop rotation method was developed in England, known as the Norfolk four-field system. It comprises of rotation of several crops such as wheat, turnips, barley, clover, and ryegrass on a yearly basis. One of the major obstacles, which was responsible for the backwardness of this region, is the slavery of the European invaders [15, 17, 37].

### *1.2.4 CONTEMPORARY PERIOD (1945 TO CURRENT)*

Similar to the modern period, farmers of Europe and the U.S. were able to produce food more than they need in the early 1900s, because of immense development in the field of agriculture science and the shift of manual power to gasoline and electrical power in late 1950. These developed countries replaced their animal power with tractors and steam power. Subsequently, machines and technology had been used at each step of the agriculture process along with livestock care. Electrical power eventually adopted by all the developed countries like the U.S., Germany, and Japan by the 1900s. By the development of advanced technology, aquaculture and hydroponics are introduced in the present agriculture system. Fish and shellfish cultivation through aquaculture was widely practiced in yearly time of China, Japan, Egypt, and India and now are being used all over the globe. During the late 1800s, the use of agriculture chemical also had been initiated, because of their capacity to increase crop yield. These agrochemicals were first manufactured in the U.S. and Europe. But irrational use of these chemicals emerged out which contributed towards the emergence of hazardous situation towards the ecosystem in terms of soil and water pollution, life hazards, and many more.

A new chapter added in the history of agriculture innovation was the Genetic Modification during the period of 1950s–1960s. High yielding varieties of various crops were evolved through this technique, particularly for wheat and rice. These varieties had been introduced in some parts of the world, including Mexico and Asian countries. Resulting into a record crop production had been achieved in these areas. This step is known as the 'Green Revolution' in the world agriculture history. Soon it was adopted by various developed and developing countries for the sake of fulfilling the food demand of rapidly growing population. Later on, Green revolution has also created a large set of problems, including environmental and socio-economic issues. Major issues are: high yielding crops required a high amount of water and fertilizers to produce high yield; lowland farmers cannot afford this technology; biodiversity or gene pool loss. Furthermore, the 'Genetic Modification' term has been coined during the 1970s in the agriculture system, including the plant and animal breeding process to get desired characteristics in plant and animal, respectively. This modification is also being used in the production of genetically modified food. Today genetically modified organisms and genetically modified food are common across the major economic region of the world. GM techniques also have its side effects as it decreases the biodiversity and natural selection of traits. GM food was also criticized for having low nutritional value. Within the recent decade, 'organic farming' occupies its value not only among the scientific gallery but also in the mind of each and every people, who want to take care of the earth. People want to consume chemical-free food, which could be possible only through organic farming techniques in agriculture. Thus, this period was full of new technologies, inventions, and research, which undoubtedly can take agriculture to new heights (Figure 1.2).

## 1.3 SOCIO-ECONOMIC IMPORTANCE OF AGRICULTURE AROUND THE WORLD

Earlier, agriculture has been only related with the cultivation of some necessary food crops, but now it is involving various other fields like milk production, fisheries, horticulture, forestry, mushroom cultivation, honey production, and many more arbitrary. Moreover, many steps after cultivation and production like post-harvesting procedure, crop, and farm product distribution, processing, along with their commerce were also supported in making of a complete agriculture system. Not only these; status of farmer, environment, and related economy were dealt with in the study of agriculture sector. Thus, agriculture acts as a crucial component of an economy since time before present to till now and in the future too. With the generation of advanced food product, it also engages people within its several activities in terms of employment generation for a huge part of

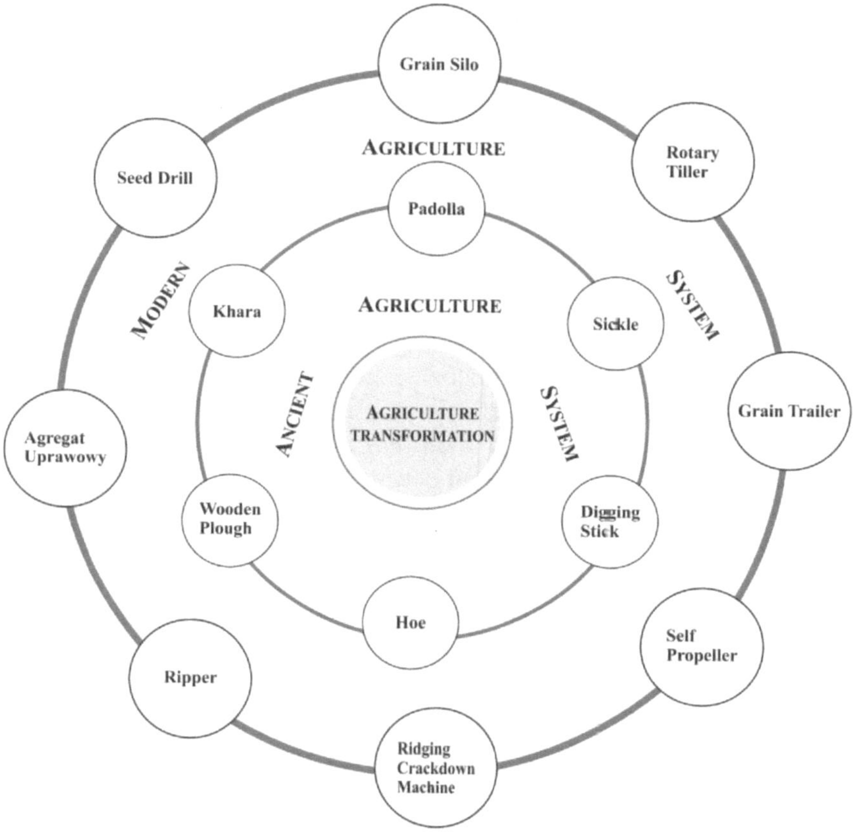


**FIGURE 1.2** Depiction of agricultural transformation in terms of equipment, tools, methods, and practices in ancient and modern era.

the population. A significant percentage of economically active population about 35–40% were observed to be engaged in agriculture, though it is varying from country to country. However, less than 10% workforce of developing countries were found to be involved in agriculture system. Now, from the detailed view of agriculture origin, development, and involvement in economy building in all parts of the globe, it is clear that agriculture is an essential part in economic development. It has already made a major contribution to developed countries economic growth, and its position in under-developed countries. It can be stated that industrial advancement is an aftereffect of the agriculture revolution in most of the developed countries like England, Japan, and U.S. Thus, industrial development is directly related to agriculture progress, which has ultimately supported the economy of a Nation. Therefore, as agriculture flourishes, the economy would automatically be developed.

Some economic impacts of agriculture around the world:

- The number of people involved in the agricultural sector contributes significantly to national GDP. When there is an increase in agricultural production and productivity, the country's gross domestic product increases. Subsequently, more wealth is created in the Nation.
- Another impact is a regional effect on GDP. Agricultural products have always been a leading contributor to globalization. Once there is a worldwide commitment to production, this gives countries a greater chance to produce more than they consume. The economic impacts of agriculture around the world are not just about farmers.
- Consumers also contribute to GDP. When consumers enjoy the same quality of food as farmers, they will spend less money to purchase it. Therefore, global agriculture system gives consumers more choices to adapt.

The economic impact of agriculture can be separated into two different components. The first component is direct agricultural production. This means that farmers can grow crops for sale and then sell their produce at local markets. The second component is indirect agriculture which contributes indirectly to GDP. A huge portion of the value of world trade is caused by exports and imports. When the export and import of goods happen in the same country, then the entire economy attains the monetary benefits out of it (Figure 1.3).

Agriculture also helps to improve the status of the global environment. Production of goods that are a part of the global market creates a demand for new products. This will improve the sustainability of the planet's environment. By improving the global environment, financial issues of farmers can be significantly resolved.

Many countries have made significant contributions to agriculture. Countries such as China, India, and Brazil made huge contribution towards GDP, but have a high proportion of farmers. Therefore, they can affect the status of their country to a greater extent than countries with lower agricultural participation. Although the economic value of agriculture can be influenced by a multitude of factors, these three factors contribute to a Nation's economic status. In this case, global agriculture has a major effect on GDP and the economic welfare of their citizens. These factors are clear evidences that the effects of global agriculture are economic in nature [38, 67].

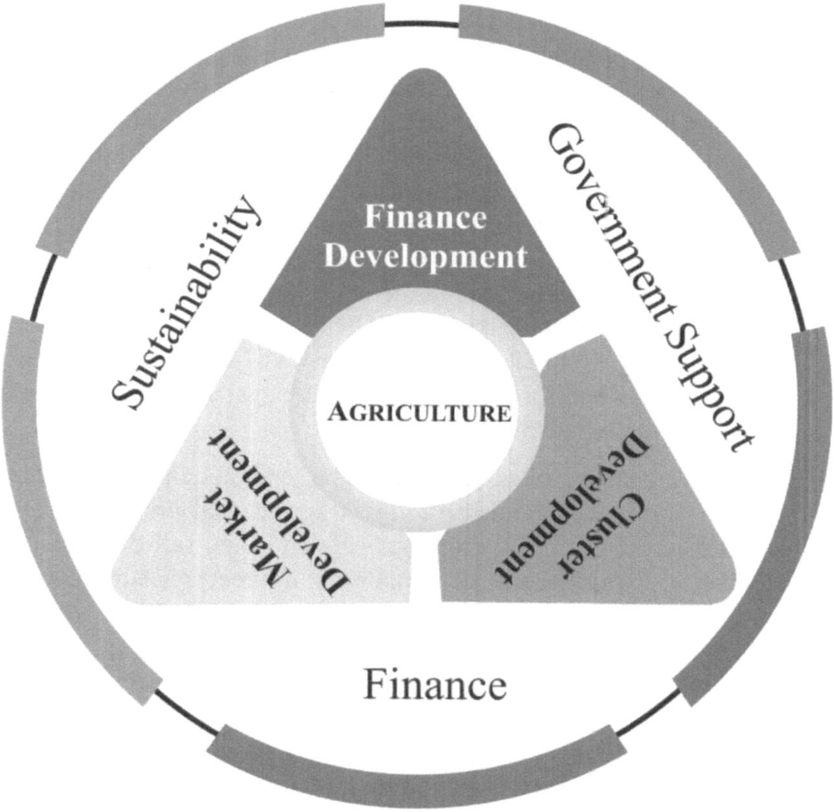


**FIGURE 1.3** Key determining factors for agricultural transformation.

Agriculture can significantly contribute to socio-economic development of any country in several ways:

- First and the foremost contribution is to ensure food availability as well as raw matter for mankind and non-farming sectors.
- Agriculture also plays a significant role in terms of investment in other sectors as well.

- The export of agriculture produce helps in exchanging the foreign currency.
- Engaged a large number of farm and non-farm labor by providing employment opportunities.
- Agriculture has the power to build an economy by contributing towards National income.
- Continuous supply of food material for the growing population.
- Make a country to become self-dependent in terms of food production for its people and economy.
- This sector is able to provide surplus production for the possibility of increase in export.
- Agriculture is directly and indirectly connected to the development of other sectors.
- It can provide job opportunities to surplus agriculture labor in other sectors to lower down the overpressure on agriculture.
- Agriculture is a basic measure to reduce poverty in Nation by generating enormous job opportunities for jobless people.
- It can uplift the life standards of rural people.
- It is a low capital investment sector, which could minimize the foreign capital growth.
- Agriculture is a tool to solve the problem of inequality among rural and urban income, by giving the high priority of agriculture development in a country.
- Agriculture is a connecting bridge between social and Government component of the economy. Thus, it is essential to develop agriculture to build a strong economy.
- It can create a significant nonstop demand, because of producing basic life necessities even at the time of economic depression too.
- Agriculture is also important for environmental protection by adopting sustainable agriculture practices.

Therefore, the growth of the agriculture sector is a necessity for improving the economy along with a tool to rectify various issues of the present era like unemployment, poverty, food insecurity, and many more (Figure 1.4). The socio-economic relation of agriculture may act as a catalyzer in the growing economy of a Nation [14, 25, 48]. Continent-wise, the socio-economic contribution of agriculture has been discussed in the upcoming sections of the chapter.

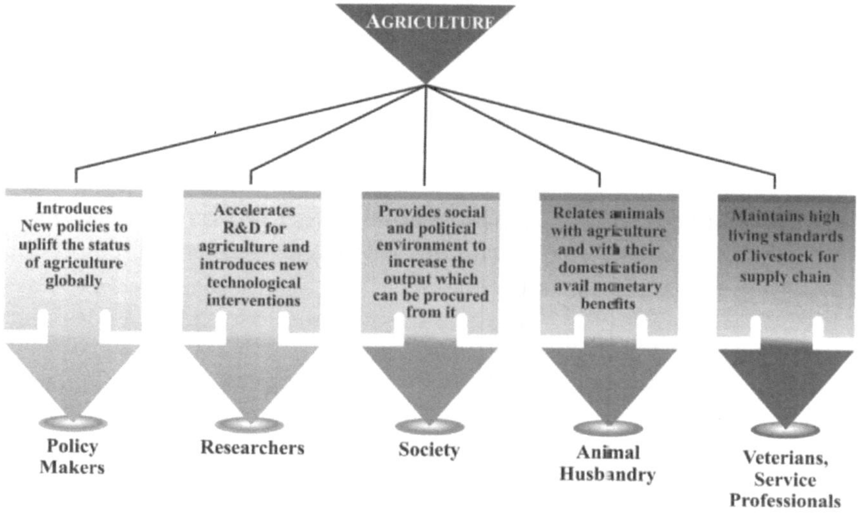


**FIGURE 1.4** Contributing elements for sustainable agriculture growth and its development.

### 1.3.1 AFRICAN CONTINENT

The economic and social consequences of Agriculture have greatly affected the African population. They harvested crops, manufactured products, and sold them at markets and storehouses. In some countries, such as Kenya, most of the income comes from the agricultural sector. Most of the land is used for crops, cattle, and sheep, while food is produced for export. The farm investment contributes significantly to developing nations. Many of them deeply rely on agriculture for their development. Agriculture helps to improve the standard of living. Improved education and employment opportunities are an outcome of agricultural development. Consumers have provided access to more products and services. The main advantage of agricultural development is easy accessibility to markets to sell their farm products and to attain monetary benefits out of it.

Agriculture products are usually more affordable than imported goods. The importance of agriculture is seen in the rural areas of Africa. The farmers in these areas tend to work harder to produce the best products. These products are often shipped to other African countries, Europe, and the rest of the world. The farmer contributes significantly to his or her community and country.

In the industrialized world, there is an increased demand for plant-based oils and fats. Nutrient content of foods has improved as a result of new technologies. One of the most important food products that is eaten in Africa is bananas. The demand for bananas is increasing due to the World Health Organization's (WHO) declaration of a low-nutrient banana as the food staple. This means that African countries can increase the production of bananas to meet demand. Agricultural economics has remained crucial for many things furthermore to improve the standard of living, to provide employment and to improve nutritional constituents [9, 11, 17, 36, 46, 64].

### 1.3.2 AMERICAN CONTINENT

The social and economic significance of agriculture in America is as great as the growth of agriculture in Asia. The arrival of European settlers made this very different from what it had been before. Agriculture provided a lot of jobs for people and provided them with money. One important fact that must be borne in mind is that agriculture began in Europe and had spread to the rest of the world. In some places, it was known as the Spinning Mule, which was used as a tool for the same purpose. This meant that people who lived in Europe were using this tool while people in Africa and other parts of the world were enjoying it. Because of this, a number of people changed their lives for the betterment of society.

Development of agriculture in America has also transformed the entire continent. For instance, those who lived in the western countries could sell the agricultural products that they had produced. This was because they could have them manufactured in the United States. The economic progress in the United States is dependent on the cultivation of crops and meat. These include clothing, foodstuffs, tools, animal feeds, chemicals, machinery, and many others. This gave the United States a much-needed edge over its competitors. There were even goods manufactured in the United States that were not traded Internationally. Things like high-tech machinery, automobiles, and computers were made in the United States. This gave the country an edge that had the potential to transform the world.

Agricultural development in the Americas also created a number of industries. The amount of agricultural land and animals doubled in just one century. The greatest of these industries was cattle ranching. This industry made sure that the needed beef products were made available for all people. It also gave the country an opportunity to produce the necessary livestock feed. Another important fact is that without agriculture, there would be no more people. Without people, there would be no opportunity to make things. There would be nothing to trade for. Thus, agriculture in America had a role in facilitating the existence of every single person living on the planet. There are many researchers who say that without agriculture in America, people would have faced a lot of problems such as poverty and hunger. To alleviate these problems, agricultural development in America have been undertaken by Governments around the world. There has been a lot of effort put up into this field.

The Government of different countries have funded research and economic development for the agricultural development. They have invested heavily in agri-businesses. By investing in agri-businesses, the Government hoped to gain some sort of control over its economy. They hoped to ensure that their countries were able to be self-sufficient and strong. This has actually worked in a number of ways. Economically, countries were able to expand their markets by getting supplies through agri-businesses. At the same time, they were able to benefit from economy of scale, allowing them to produce more products than they could afford to make.

In rural areas, agricultural development has allowed the poor to have jobs. More importantly, farmers were given an opportunity to produce enough food to eat. This ensured that their children grew up with surplus availability of food to eat and were not dying of hunger anymore. They were able to feed their families and have enough money [6, 39, 51, 65].

### *1.3.3 ASIAN CONTINENT*

In this section, we analyzed the socio-economic impact of agriculture in Asia. Food has always been a very important economic factor in any country. If a country lacks food, then they were forced to use up their resources to create more food in order to feed themselves. Not only does this make them dependent on foreign countries for their food, but the idea of world hunger has also made an enormous psychological impact on the world. It is highly important to understand that agriculture is a broad-ranging industry, with many different types of products that are produced around the world. With this, it is important to look at some of the kinds of agricultural products that are being developed in the agriculture sector around the world. These products have an incredible impact on how well developed a country is. With agriculture around the world, you can see that there are many different areas that produce products that would be considered part of the food chain in the western world. So many of the items that you buy on a daily basis are actually derived as an output of agriculture.

There are many reasons why agriculture in Asia has developed differently than food products that are imported from outside of Asia. One reason is that the methods of farming that are used in Asia were not very common or acceptable in the past. As a result, the amount of people who have had access to the products that they produce and process was very small, even though the products that they produced and processed did a lot of good for the world. Another thing that we need to understand about agriculture in Asia is that the products that are harvested and processed by the farmers were made from items that were grown and produced locally. Agricultural products of Asia are going to make a positive impact on the world. Because of the things that are put into these agricultural products, the cost is usually much less than it would be if these products were imported. This is because the processing of these products allows for the increased production and lower costs of these products as well. This is because of the fact that agricultural products are so cheap to produce, so it becomes much easier for many people to afford them, and for the local economy to flourish. It is not uncommon for the economies to boom as the number of farmers increases, and the amount of agricultural products that are produced goes up as well. With the way that the agricultural products are processed, the prices go down so that everyone can afford them. Therefore, when we think about how agriculture can shape the world as well as our own life, we can begin to see that this is an industry that is very important to the world.

However, Asia faces a daunting problem from a global viewpoint when it comes to solving emerging food demand and supplied imbalance in the future. Although global food demand is projected to double by 2050, there is some doubt as to whether or not our agriculture sector will be able to satisfy the growing demand for food, arising from population growth and employment. In particular, considering the anticipated scarcity of water supplies, soil degradation, and adverse effects of climate change, concerted efforts to address the challenge for improving the agricultural production capacity and are vitally necessary for our global civilization. In the case of India and China, they solely constituted more than 30% of the world's population and are witnessing sustained double-digit economic growth rates coupled with increasing urbanization. There are different sub-regions in Asia dealing with special and heterogeneous problems related to agriculture, meals, and the environment. Three industrialized Far Eastern countries, which includes Korea, Japan, and Taiwan, have intrinsic comparative disadvantages in agricultural manufacturing and have appreciably decreased meal self-sufficiency costs when in contrast to different industrialized Nations. It stays to be seen how these Nations would cope with the developing dependence on overseas imports for food. On the different hand, with wealthy herbal and forestry resources, South-eastern Asian countries have been developing speedily in the latest decades, struggling noticeably from high incidences of poverty. While agriculture is regarded as the major engine for reducing poverty and fueling, the increase of standard economy, sustainable improvement balancing industrialization and environmental conservation stays a foremost challenge. In brief, Asia faces numerous challenges ranging from poverty which is expected to be a boom in demand for meals and herbal resources (particularly energy sources) [7, 38, 40, 47, 66, 70, 71].

### 1.3.4 AUSTRALIAN CONTINENT

The socio-economic impact of agriculture in Australia has remained to be dramatic. These days, agriculture has a lot to do with how well the economy flourishes. This is largely due to the economic dependence that our country had on agriculture. Agriculture provides many jobs for the people in Australia. It helps the country's economy to develop because of the high demand for the products that are being produced. One of the economic benefits of agriculture is the revenue that it generates. This revenue goes into the government coffers, which in turn is used to build infrastructure for the betterment of the people.

As you can imagine, the food that comes from agricultural commodities have a lot of nutritional value. This contributes to the overall health benefits of the country, as well as to the economy's welfare. The growth of agriculture in Australia has also brought about an increase in the price of agricultural commodities, as there is a high demand for these products. So, the importance and growth of agriculture in Australia are also important for the people in the country.

The growth of agriculture in Australia also brings a wider variety of food in the marketplace. With an increased amount of population, the amount of food that gets consumed by people is also expected to increase. But as Australia is a large country, it is more convenient for people to purchase their food from farms. And since the income of people in agriculture-based countries is not usually very high, they do not have much of a need to purchase food that is far away from their homes. Because of the wide range of agricultural commodities that are available in Australia, it is also easy for people to purchase products that are considered to be good for health. The significance and growth of agriculture in Australia are important for the people in the country, because they can always count on these kinds of foods whenever they need them.

The economic growth of the Nation also has a lot to do with the growth of agriculture in Australia. The livelihood of people in agriculture-based countries are directly tied to the amount of people that the agriculture sector employs. In addition to this, people also want to get employed in a particular demographic that will have a greater job possibility. The population of one region or another will affect the growth of agricultural commodities. So, these factors also play a part in the significance and growth of agriculture in Australia.

Another contributing factor to the importance and growth of agriculture in Australia is the impact that unemployment rate has on the value of the products that are getting sold in the market. People who are jobless in Australia may end up finding that they do not have enough money to buy the products that they want to buy. The recession in Australia has also made it difficult for people to get work. Not only is the food that we eat were important for our economic development, but also the foods that we grow by ourselves. The significance and growth of agriculture in Australia cannot be overlooked, and if we want to contribute to the economic growth of our country, then we need to have a hand in developing the agriculture sector [24, 29, 33, 59].

### *1.3.5 EUROPEAN CONTINENT*

The economic and social benefits of agriculture in Europe are profound. This economic and social value is a result of the agricultural sector's contributions to the economy, as well as its contribution to society. It is an example of the fact that we can take advantage of nature by growing crops, while increasing the ecosystem that supports us. There are many contributing factors to the economic and social value of agriculture in Europe. Agriculture has provided several benefits, including feeding the population, rural development, infrastructure, and recreation. In addition, these benefits have helped countries in Europe and other regions of the world to become more productive and self-sufficient.

Economic and social benefits of agriculture in Europe are also not limited to food production alone. The economy and the social benefits of agriculture are multiplied due to the overall benefits it brings to the local economy. In addition, the impacts of agriculture on the environment are also very positive. Growth is seen when we invest in various sectors of the economy, including agriculture. Farms, for example, provide employment for many people. At the same time, more services are provided to satisfy the needs of the population. Agricultural growth can also lead to economic growth by supporting International trade and attracting tourists.

Many countries have developed policies to promote agriculture, but the agricultural policy of Europe is the most developed. It has implemented several programs to promote agriculture. These programs are aimed at boosting production, improved the environment, and it has also promoted International trade and tourism. Because of these benefits, many European countries have developed policies that are aimed at promoting and developing agriculture. These policies have become stricter, but they have been able to achieve positive results due to the efficiency of these policies [8, 43–45, 58, 64].

## 1.4 THREATS AND OPPORTUNITIES

Agricultural activities in the Global South contributes to global threats such as soil degradation, air and water pollution, and water scarcity. In addition, growing population and consumption of food are driving some countries towards the risks of climate change, unsustainable practices, and even famines. This means that there are many areas where people in some regions of the world are not being well-served by today's food production systems. These issues have been around for a while, but as farming becomes more industrialized and rural, farmers struggle to handle them. As a result, the existing system is finding it hard to adjust to these new challenges.

We believe that several developed countries like the United States, China, Germany, etc., will take over the role of global agricultural leader, but there are other experts that think, that it will not be able to compete with other countries on an equal footing for long. However, the next century will see many changes for the betterment of agriculture at the global level. For instance, the average amount of land needed to grow a certain crop is going to decrease. Several other countries are looking for alternative sources of energy to supplement their current sources so that they can sustain the energy which they need to continue to uplift their economy. They are also looking at new power plants that will allow them to generate the energy which they need. The question is, will they be interested in powering farms, or will it be more the case that the power their homes with solar energy? This means that they will be using land for agricultural purposes but will it be enough? Of course, one of the biggest challenges will be feeding people if we are going to continue to farm agricultural crops.

In today's world, everything has to come from somewhere, and our current sources of energy are not up to the task. There will be a period of adjustment, but it will happen sooner rather than later. If we continue to rely on oil and coal for our fuel sources, we will eventually run out of those sources, and if we were to rely solely on solar energy and wind power to power our farms, we would be running out of energy as well. This is why there will be a need for new types of energy sources. So far, these alternative sources of energy have only been in existence for a few decades, but they will become more widespread in the coming years.

In fact, we already have several Global Agricultural Leaders that has adopted the use of wind turbines and solar energy. In Canada, the Government provides incentives to companies to use alternative energy sources such as solar panels, wind turbines, and bio-diesel that can be potentially used for fuel. The United States, on the other hand, is on the opposite end of the spectrum in the Global Agricultural Industry. They still rely on traditional sources of energy. They are in the process of changing from fossil fuels to green energy sources such as wind, water, and solar energy. Other countries, such as China, India are investing billions into research and development to develop new types of energy. As more people turn to green energy sources, we will see new advances in both technology and process. It is imperative that we get on the same page as the rest of the world and work together to make our agricultural industry more efficient. Right now, we are fighting against a number of challenges, but we can address them one by one, and soon we will have solutions that no one in the world has before.

As per FAO 2018, the present trend in agriculture reconfirms the hypothesis that "*with the development of a country, the share of agriculture in GDP goes down, which usually is accompanied by migration of the labor force from agriculture to non-agricultural sectors. In order to keep agriculture playing its vital role as provider of food, fiber, and fodder, it is important to increase per-worker productivity of the agricultural sector, which can come there exists through greater investments in agriculture and use of improved technologies*" [8]. Therefore, an urgent need to mitigate hurdles and obstacles lying in the path to attain sustainable growth of this sector.

For example, in 2016, the world total usage of chemical or mineral fertilizers was 110 Mt nitrogen (N), 49 Mt phosphate ($P_2O_5$), and 39 Mt potash (K). With respect to 2002, this represents increases of 34, 40 and 45%, respectively. According to the FAO report of 2018, for each nutrient, the top three countries, such as China, India, and the USA, together reflect half of the total. Agriculture is primarily driven by the increasing requirement of fiber and food at the global level, due to the rapid growing population. This situation leads to create a pressure on environmental resources such as land and water across the globe. This further increases the problem of degradation and spoiled the environmental factors, which were directly or indirectly concerning to the global agriculture system. Not only this, in the present era, agriculture has emerged out as a major contributor of GHGs into the environment.

It should also be taken into consideration that the food system adopted in present days also debilitates the wellbeing of individuals as well as the planet. As agribusiness accounts for 70% of water utilization alone and creates unsustainable levels of contamination and squander. Dangers related to malnutrition in diets are too the driving cause of deaths around the world. Millions of individuals are either not eating sufficiently or eating the off-base sorts of nourishment, coming about in a twofold burden of lack of healthy sustenance that can lead to sicknesses and wellbeing emergencies. As per a report in 2018, most of the number of hungry and undernourished individuals expanded to about 821 million in 2017, from around 804 million in 2016. Not only hunger, but obesity is additionally expanding [6, 7, 16, 23, 30].

Agriculture has the biggest physical impression on any human enterprise; thus, it also showed some impacts. About 10 billion people in the world were fed by means of agriculture along with conserving the environment. Apart from all the contribution, agriculture has posed a series of threats:

- Agriculture faces a scarcity of public fund investment at regional and global level.
- Among the globe, farmer's census showed major contribution of older aged people in agriculture. Average of young people does not involve in farming, maybe due to its requirement of hard-work and low-profit tendency. It can only be achieved by raising the profit and distribution of labor through the increasing mechanization and robotic farming.
- Agriculture has a big challenge in the form of Climate change. There is variation which is occurring in the form of high temperature and a variable amount of precipitation, responsible for the low productivity. Danger from the emergence of resistant strains of pest. It drives to incorporation of new techniques and plans to achieve maximum output from farming. About 25% of GHG contribution in the global environment from the farming sector. This urges an immediate action to stop or minimize the activities related to the problem.
- An accelerated depletion of water resources also seeks a great concern globally, due to not only by the climate change but also through the overexploitation and unsustainable irrigation practice. The alarming rate of depletion of aquifers, water pool, rivers with increasing salinity level are some of the waters related problems, which were emerged across the world. Mitigation measure should be adopted by the Government and farmers in the form of sustainable irrigation practices and restoration of above ground and underground water reservoirs.
- Shortage of skilled labor is a big challenge in the agriculture sector.
- The shortage of contractual labor is also a big challenge for a farmer, who need them particularly at the time of hand labor operations in farming. Any crop that requires hand labor to plant, tend or harvest is at risk.
- After labor, land, and often water, the upcoming largest cost is on the energy. Now, most of the agriculture activities totally depend upon the energy in terms of the use of machines like tractors, generators, etc.; manufacturing and transporting the goods or products used in cultivation as well as production. Eventually, emerging cost from these expenditures ultimately becomes an overburden for the last

man of the society. Therefore, progressive and optimistic steps should be taken by the Nations to rectify all these problems. For example, waste heat and LED lighting may be used to fulfill energy requirements.

- It is a big threat for attaining a real sustainable farming that, a high number of farmers does not have their own land. They rented it and doing farming on cost or rent basis.
- Investment uncertainty in global agriculture due to the fragility and vulnerabilities involved in this sector at public as well as political level.

With ample time, resources, and imagination, each of these problems can be resolved, but concurrently, global agriculture faces them all. That is a big obstacle. The nations that are largely dependent on the export market for their food supply have little knowledge or ability to help the local population to adapt to changing climate and a changing world. However, as the benefits of a global agricultural community become apparent, more Nations will begin to realize the benefits and begin to support sustainable agriculture [1, 18, 47].

## 1.5 SUGGESTIVE MEASURES

1. Global agriculture is dominated by nations that are largely dependent on the export market for their food supply. However, as the benefits of a global agricultural community become apparent, more Nations will begin to realize the benefits and begin to support sustainable agriculture.
2. As food production becomes more efficient, food security and stability become a necessity in the 21$^{st}$ century. The ability to increase food production will require that we reduce our dependency on imports. We must change the way we do farming, harvesting, and store our food so that we will be better able to meet the demands of a global community (Figure 1.5).
3. One of the most obvious changes that will need to occur is the conversion of croplands to more resilient crops that will be less impacted by climate change. Many farms are currently being used as wind turbines. However, these wind turbines will only help to slow down climatic changes.

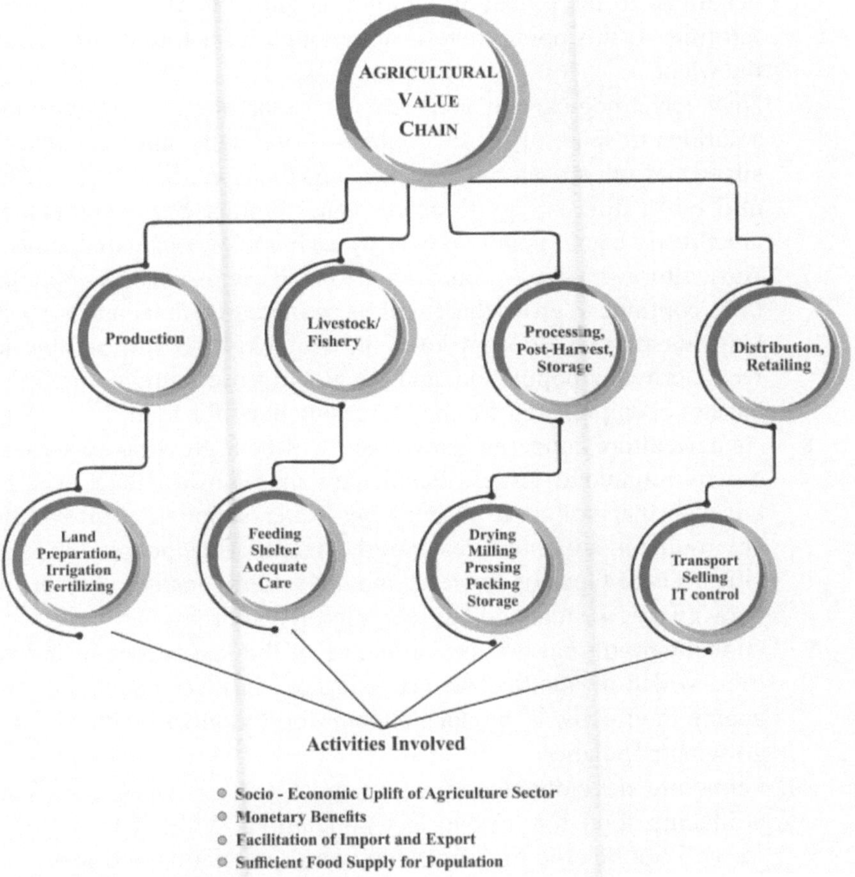


**FIGURE 1.5** Illustration of different segments of supply chain in agriculture sector and their benefits.

4. Global agriculture is all about enhancing efficiency. The energy necessary to produce crops, transport, and store food will increase at a rapid rate. We need to convert more of our farmland to crops that are more resistant to extreme weather conditions and climate change.
5. One of the largest threats to the sustainability of the global agricultural community is water pollution. We are not utilizing all of the water that is available on the planet. There is an enormous amount of water that is locked up underground in aquifers that would be

beneficial to the global agricultural community if we used it for farming. At this point, there is not enough infrastructure to get all the water.

6. New technological advancements in solar energy may provide solutions to some of these problems. Solar cells that are made of silver and other materials are very inexpensive, and they can be utilized to store energy to power water pumps. Some companies are already using them to power water pumps for irrigated areas.
7. Agriculture is a major contributor to GHG emissions. As populations continue to grow, there will be an increased demand for food. Unfortunately, we do not know how quickly we will be able to feed a growing population, and the global agricultural community has not been prepared for the increasing need for food.
8. As agriculture concerns grow, there will be a growing awareness of environmental issues. People are more aware than ever of the way that pollution pollutes our water supplies, destroys the environment, and interferes with the health of the people living in cities. These types of concerns may eventually lead to changes in the way that we manage our agricultural resources.
9. Global agricultural production is one of the least efficient industries, which means that we are using our best resources, but are unable to effectively produce enough food to meet the needs of a growing population.
10. Economic development is an essential part of what goes into producing food for the world's population. Without these jobs, we will not be able to sustain the world population and meet the needs of a growing population. The fact that the world is growing by leaps and bounds means that there will be millions of hungry people in the future.
11. We need to build an agricultural community where farmers and ranchers can utilize technology and tools that will allow them to convert more of their land to crops that are not affected by climate change or water pollution. The key to ensuring that we continue to be a global agricultural community will be understanding how to grow crops that are more resilient and self-sustaining. Understanding how to develop sustainable agriculture will be the first step in creating a sustainable global agricultural community. As more people become more aware of the need for better resources and a global agricultural

community and how to make it work, we will find that we have a chance to eliminate global hunger once and for all.

12. One of the emerging powerful tools is the digitalization of the agriculture sector, which has immense potential to rectify and minimize the risks associated with it. All the related fields such as cultivation, animal husbandry, fisheries, food processing, and management were benefitted with the digitalization technology and with the use of various advanced analytical tools for generating the easy data interpretation, involvement of satellite information, advance equipment along with artificial intelligence (AI). These measure helps to maintain sustainability of resources, increment in production with the ability to ensure quality food accessibility for mankind. Overall, through digitalization, farmer or grower could be able to evaluate and monitor their produce and livestock in a well-organized manner. He also can judge his efforts towards inputs and were able to estimate the output from the farm. Moreover, grower has a market connection or accessibility on his digital device. Digitalization should comprise of a traceability network to monitor each and every steps of this sector to provide value chain assurance to consumer. The use of digitalization, the agriculture sector could sustain its originality and enter into a new horizon of sustainable agriculture with the involvement of environment-friendly technology and policies.
13. To mitigate all these challenges and opportunities, Governments have to reframe and update their policies and schemes related to the holistic development of the sector. A regular realistic audit must be undertaken for the betterment and utilization of these policies at the ground level. New schemes also have to be incorporated in terms of Agro-economic, Agro-social, and Agro-environmental aspects [4, 71].

## 1.6 SUMMARY

This chapter reflects a journey of agriculture in a global perspective with an adaptation of various innovations on a regular basis. Agriculture based on the domestication of wild plants and animals arose independently in several parts of the world between the period about 10,500 and 4,500 years ago. This led to the emergence of the first economy in human history. In spite, there are still some variations in opinion among archaeologists as to the correct timeline of agriculture. It is started independently at diverse times in a few places around the world. The earliest agriculture within the Fertile Crescent of the Middle East dated around 11,000–10,000 years back. Another was the valleys of the Indus, the Yangtze, and Yellow rivers in India and China, between 10,000 and 9,000 years ago from the present. A proper agriculture system could ensure a plentiful supply of food while being ecologically sustainable. However, we believe that the best way to approach sustainability is to see it from a global perspective. Since sustainability involves not only producing the best quality products, it is also important to ensure that the products are affordable for mass and also to ensure that they are of high quality. However, the primary focus should be on finding the best kinds of products which can be produced without using a lot of inputs and without causing environmental damage. In this global perspective, we need to ensure that the use of our resources is also limited. We cannot expect that all resources will be used efficiently and effectively without any unwanted side effects.

## ACKNOWLEDGMENTS

The authors are grateful to Dr. Anurag Varshney, Vice-President of Patanjali Research Institute, for his tremendous support. The authors are highly thankful to our designer Mr. Ajeet Chauhan for his phenomenal efforts in designing the graphics of the chapter.

## KEYWORDS

- **ancient era**
- **classical era**
- **modern era**
- **Neolithic revolution**
- **potash**
- **sustainability**

## REFERENCES


1. *Agriculture Goods: Why Is Agriculture Important and its Role in Everyday Life*, (n.d.). Retrieved from: https://agriculturegoods.com/why-is-agriculture-important/ (accessed on 9 April 2021).
2. *AGWEB: A Brief History of Ancient Agriculture*, (n.d.). Retrieved from: https://www.agweb.com/blog/straight-from-dc-agricultural-perspectives/a-brief-history-of-ancient-agriculture (accessed on 9 April 2021).
3. *Ancient Civilization World: Agriculture*, (2017). Retrieved from: https://ancientcivilizationsworld.com/agriculture/ (accessed on 9 April 2021).
4. Aryal, J. P., Sapkota, T. B., Khurana, R., Khatri-Chhetri, A., & Jat, M. L., (2019). Climate change and agriculture in South Asia: Adaptation options in smallholder production systems. *Environment, Development and Sustainability*, 1–31. Retrieved from: https://link.springer.com/content/pdf/10.1007/s10668-019-00414-4.pdf (accessed on 9 April 2021).
5. Atkins, P. J., Simmons, I. G., & Roberts, B. K., (1998). *People, Land and Time London.* Hodder Arnold. Retrieved from: https://www.researchgate.net/publication/271841247_The_origins_and_spread_of_agriculture (accessed on 9 April 2021).
6. Barrientos-Fuentes, J. C., & Torrico-Albino, J. C., (2014). Socio-economic perspectives of family farming in South America: Cases of Bolivia, Colombia, and Peru. *Agronomía Colombiana, 32*(2), 266–275. Retrieved from: http://www.scielo.org.co/pdf/agc/v32n2/v32n2a14.pdf (accessed on 9 April 2021).
7. Bashir, Z., & Ahmad, M., (2000). The role of agricultural growth in South Asian countries and the affordability of food: An inter-country analysis [with comments]. *The Pakistan Development Review*, 751–767. Retrieved from: https://www.jstor.org/stable/41260296 (accessed on 9 April 2021).
8. Brisson, N., Gate, P., Gouache, D., Charmet, G., Oury, F. X., & Huard, F., (2010). Why are wheat yields stagnating in Europe? A comprehensive data analysis for France. *Field Crops Research, 119*(1), 201–212. Retrieved from: doi: 10.1016/j.fcr.2010.07.012.
9. Clute, R. E., (1982). Clute the role of agriculture in African development. *African Studies Review, 25*(4), 1–20. Retrieved from: doi: 10.2307/524398.
10. Dennell, R., (1992). In: Watson, P. J., & Cowan, W., (eds.), *The Origins of Crop Agriculture in Europe: The Origins of Crop Agriculture in Europe* (pp. 71–100) Publisher: Washington: Smithsonian Institution. Retrieved from: https://www.researchgate.net/publication/271852777_The_origins_of_crop_agriculture_in_Europe (accessed on 9 April 2021).
11. Diao, X., Hazell, P., Resnick, D., & Thurlow, J., (2007). *The Role of Agriculture in Development Implications for Sub-Saharan Africa*. International Food Policy Research Institute. Retrieved from: doi: 10.2499/9780896291614RR153.
12. Doughty, C. E., (2010). The development of agriculture in the Americas: An ecological perspective. *Ecosphere, 1*(6), 1–11. Retrieved from: https://esajournals.onlinelibrary.wiley.com/doi/pdf/10.1890/ES10-00098.1 (accessed on 9 April 2021).
13. Eagri, (2011). *AGRO102-Introductory Agriculture- Agricultural Heritage of India (1+0) (Lecture 01)*. Tamil Nadu Agricultural University.
14. *Economic Discussion: Role of Agriculture in the Economic Development of a Country*, (n.d.). Retrieved from: http://www.economicsdiscussion.net/economic-development/

role-of-agriculture-in-the-economic-development-of-a-country/4652 (accessed on 9 April 2021).

15. Encyclopedia Britannica, (2020). *Origins of Agriculture*. Retrieved from: https://www.britannica.com/ (accessed on 9 April 2021).
16. *FAO: GIEWS-Global Information and Early Warning System* (n.d.). Retrieved from: http://www.fao.org/giews/country-analysis/external-assistance/en/ (accessed on 9 April 2021).
17. Fischer, G., Shah, M., Tubiello, F. N., & Van, V. H., (2005). Socioeconomic and climate change impacts on agriculture: An integrated assessment, 1990–2080. *Philosophical Transactions of the Royal Society B: Biological Sciences, 360*(1463), 2067–2083. Retrieved from: https://www.ncbi.nlm.nih.gov/pmc/articles/PMC1569572/pdf/rstb20051744.pdf (accessed on 9 April 2021).
18. Flannery, K. V., (1973). The origins of agriculture. *Annual Review of Anthropology, 2*(1), 271–310. Retrieved from: https://www.annualreviews.org/doi/pdf/10.1146/annurev.an.02.100173.001415 (accessed on 9 April 2021).
19. Fuller, D. Q., & Murphy, C., (2014). Overlooked but not forgotten: India as a center for agricultural domestication. *General Anthropology, 21*(2), 1–8.
20. Fuller, D. Q., Kingwell-Banham, E., Lucas, L., Murphy, C., & Stevens, C. J., (2015). Comparing pathways to agriculture. *Archaeology International, 18*, 61–61. Retrieved from: doi: http://dx.doi.org/10.5334/ai.1808.
21. Fussell, G. E., (1967). Farming systems of the classical era. *Technology and Culture, 8*(1), 16–44. Retrieved from: doi: 10.2307/3101523.
22. *Geography and History: Prehistory Timeline*, (n.d.). Retrieved from: https://slideplayer.com/slide/9753724/31/images/2/Prehistory+Timeline.jpg (accessed on 9 April 2021).
23. Godfray, H. C. J., Beddington, J. R., Crute, I. R., Haddad, L., Lawrence, D., Muir, J. F., & Toulmin, C., (2010). Food security: The challenge of feeding 9 billion people. *Science, 327*(5967), 812–818. Retrieved from: doi: 10.1126/science.1185383.
24. Gunasekera, D., Kim, Y., Tulloh, C., & Ford, M., (2007). Climate change impacts on Australian agriculture. *Australian Commodities: Forecasts and Issues, 14*(4), 657. Retrieved from: https://www.researchgate.net/publication/289781717_Climate_change_Impacts_on_Australian_agriculture (accessed on 9 April 2021).
25. *High Level Expert Forum Global Agriculture Towards 2050*, (2009). Retrieved from: http://www.fao.org/fileadmin/templates/wsfs/docs/Issues_papers/HLEF2050_Global_Agriculturepdf (accessed on 9 April 2021).
26. *History Discussion: Economic Conditions During Maurya-Scythian Era*, (2016). http://www.historydiscussion.net/history-of-india/maurya-scythian-era/economic-conditions-during-maurya-scythian-era/5727 (accessed on 9 April 2021).
27. Hubbard, R. N., (1980). Development of agriculture in Europe and the Near East: Evidence from quantitative studies. *Economic Botany, 34*(1), 51–67. Retrieved from: https://link.springer.com/article/10.1007/BF02859554 (accessed on 9 April 2021).
28. *Johns Hopkins-Center for a Livable Future: History of Agriculture* (n.d.). Retrieved from: http://www.foodsystemprimer.org/food-production/history-of-agriculture/ (accessed on 9 April 2021).
29. Keating, B. A., & Harle, K., (2004). Farming in an ancient land-Australia's journey towards sustainable agriculture. *Paper no 102, Presenting in ISCO 2004-13th Inter National Soil Conservation Organization Conference-Brisbane*. Retrieved from:

https://tucson.ars.ag.gov/isco/isco13/PAPERS%20F-L/KEATING.pdf (accessed on 9 April 2021).

30. Kirchmann, H., & Thorvaldsson, G., (2000). Challenging targets for future agriculture. *European Journal of Agronomy, 12*(3, 4), 145–161. Retrieved from: doi: 10.1126/science.1185383.
31. Kumar, D., (2017). River Ganges-historical, cultural, and socio-economic attributes. *Aquatic Ecosystem Health and Management, 20*(1, 2), 8–20. Retrieved from: doi: 10.1080/14634988.2017.1304129.
32. Jennifer, W., (2012). *LiveScience: How European Farmers Spread Agriculture Across Continent.* Retrieved from: https://www.livescience.com/19924-agriculture-move-north-europe.html (accessed on 9 April 2021).
33. Lockie, S., (2015). *Australia's Agricultural Future: The Social and Political Context.* Report to SAF07-Australia's agricultural future project. Retrieved from: https://acola.org.au/wp/PDF/SAF07/social%20and%20political%20context.pdf (accessed on 9 April 2021).
34. *Lumen Sociology: Reading: Types of Economic Systems*, (n.d.). Retrieved from: https://courses.lumenlearning.com/alamo-sociology/chapter/reading-economic-systems/ (accessed on 9 April 2021).
35. Dastagiri, M. B., (2017). Global agriculture: Vision and approaches. *European Scientific Journal, 13(*21). Retrieved from: doi: 10.19044/esj.2017.v13n21p312.
36. Mazvimavi, K., (2010). *Socio-Economic Analysis of Conservation Agriculture in Southern Africa.* FAO Regional Emergency Office for Southern Africa (REOSA). Retrieved from: http://www.fao.org/3/i2016e/i2016e00.pdf (accessed on 9 April 2021).
37. Mclntosh, J., (2008). *The Ancient Indus Valley: New Perspectives* (p. 387). ABC-CLIO. ISBN 978- 1576079072.
38. Mendelsohn, R., (2014). The impact of climate change on agriculture in Asia. *Journal of Integrative Agriculture, 13*(4), 660–665. Retrieved from: doi: 10.1016/S2095-3119(13)60701-7.
39. *National Geographic: Agriculture*, (n.d.). Retrieved from: https://www.Nationalgeographic.org/encyclopedia/agriculture/ (accessed on 9 April 2021).
40. National Research Council, (2015). *A framework for Assessing Effects of the Food System.* National Academies Press. Retrieved from: https://www.ncbi.nlm.nih.gov/books/NBK305181/pdf/Bookshelf_NBK305181.pdf (accessed on 9 April 2021).
41. NEPAD, (2013). *Agriculture in Africa: Transformation and Outlook.* Retrieved from: https://www.un.org/en/africa/osaa/pdf/pubs/2013africanagricultures.pdf (accessed on 9 April 2021).
42. *New World Encyclopedia: Indus Valley Civilization*, (n.d.). Retrieved from: https://www.newworldencyclopedia.org/entry/Indus_Valley_Civilization (accessed on 9 April 2021).
43. Olesen, J. E., (2016). Socio-economic impacts-agricultural systems. In: Quante, M., & Colijn, F., (eds.), *North Sea Region Climate Change Assessment.* Regional Climate Studies. Springer, Cham Retrieved from: https://link.springer.com/chapter/10.1007/978-3-319-39745-0_13 (accessed on 9 April 2021).
44. Olesen, J. E., & Bindi, M., (2002). Consequences of climate change for European agricultural productivity, land use and policy. *European Journal of Agronomy, 16*(4), 239–262. Retrieved from: doi: 10.1016/S1161-0301(02)00004-7.

45. Olesen, J. E., Trnka, M., Kersebaum, K. C., Skjelvåg, A. O., Seguin, B., Peltonen-Sainio, P., & Micale, F., (2011). Impacts and adaptation of European crop production systems to climate change. *European Journal of Agronomy, 34*(2), 96–112. Retrieved from: https://pure.au.dk/portal/files/44136103/Bilag1_JEOA105.pdf (accessed on 9 April 2021).
46. *Organization for Economic Co-Operation and Development: Official Website*, (2019). Retrieved from: https://www.oecd.org/india/ (accessed on 9 April 2021).
47. *Organization for Economic Co-Operation and Development: Opportunities and Threats for Agriculture-Challenges and Opportunities for the Global Food System*, (n.d.). Retrieved from: https://www.oecd.org/agriculture/understanding-the-global-food-system/opportunities-and-threats-for-agriculture/ (accessed on 9 April 2021).
48. Paniagua, A., & Baker, K., (n.d.). *The Socioeconomics of Agriculture* (Vol. I.). Social and economic development. Retrieved from: https://www.eolss.net/Sample-Chapters/C13/E1-20-03-01.pdf (accessed on 9 April 2021).
49. Pinhasi, R., Fort, J., & Ammerman, A. J., (2005). Tracing the origin and spread of agriculture in Europe. *PLoS Biology, 3*(12). Retrieved from: https://www.ncbi.nlm.nih.gov/pmc/articles/PMC1287502/pdf/pbio.0030410.pdf (accessed on 9 April 2021).
50. Pokharia, A. K., Sharma, S., Tripathi, D., Mishra, N., Pal, J. N., Vinay, R., & Srivastava, A., (2017). Neolithic-early historic (2500–200 BC) plant use: The archaeobotany of Ganga plain, India. *Quaternary International, 443*, 223–237.
51. Rodríguez, A., Rodrigues, M., & Salcedo, S., (2010). The outlook for agriculture and rural development in the Americas: A perspective on Latin America and the Caribbean. Retrieved from: http://www.fao.org/3/i8048en/I8048EN.pdf (accessed on 9 April 2021).
52. Roy, M., (2009). Agriculture in the Vedic period. *Indian Journal of History of Science, 44*(4), 497–520.
53. Sameer, M. A., & Zhang, J. Z., (2018). A case study on intangible cultural heritage of Indus valley civilization: Alteration and reappearance of ancient artifacts and its role in modern economy. *Int. J. Recent Sci. Res., 9*(12), 29898–29902. doi: http://dx.doi.org/10.24327/ijrsr.2018.0912.2955.
54. Scienmag, (2016). *Rice Farming in India Much Older Than Thought, Used As 'Summer Crop' By Indus Civilization*. Retrieved from: https://scienmag.com/rice-farming-in-india-much-older-than-thought-used-as-summer-crop-by-indus-civilization/ (accessed on 9 April 2021).
55. Shereen, R., (1986). An aspect of Harappan agricultural production. *Studies in History, 2*(2), 137–153. Retrieved from: https://doi.org/10.1177/025764308600200201.
56. Study.com. (n.d.). *Periods of World History: Overviews of Eras from 8000 B.C.E to the Present*. Retrieved from: https://study.com/academy/lesson/periods-of-world-history-overviews-of-eras-from-8000-bce-to-the-present.html (accessed on 9 April 2021).
57. Svizzero, S., & Tisdell, C., (2014). Theories about the commencement of agriculture in prehistoric societies: A critical evaluation. *Rivista Di Storia Economica, 30*(3), 255–280. Retrieved from: https://www.researchgate.net/publication/265972354_Theories_about_the_Commencement_of_Agriculture_in_Prehistoric_Societies_A_Critical_Evaluation (accessed on 9 April 2021).
58. The Conversation, (2014). *Agriculture in Australia: Growing More Than Our Farming Future*. Retrieved from: https://theconversation.com/agriculture-in-australia-growing-more-than-our-farming-future-22843 (accessed on 9 April 2021).

59. The World Bank Group, (n.d.). *Agriculture and Food.* Retrieved from: https://www.worldbank.org/en/topic/agriculture/overview (accessed on 9 April 2021).
60. The World Bank Group, (2019). *Agriculture, Forestry, and Fishing, Value Added (% of GDP).* Retrieved from: https://data.worldbank.org/indicator/NV.AGR.TOTL.ZS?end=2018&start=1960&view=chart (accessed on 9 April 2021).
61. The World Bank Group, (2017). *World Development Indicators: Structure of Output.* Retrieved from: http://wdi.worldbank.org/table/4.2 (accessed on 9 April 2021).
62. Thrall, P. H., Bever, J. D., & Burdon, J. J., (2010). Evolutionary change in agriculture: The past, present and future. *Evolutionary Applications, 3*(5, 6), 405. Retrieved from: https://www.ncbi.nlm.nih.gov/pmc/articles/PMC3352499/pdf/eva0003-0405.pdf (accessed on 9 April 2021).
63. Vikrama, B., & Chattopadhyaya, U., (2002). Ganges neolithic. In: *Encyclopedia of Prehistory* (pp. 127–132). Springer, Boston, MA. Retrieved from: https://link.springer.com/chapter/10.1007%2F978-1-4615-0023-0_14 (accessed on 9 April 2021).
64. Vink, N., (2014). Commercializing agriculture in Africa: Economic, social, and environmental impacts. *African Journal of Agricultural and Resource Economics, 9*(311-2016-5574), 1–17.
65. Walthall, C. L., Anderson, C. J., Baumgard, L. H., Takle, E., & Wright-Morton, L., (2013). *Climate Change and Agriculture in the United States: Effects and Adaptation.* Retrieved from: https://core.ac.uk/download/pdf/12897964.pdf (accessed on 30 April 2021).
66. Wanki, M. W., & Lee, J., (2010). *Economic Development and Agricultural Growth in Asia.* Retrieved from: https://www2.gsid.nagoya-u.ac.jp/blog/anda/files/2010/06/4_wankimoon.pdf (accessed on 9 April 2021).
67. Weebly, (n.d.). *The Maurya and Gupta Empires.* Retrieved from: https://globalprogect.weebly.com/economic.html (accessed on 9 April 2021).
68. Wik, M., Pingali, P., & Brocai, S., (2008). *Global Agricultural Performance: Past Trends and Future Prospects.* Background paper for the world development report. Retrieved from: https://openknowledge.worldbank.org/bitstream/handle/10986/9122/WDR2008_0029.pdf?sequence=1&isAllowed=y (accessed on 30 April 2021).
69. *World Food and Agriculture Statistical Pocketbook,* (2018). Retrieved from: http://www.fao.org/3/CA1796EN/ca1796en.pdf (accessed on 9 April 2021).
70. Ye, J., & Pan, L., (2016). *Concepts and Realities of Family Farming in Asia and the Pacific (No. 139).* Working paper. FAO-UNDP. Retrieved from: http://www.fao.org/3/a-i5530e.pdf (accessed on 9 April 2021).
71. Zhan, J., Mirza, H., & Speller, W., (2016). The impact of large-scale agricultural investments on communities in south-east Asia: A first assessment. In: *Large-Scale Land Acquisitions* (pp. 79–107). Retrieved from: https://doi.org/10.4000/poldev.2029.

# CHAPTER 2

# Rainfed Agriculture: Giant Strides of Current Farming Practices

JAWAHAR LAL DWIVEDI, RITIKA JOSHI, VEDPRIYA ARYA, and VEMURI RAVINDRA BABU

## ABSTRACT

Rainfed farming is mostly found in the Middle East, Asia, Africa, and Latin America. The harsh environmental conditions of West Asia, North Africa, and Sub-Saharan Africa (SSA) are the main factors limiting crop production. Demand and problems of rainfed arable land are increasing in demand for increasing grain production due to the problems and challenges of the fast-growing world population, global climate change, lack of suitable water for irrigation, and erosion of agricultural land. Rainwater harvesting methods are commonly practiced, and it is possible to improve water access for domestic and agricultural use in arid and semi-arid regions. New funds are becoming available to encourage the implementation of water harvesting methods. This chapter deals with crops and management, research that can improve water efficiency, water harvesting methods, and opportunities in rainfed farming for better working as well as commercial development. The people of the semi-arid region are constantly fighting problems related to water in farming methods. This can only be achieved by proper agricultural methods and can provide good technology by governance.

## 2.1 INTRODUCTION

The recent increase in food prices is necessary to increase the production of rainfed crops through excellent water management. In 2008, the cost of food products increased rapidly due to several environmental factors and was predicted to remain somewhat high. Already, this increase is showing significant adverse changes, threatening existing benefits in malnutrition and poverty [40]. Technical and financial support has to be provided to motivate the agriculture system to overcome the food crisis under moderate conditions, as water-related funds will be the pillars of a unified global response [41]. The World Bank's Global Food Price Response Program in 2008 was organized to deal with supply and farmers issues. Rain-based sources of food have little importance on women, and 70% of it is given to poor women around the world. Agriculture plays a key role in poverty reduction and economic development, there are indications of a 1% rise in agriculture production and 0.6–1.2% reduction in the farmer's condition [33].

Rainfed agriculture is the technique based on rainwater. It is the most valued and highly productive agricultural method for the production of food in developing countries. For example, in Europe, in many arid sub-humid, semi-arid, and tropical regions where the production rate was very slow, rainfed farming is applied [4]. In rainfed areas, farming critically depends on the spatial and temporal movement of rainfall, which should be adequate for soil moisture distribution and evaporative demands [17]. Cropland rains account for about East Asia (65%) and South Asia (58%), Latin America (87%), Middle East, and North Africa (67%). In sub-Saharan Africa (SSA) where 96% of agriculture is the most prominently based on rainfed agriculture [8]. In 2008, World Development Report on Agricultural Development suggested that to complete the food demand in the future by agriculture productivity should be developed in both irrigated and Rainfed areas [38]. There are two core motives for rainfed production; first, around 82% of total cropland (1.5 billion hectares) in worldwide is rainfed [8] out of which 60% agriculture products is generated by rainfed agriculture in developing countries [37]. Secondly, the chances of development in irrigation are very limited due to investment cost and high-water demand with some exceptions, for example, in SSA. The evaluation was found in rainfed mixed farms about 70 to 90% of ruminant's livestock. The description for food production is instant for aiming rainfed areas and is connected to the following [27]:

- Intense poverty and push for survival due to hash and Human-induced climatic changes;
- The profitability and potential output are production objectives;
- Interaction between crop, soil, and animals;

- Ensuring the benefit of environmental integrity, production yield, and renewable development.

## 2.2 IMPORTANCE OF RAINFED FARMING

Crop production in developing countries depends entirely on irrigation and 60% of the crop it produces but rainfed farming is also very essential in developing countries. In developing countries 2021–2025, rainfall production probably increases from 1.5 to 2.1 metric tons per hectares. The area of rain will be 43% of the total grain area, and 40% of the area of rain will be responsible for the increase in grain production. In various outlines, it is suggested that rainwater yields increase more quickly and may offset investment in irrigation to reduce groundwater pumping, but the necessary improvement in achieving rainfall production is an important improvement. For example, an outline of 2021–2025 states that 20.1 million metric tons of irrigated grain production during groundwater mining decreases in China, such as in WANA and India, it is 18.4 and 18 million metrics tons, which is the case for developed countries and developing countries. In these countries, it is 1.6 and 53.0 million metric tons. These can be offset by the decrease in area and yield of rain, but need to increase the production rapidly. Compared to the standards, in China average rainfall-based cereal production (0.6 metric tons per hectare) would require 13%, 20% production in India (0.30 metric tons per hectare) and in WANA (0.3 metric tons per hectare); the rainfed production areas in China, India, and WANA is 0.6, 0.8 and 0.10 million hectares. In the world, about 71.74 million hectares is rainfed agricultural land and 68.38 million hectares is the irrigated land, characterized by low levels of productivity and low input usage, complex, and risk-prone. In the country, rainfed farming shares a large number of food production if it managed properly because the government of India gives priority to the sustainable development and provides proper water management in the rainfed areas [32].

## 2.3 STATUS OF RAINFED AGRICULTURE AROUND THE WORLD

In developing countries and worldwide, the crops produced by rainfed agriculture, more is consumed by poor communities. Water productivity, 'crop per drop crop,' is reducing rainfed farming methods due to high evaporation rates, land degradation, floods, or droughts leading to crop die so methods should be applied for more water management. Production rate is very slow in SSA and South Asia due to lack of food and poverty for rural areas. In the world, most countries rely mainly on rainfed farming for their crops. In many developing countries, main efforts to improve environmental conditions and production rate because in Asia and Africa, there is a huge ratio of poor families which are still facing food insecurity, starvation, poverty, and malnutrition were rainfed farming is major activity. Though, in rainfed areas, many farmers harvest water for irrigation methods, but millions are completely depending on rainfall. Rainfed system presents research quires related to irrigation farming that inherent poverty and uncertainty. Researcher should understand risk and trade-offs that are domestic in rainfed settings and search several methods have not been adopted to increase the water and soil management, find out the reasons for this, given the institutions and maintenances, socioeconomic, and political limitations. Researcher should have a deep understanding of water scarcity in the environment and improvement of livestock. Increasing population has exerted substantial pressure on land and water resources in many areas, which was used by livestock and rainfed cropland. In the ecosystems, soil contains an inadequate amount of organic matter and nutrients that are loses a part of its underlying biodiversity [16].

Management of soil moisture and rainwater effectively and supplementing with the increased use of organic and inorganic fertilizer and using small-scale irrigation, improved entree to markets and increased protection on soil and water resources for rainfed farmers Areas will be necessary to improve livelihood. In SSA, for reducing dry spells in the soil, farmers use more and more water to maintains the soil moisture contents. Farmers need to invest on productivity-enhancing technologies such as better-quality seeds and good fertilizer with good protection. They are allowing crops that are high in value and highly sensitive to water stress. When farmers grow more crops, they will be on the path of income and food security [35].

In different regions of the world, rainfed agriculture produces the largest number of crops. Productive soils and relatively frequent rainfall occur mainly in temperate regions. However, in tropical regions, mainly in humid and sub-humid regions, crop production exceeds 5–6 tons per hectare in marketable rainfed agriculture. Meanwhile, the semi-arid and sub-humid areas have experienced the lowest production per unit of land and poor production improvement. Here, production can be as low as 1/10 of more production areas. In rainfed agriculture, awareness already exists for at least double production, even in places where water constitutes a special challenge: economic and environmental institutions are ready to promote and adopt the methods that were facilities. It is essential to the human success to build strong institutions. Rainwater-based agriculture in the catchment area requires investment in institutional and human capabilities for water planning and management, where rainwater harvesting techniques can be applied properly. Soil and water management requires a significant need to create a new paradigm in rainfed systems, to enhance the production and to reduce poverty a holistic method is needed for the management. So that no further deterioration occurs in the natural resource.

Complementary irrigation plays a key role in minimizing the risk of production optimization and crop failure. Storage structures or groundwater have the potential to carry additional rainwater, there is loss of rainfall 12 to 30% as a runoff under improved systems. The division between irrigated and rainfed system alike artificial and ineffective consider increasingly in many observation [33].

Main problems in rainfed agriculture include:

- In rainfed system improved rainwater management can enhanced the production by creating excess soil erosion and runoff. The water holding capacity of soil and high rainfall infiltration decrease soil erosion and in soil increased availably of water is helpful in crop production. Due to infiltration, improvement in the natural and aquatic ecosystems water scarcity.
- In agricultural policy irrigation, the management of water resources is essential, while the distribution and management of groundwater, lakes, and rivers primarily for joint water resources. An effective integration is required that emphases on water management investment options between continuity for rainwater-fed agriculture and which includes a multilateral approach.
- Typically, rainfall-based agricultural systems receive sufficient rainfall for double and often quadruple yields, also in water-limited areas. But obtain at incorrect time which causes dry spells and huge damage. In addition, upgrading to water and rainfed farming, crop management and good set-up, market investments and excellent and reasonable access to soil and water resources. In rainfed areas, to improve yield and rural living, there is a need to reduce the risks related to rainfall, meaning that investment in management of water and unlock possibilities of the entry point in rainfed agriculture.

Asian rainfed rice farmers, like other farmers alike, are faced with a choice of techniques best suited to the family through tasks, lifesaving management, as well as varied tasks of resource custody, especially land resources enter through and increase income for domestic economic development, especially to promote diversification and better family welfare. The nature of rainforest farming, which is being done, all these different burdens are also creating more complex trends of uncertainty by the plex. Through uncertainties, the most important thing the arises from the natural environment is unexpected changes in rainfall, radiation, other atmospheric events. Overflows such as events, and floods, such as strong winds, hail, extremes, and drowning, but uncertainty are not limited to the natural environment. They allow for an economic environment [3] and include unpredictable price levels for both input and output and in a policy environment. The environment, which in some cases may be linked to market events or through changes in the tenure of the system, can directly influence the decisions of the farm and marketing the arrangement.

## 2.4 TYPE OF RAINFED FARMING IN THE WORLD

Several attempts have been made to classify farm systems for different resolves using different standards Grigg [13], investigated globally and recognized the different types on farming methods: Wet-rice farming (Asia); shifting agriculture; Mixed farming (Western Europe and North America); Rustic nomads; Mediterranean Agriculture; tree planting; large-scale crop production and dairy; Animal husbandry. This class ignores the options that rainfed farming methods may link with elements from different listed systems. Duckham and Mesfield [9] studied a classification on the basis of two factors: firstly, the intensity, which consists of widespread intensive and semi-intensive practices and secondly, major land use, which consists of pasture, perennial tree, and annual crop practices. Another alternative practice is plowing, which contains grassland and shrubs. These systems were practices in temperate climate regions. In the rainfed system taxonomy of the system is ignored due to the fact their integration with livestock. Dixon and Gibbon [7], in a recent FAO publication discussing farming systems in developing countries, favored a larger range of groups. Around the world, researchers have deliberated different type farm systems, each were explained in terms of the region where it dominates. These systems were divided into wide groups, and they finished with the following assembly: coastal artisan fishing; small-holder irrigated schemes; Wetland rice-based; Humid from rain; Dualistic; Rainy Highland; and dry or cold from rainfall. In many rainfed systems, crop-livestock integration is identified as very important; many developed countries were excluded from these classifications and excuses from commercial production opportunities. Yet, it delivers a sound basis for more development of an arrangement.

Rainfed farming is divided into four main groups, which are given as follows:

- High-latitude rainfed systems (cold winters);
- Mid-latitude rainfed systems (mild winters);
- Subtropical and tropical highland rainfed systems; and
- Semi-arid tropical and subtropical rain-fed systems.

## 2.5 WATER HARVESTING TECHNIQUES FOR RAINFED FARMING

The most significant problem is the lack of water for farmers in Africa, Asia, and the Near East, where 80–90% of the water is diverted for agricultural purposes [11]. The very basic unit of life and most essential components in our planet is water, a finite resource and at the present time, the most endangered natural resource. According to millennium development goals, water is the most significant driver, as shown in Figure 2.1. in the framework of the four-goal of the millennium, in 2015, the involvement of water resources management has achieved the milestones. Yet, in many sets of conditions, water content per productivity is not the limiting factor for increased productivity, but its management and efficient use are the main yield determinants. Alternatively, rainfed agricultural areas like semi-arid and dry sub-humid regions suffer major water-related challenges to deal with variable rainfall, considered by high-intensity storms and high frequency of drought [20], as shown in Figure 2.1.

Previously, extensive duration crop water harvesting has been used more frequently than water crop rainwater harvesting. Many authors define water harvesting and rainwater harvesting as interchangeable, as the collection and storage of any type of water from runoff or drainage for irrigation use. Ancient methods were commonly used in all agricultural purposes such that they were designed to meet the basic needs of water and technologies.

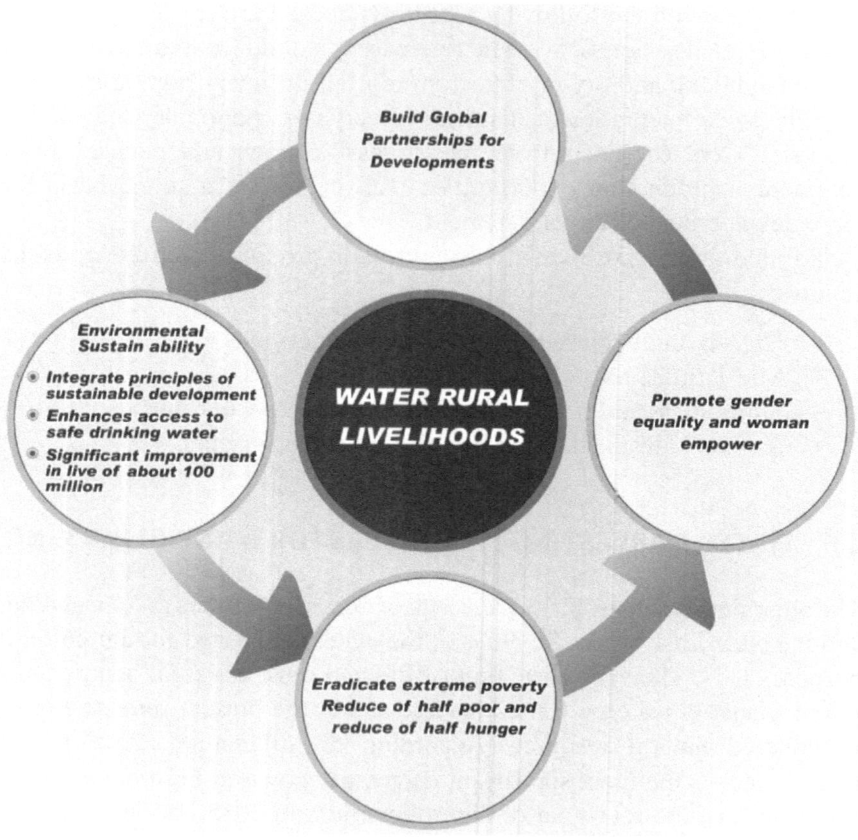


**FIGURE 2.1** Water is an important driver for achieving the millennium development goals.

In modern periods, in the Middle East and Southeast Asia, researchers have made test for ranging different technologies of rainwater harvesting for agricultural purposes, which includes farming, crop production, livestock water, and supply of water for fishes and ducks in ponds. Currently, an idea has been combined appropriate land management methods and in situ techniques that will enhance infiltration and minimize soil evaporation and runoff. This chapter will deal with comprehensive and modern term of rainwater harvesting and management '(RWHM), which are used in all methods of rainwater harvesting techniques [20], as shown in Table 2.1.

**TABLE 2.1** Description of Water Harvesting Techniques Implemented Worldwide [19]

| Water Harvesting Techniques | Methods | The Procedure for water harvesting |
|---|---|---|
| **Small storage in container** | Rain jar | Small tank (0.2 $m^3$–6 $m^3$), deposited water cut from the roof, made of reinforced cement. Water from the tap is drained from the tank. Larger tanks require a foundation. |
| | Ferro cement tank | Above ground, including a tap, capacity 3 $m^3$–15 $m^3$ water stored from a catchment area like a terrace. Made of ferro-cement consisting of a thin sheet of reinforced cement mortar with wire mesh and steel bars. It can be cast into any shape and can be manufactured by a semi-skilled worker. |
| | Stone masonry tank | The tank capacity is similar to that of rain jars and ferro cement tanks, but the structure is built using stone masonry and water-tight using cement mortar; Can be constructed by semi-skilled workers. |
| Larger storage in container | Open reservoir | Natural or (hand) open reservoir to dig water collected from elsewhere. The permeability of the pond can be reduced by using lining (concrete or plastic). Sizes vary from 30 $m^3$ (individual household use) to 20,000 $m^3$ (community use). Simple structures that can be built by non-trained workers. When the lining is used, some expertise is required. Water is extracted using a bucket, foot pump or motor pump. |
| | Cistern | Man-made underground reservoirs of various shapes and geometry, storing water collected from a surface plot, capacity 5 $m^3$ e100 $m^3$. Construction materials are concrete blocks or stone masonry for the walls and floors, cover materials can be corrugated iron plates, mortar or concrete. Water is drained using a hand pump or bucket. |
| **Small in-soil storage measures** | Contour trenches or ridges | Small trenches or ridges 1.5–3 m apart, following the contours of the landscape. The purpose of increasing infiltration. Requires less investment, less stability, annual maintenance. |
| | Terraces | Unit consisting of a relatively steeply faced structure across the slope (a riser, bund, bank, dyke, ridge, wall, embankment, etc.) supporting a relatively flat terrace bed. Aims to increase infiltration, storage in the soil and reduced erosion. A subdivision is based on the material used for the bund, as this influences the construction costs. |

**TABLE 2.1** *(Continued)*

| **Water Harvesting Techniques** | **Methods** | **The Procedure for water harvesting** |
|---|---|---|
| **Large in-soil storage measures** | Spate irrigation | Pre-planting, diversion of floodwaters (spate floods) from beds of ephemeral rivers by free intakes, by diversion spurs or by bunds, that are built across the river bed, to spread over large areas as irrigation water and to be partially stored in the soil. An uncertain method, as it is dependent on the occurrence of floods. The structure often needs to be rebuilt after a flood. |
| | Sub-surface dam | Dam built in a river bed of seasonal river. The dam is based on an impermeable layer to create an artificial aquifer, to be filled by intercepted groundwater. The dam can be constructed of clayey soil, stone masonry or concrete. Wells can be used to abstract the stored water from the aquifer. Instead of storing the water in surface reservoirs, water is stored underground. The main advantage is that evaporation losses are much less for water stored underground and the risk of contamination is reduced. |
| | Sand dams | Impermeable concrete or stone masonry structures constructed across seasonal rivers. Increasing water storage capacity, by enlarging the aquifer above the original river bed, through accumulation of sand and gravel particles against the dam. The sub-surface reservoir is recharged during flash floods and when the reservoir is filled surplus water passes the dam. The stored water is captured for use through digging a scooping hole, or constructing an ordinary well or tube well. By storing the water in the sand, it is protected against high evaporation losses and contamination. |

## 2.6 FACTORS AFFECTING THE RAINFED FARMING

The concepts and general indications on the significance of biotechnology and biological practices for production and sustainable agricultural methods given in this chapter are derivative from research in the world that are characterized. Temperature and soil moisture content in rainfed agriculture can be determined by the activity of soil microflora, micro, and macrofauna and population as shown in Figure 2.2. Carbon as a source of energy required by the soil microorganism; consequently, the main impact on biological methods and populations are carbon inputs through roots and shoots of the plants. The main reason is the carbon requirement in the concentration of carbon-enriched microsites such as detritusphere (decomposing residues) and rhizosphere (soil root). Organic carbon content in soil is very low in rainfed areas (> 60%) concentration of biological activity carbon in the rhizome near soil surface and crops residues [14]. The biological and structure-activity is beneficial to pathogenic microbiota of the plant. Benefit management is crop-specific and in the Mediterranean environment, biological function is highest and water-limited. During summers, drying and wetting is very frequent and potent towards stability of biological activity, especially under low carbon content as shown in Figure 2.3.

The characterization of plant-biota communications are:

- Barriers to root and shoot growth (such as root diseases caused by soil-borne pathogens may limit yield potential in cereal crops [26];
- Biological interactions linked with nutrient and plant health (such as plant nutrition and disease suppression).

In crop management system, it is successful and most essential plant performance. Such as mineralization-stabilization processes to availability of synchronize nutrient, plant requirement to loss of minimize nutrient through leaching researcher needed to understand. Likewise, optimization of suppression of disease in organisms by other soil biota there is needed to recognize the pathogen in roots-soil biota-plant interactions.

In the temporal dynamics and regulation of factors for controlling acute periods of biological activity must be understood to optimize crop management benefits. Such as during winter, limited non-symbiotic nitrogen fixation by lower temperature than soil in winter-rainy Mediterranean environments and in summers, nutrient mineralization is associated the stabilization and intermittent precipitation, it very slows accumulation of N in the soil. A conservation agricultural system occurs in southern Australian rain fields which includes retention and reduced tillage of crop residues.

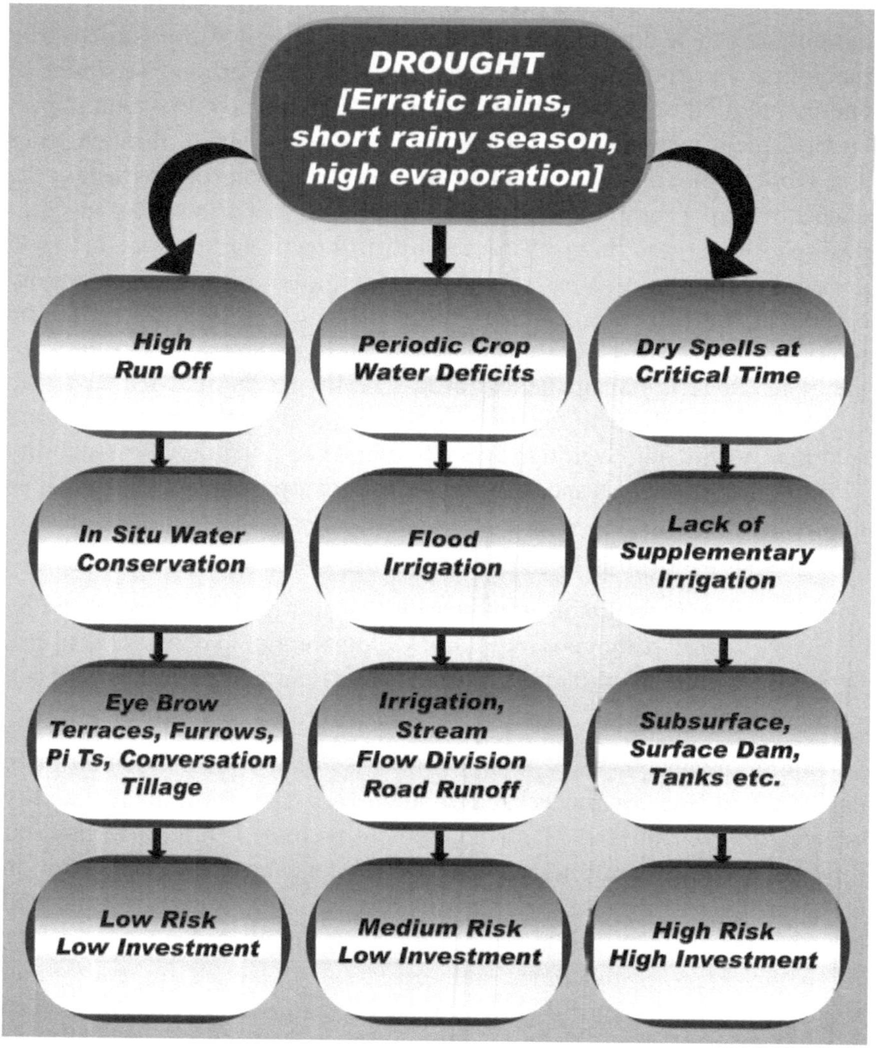


**FIGURE 2.2** Methods of dealing with arid conditions.

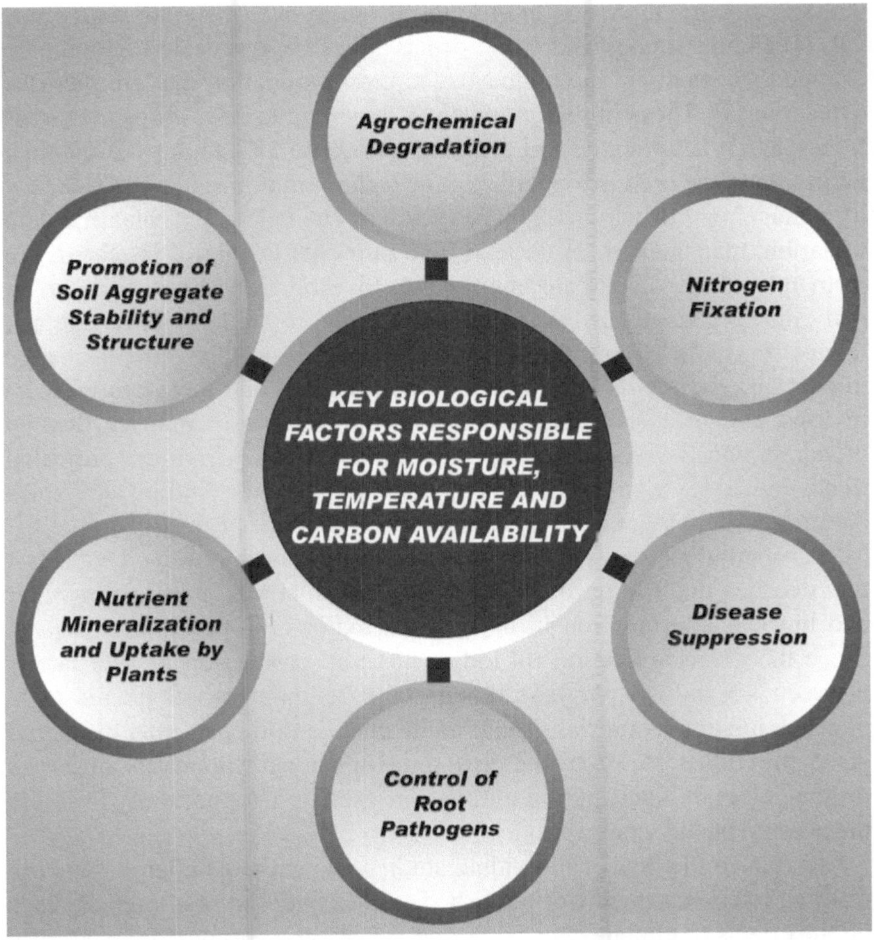


**FIGURE 2.3** Key biological factors responsible for optimal performance of moisture, temperature, and carbon availability.

## 2.7 TECHNOLOGY ADVANCEMENT AT GLOBAL LEVEL

World future food security concern with rainfall, rise in temperature and $CO_2$ changes which affects the crop's susceptibility. Agricultural impact has been evaluated to inform policy-makers, climate change has potent impact and adaptation of option to mitigate [1]. These evaluations use biochemical process-based models and statistical methods [42] to show climatic change impact on crop yield in future over different time intervals. Different studies and IPCC observations have revealed the climate change that results in the major loss in crop production that are increase over time [5]. These observations have shown up to 25% crops and areas losses, and it is estimated that these losses will be 50% high by 2080 alike harm trends are used irrespective of the technologies used [21]. Overview of management of various methods for water management, planting date variation, management of varieties and nutrients is helpful in recovering from the effects of climate change [2]. Investigations have also shown that climate change has started affecting crop production and its damage ranges from 1 to 10% in the event of climate change. The climate change impact on crop yields is estimated to be affected (and therefore weaker) in low- and middle-income tropical and developing countries than in moderate and developed countries. South Asia and Africa are projected to have yield losses of up to 25% by 2020 compared to Europe and North America, where there may be some production gains [19].

Substantial increases in crop productivity have been observed in the last few decades due to improvement in crop productivity, increase in water and nutrient consumption in the world's major crop-producing regions. Since the green revolution till today, the crop production growth rate is between 1.5 and 2% per year, though huge regional variations and yield plateaus in some countries. The climate change issue has been measured since the 1980s to 2020, the first time in impact valuations observed production in the same period of time, and there is an opportunity to assess these estimates by comparison.

To reduce the bias of individual studies, we gathered a large database from a systematic review of published studies over the past three decades on the effects of climate change on rice, wheat, and corn yields, especially in the period 2010–2020 and averaged them nationally. Some observations have been completed in the past, but emphasis has been placed on crop production analysis of climate change, different time intervals, and adaptation benefits.

### *2.7.1 RESEARCH ON RAINWATER HARVESTING IRRIGATION*

The research shown by the countries involved used a keyword check to identify preferences. A group of words has been extracted and grouped in a line from different topic. Groups of developed countries have established

joint relations which are documented here. Thus, some observations of Brazil, India, and China analysis in developed economies. The most concern for food safety and water availability for human consumption are India, United States of America (USA), Germany, Canada, and Australia. Groundwater degradation in areas where these types of resources are practically the only source is the most serious issue faced by Australia and India. Therefore, both countries are recovering, recharging, and managing the water bodies on their priority basis and RWHI is contributing positively. The development and innovation of rainwater harvesting and highly effective irrigation systems is an excellent step forward in developed countries and China. A combination of remote sensing and hydrological models has been used in the case of India. Chemistry is seen as the main subject in this country, which is mainly based on conservation of soil and poor water structure due to saltwater irrigation. China made an investigational study and analysis of different agricultural practices and irrigation methods. In United States, an investigation has shown various characteristics regarding economic and financial problems resulting from hurricane and irrigation projects. The main topics included in of South Africa are water-soil participatory processes (like infiltration and evaporation levels; modernization processes; agricultural adaptation). Meanwhile, Germany has most of the documentation on climate models. The main topic of observation on aspects of sustainability was conducted by Kenya and UK [34] shown in Figure 2.4.

## 2.8 CROPS SUITABLE FOR RAINFED AGRICULTURE

Farmers have two options for increasing crop production. First is the use of extensive systems (which will increase the area of plants) and secondly, is intensive farming (which will increase both the area of plants and production rate). To fulfill the need of food, in many rainfed areas, farmers must increase the production rate. These eco-sensitive zones are vulnerable to environmental degradation, especially due to overcrowded farming, soil erosion, and over-grazing. This issue is observed in different areas in Africa, where the use of extensive to intensive systems very rare than in other areas. Health and economic issues are occurring due to these environmental impacts, mainly in poor people who regularly live-in marginal regions.

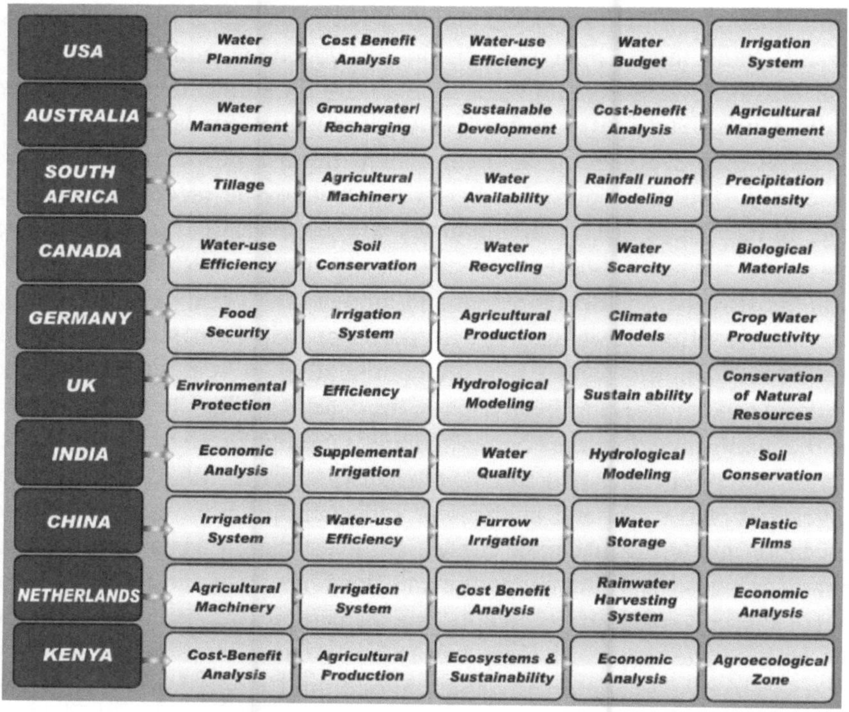


**FIGURE 2.4** Keywords for the most active countries in RWHI (research on rainwater harvesting irrigation) research.

The poor are severely affected by these problems compared to other groups of the population because they do not have sufficient assets to mitigate the effects of environmental impacts [29]. Environmental impacts can have far-reaching effects in poor communities through a reduction in agricultural yield potential, which may further exacerbate poverty, leading to malnutrition and health deterioration. An increase in the production by increasing the area planted in marginal areas may have more adverse effects on migrating people of these regions, as survival situations in more productive areas may become tighter. Because of this environmental status of field development, production of crop cultivation is a better measure than increasing the number of plants in rainfed farming. Investigating some condition, increase the production land in the irrigated areas can decrease the pressure of using marginal land for crop production and preservation natural resources from destruction [28]. But as given below, in different

areas in the world possibility of irrigational areas is limited. Consequently, intensive cropping system incorporating inputs such as manures, pesticides, labor, or better-quality varieties to increase production will be necessary for rainfed farming. In rainfed agriculture, crop growth can be increased by Sustainable intensification but limiting environmental issues as given in Figure 2.5.

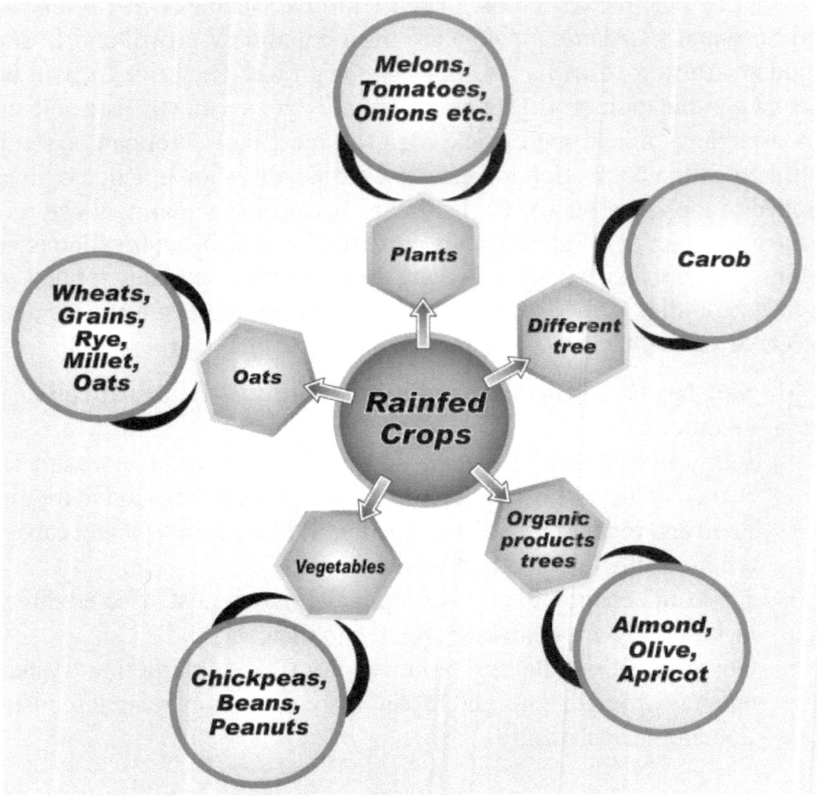


**FIGURE 2.5** Crops grown in rainfed agriculture.

There are three basic ways to increase rainfed farming through advanced crop production:

- Uses good water harvesting techniques to increase crop production;
- Uses agriculture research idea to increase crop production; and
- Renew agricultural policies and investment in rainfed areas.

Coarse grains (85%), pulses (83%), oilseeds (70%), and cotton (65%) are the main crop grown through rainfed agriculture. The single-crop farming methods are used in the long term is a basic protocol from October to June rather than an exception in the arid areas. In the means of protection and low risk, intercropping, and mixed farming are the most common practice. A large part of Vertisol is left in the semi-arid region during rainfall due to waterlogging and drainage problems. In the soil profile moisture content is stored by post-harvest crops. In central India, Chhola (*Cicer arietinum*), and Sorghum (*Sorghum bicolor*) are most commonly grown and in small amount safflower (*Carthamus tinctorius*) is grown. They can grow in both mutual combination or as sole crop. Wheat is grown mostly as a sole crop but sometimes mixed with chickpeas. The most basic cropping system is based on cotton cultivation in Vertisol. Cotton cultivation practice is always applied in the upper part of the landscape because soil quality is very good in the upper part of the slope. In the rainy season, Alfisols cultivation is very common except inside deep soil where dual cropping method is applied with excellent rainfall [26]. Cropping patterns changes are result in interactive effects of several factors which mainly classified in five groups [23]:

- Soil fertility, irrigation, and rainfall are the factors affecting the resources;
- Rainwater harvesting, seeds, and fertilizer are the factors not only related to technology but also to storage, processing, and marketing;
- Food and fodder independent need in addition to investment capacity are the factors related to household;
- Trade and economic policies input and output cost affected directly or indirectly are the factors related to the cost; and
- Government regulatory policies, farm size, marketing systems, tenancy arrangements and research are the factors related to institution and infrastructure.

## 2.9 RESEARCH ADVANCEMENTS ON RAINFED FARMING

Comparison of potential yield with actual is the most significant field in the future research and refinement model achieved by the analysis of the best farmers in the different regions of the countries. For example, in the rain-based regions of the United States do not good potential of rainfed farming. Obviously, these areas have highly benefited from crop diversity,

management, pesticides, and fertilizer rather than optimal climatic conditions. Other regions all over the world lead to the well-known conclusion of increase substantially production as socio-technically and socioeconomically improve conditions. This observation leads to three major points. Firstly, in many regions of the world of rainfed agriculture consider the potential of socio-technical and socioeconomic barriers to overcome. Secondly, more irrigation is needed for population growth in the world and increase with food production increase international trade. Thirdly, through small-scale rainwater harvesting systems will increase the production of food, partial irrigation will fulfill the water requirement of the crop [8]. The population of the world is expected to reach up to 9.1 billion in 2050, which is about 50% more as compared to 2000. The agriculture sector has an immense burden to fulfill the food demand of this population, competition of land, availability of water for non-food crops and urbanization or industrial development. Most of the production of crops is rainfed, and crop yield varies with the geographic locations of the world. Rainfed farming can also improve crop yield up to double or quadruple [6]. However, there are various gaps in the rainfed system, such as the insufficient fund for adopting yield-enhancing seed or crop techniques by farmers, lack of access to information, extension services and technical skills, deprived infrastructure, fragile institutions, and discouraging agricultural policies.

Changes in crop management techniques can also help close the yield gap. Plant breeding plays an important role in preventing yield gaps by adapting cultivars to local conditions and making them more resilient to biological (such as insects, diseases, viruses) and abiotic stresses (e.g., flooding). The first step is to target water as people without water face crop failure and hunger [11]. To meet the requirements of this goal, the productivity of rainfed farming systems must be increased. Efforts under this goal will focus on opportunities to improve the efficiency of rainfed farming to boost yields and incomes, especially in low-productivity areas.

Land and water productivity improvements measures which may include:

- Provide rainwater according to the crop requirement (conversation of soil and water, capturing of rainwater harvesting) and its irrigation uses, etc.;
- Loss of water by evaporation in agricultural water management;
- Improvement in crop varieties;
- Adoption of new technologies in the development of economic framework;

- Non-conventional forestry such as use of poor quality of water;
- Assessment of rainfall patterns.

Increasing land and water productivity in rainfed systems will promote technologies and practices, but also need to be supported by capacity building, financing, marketing systems, and substantial policies and institutional changes. Key stakeholders in the improvement of rainfed systems will include farmers, landowners, extension services in agriculture and rural development, local governments, regional/state governments, and federal governments. Costs for farmers will necessarily increase, but will be offset by increased yields, and thus total revenues will increase. Individual countries will have to assess their needs and develop strategies that are economically and technically feasible to address the constraints of the region. Social, cultural, and environmental issues surrounding the use of land and water in the area will also need to be considered. Individual countries will need to develop measurement tools to assess progress toward targets. Once these tools are developed, they can be used to determine the appropriate percentage increase compared to the 2005–2007 baseline [11].

## 2.10 CHALLENGES AND OPPORTUNITIES

Although many investigations of water harvesting approaches have yielded positive results, these techniques are widely applied in farming methods. Many journalists have mentioned the importance of socioeconomic status of farmers in the field of technology. Some plan is too costly for farmer so they required input for some resources. Besides, many farmers in arid or semi-arid regions do not enough manpower there to move huge amount of the land which are needed in huge rainwater harvesting systems. Further, organic farming methods having necessary concern in the areas where practice is applied. Such as, animal tillage project is required where farmers are using plow by hand methods. The primary decision in rainwater harvesting is risk concerns that are very essential at the farm level. Many farmers due to low financial condition basically in the regions water harvesting is used, the prediction is that before approving new technology government should exceed the implementation cost of it. At the investigational level, various rainwater harvesting techniques methods are working well, confirming the adoption of agriculture proves to be an additional task in broad-scale testing methods. Joint action difficulties can prove when demanding to water harvesting methods [25]. Meanwhile, many rainwater harvesting structures are very large and implement of considerable amounts of land and labor; these structures are created at the communal level. Difficulties development in this stage, though, unwilling farmers must give implement to the voluntary labor, and maintenance of water harvesting systems at communal level. According to Reij, Mulder, and Begemann [24], very few water harvesting projects at the farmers' family level have attempted to incorporate methods that can be implemented easily. Some researchers precisely point to concepts that are essential when trying to the promise widespread agriculture adoption. At the farm level, many factors are guarantee acceptance of various methods, including farmers' participation from the initial stages, the use of data collection and farmer's maintenance that can take ownership of the project. Further, the endowment of suitable educational and extension support to ensure farmers for early water harvesting techniques in the adoption procedure provides necessary knowledge to apply the chosen technology. It is important information about the advantages.

The environmental impacts of rainwater harvesting for farming should also be considered to determine whether to adopted certain technology. Water harvesting caused certain environmental damages which include soil erosion, sedimentation, salinization, and water-logging. Furthermore, excess use of rainwater harvesting methods will affect the use of water which depends upon the water supply in the crop production. According to Gowda [12], the present state of observation is about the seven aspects of rainwater harvesting, and additional irrigation are included in the following points:

- In semi-arid regions, organic tank is used for irrigation;
- Assessment of suitable tank water availability for crop growth stages;
- Optimization of design parameters and tank size;
- Supplemental irrigation water for proficient technique;
- Supplemental irrigation for Crop responses and Cost-effective valuation of runoff storage;
- Groundwater recharging, efficient water utilization and watershed-based water harvesting.

In rainfed agriculture, for enhanced productivity and resilience against tough the climate, soil is a very important component. Soil erosion, nutrient deficiencies, SOM depletion, soil compaction, and soil biodiversity loss are the main factors that involve in soil degradation. Thus, soil management systems must not only take into account the constraints associated with the farm conditions and topography but also consider production objectives, choice of crops, methods of cultivation, and stocking levels.

In rainfed farming systems, soil-related factors are major barriers to sustainable production, followed by water scarcity. Soil-related constraints refer to a situation where the soil environment is sub-optimal to produce high yields. Soil-related barriers can be physical, chemical, or biological. Two processes that lead to the loss of the soil's ability to perform its functions: those that change their physical, chemical, and biological properties (internal processes), and those that inhibit their use (external processes) for other reasons.

Soil chemical degradation refers to any undesirable changes in chemical properties (e.g., pH, magnitude, and ration exchange complex, SOM concentration, mineral nutrients, and composition of soluble salts). Changes in one or more of these properties often adversely affect the chemical quality of the soil, directly or indirectly, which can lead to a decrease in productivity [31].

### *2.10.1 SOIL QUALITY RESTORATION*

In rainfed and irrigation farming, the major cause of stagnation in agriculture production is regressive decline. For restoring the Soil quality RMP (recommended management practices) is applied. There RMPs are most important among rainfed agriculture include:

- To optimize the moisture content, timely plowing is needed to decrease the formation of large clods and improvement in the soil tilt;
- Decreasing the secondary tillage methods and adopting no-tillage methods;
- Legumes and cereals regular crop rotation;
- Rotation cycle of cover crops;
- Enhanced SOM concentration by using manure;
- In the crust surface, use a strong hoe for the improvement of seedling and crop stand.

The procedure of selecting RMP depends upon the type of soil and other condition-related factors. Alfisols identification is essential for dryland farming. According to Sharma et al. [30], the impacts of appropriate land management measures were developed to develop the Composite Land Quality Index (SQI) that is meaningful for the arid railway system, as given in Figure 2.6.

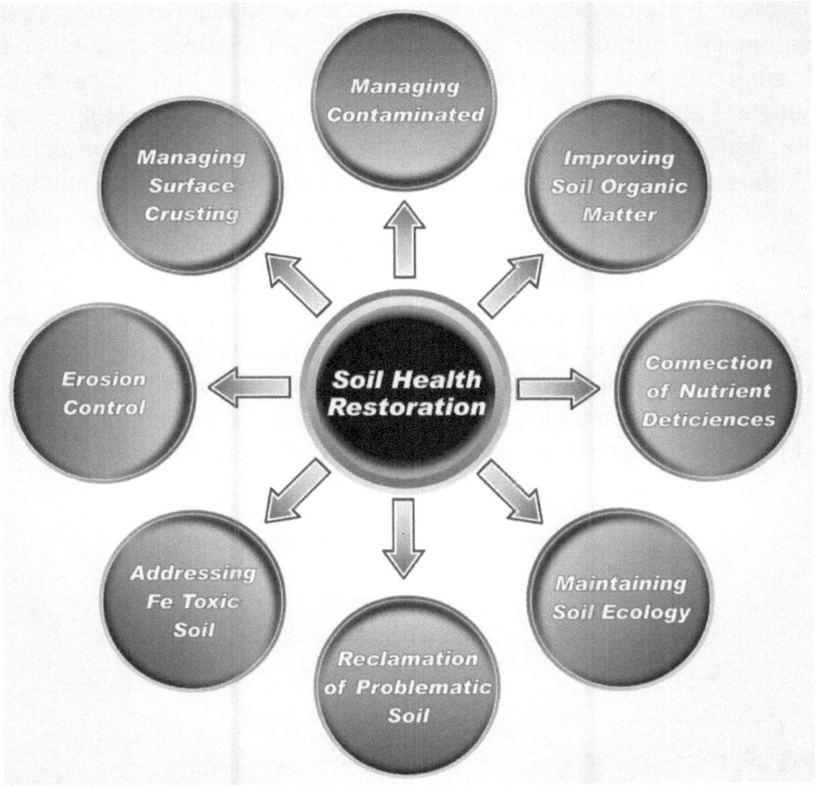


**FIGURE 2.6** Factors responsible for soil health restoration.

## 2.11 WATER HARVESTING AND MANAGEMENT

In the different stages of crop production, crops always suffer moisture stress due to rainfall distribution and spatial variability in the rainfed areas. Moreover, the agriculture sector is facing pressure on the availability of water resources and to meet the demand of water of different sectors, for example, household, drinking, industrial, and energy. Since rain is the biggest source of water and essential input for the agriculture system, effective rainwater harvesting methods should be applied for excellent rainwater farming. In arid and semi-arid areas, the approach of rainwater harvesting mainly involves selecting short-duration and low water requirement crops so that moisture stress can avoid by the crop during the period of growth. Furthermore, to in situ conservation, to meet critical irrigation requirements, surplus water is converted into storage structures, which can be utilized in both ways with groundwater (conjunction or stand-alone resource). In regions with moderately high rainfall, harvesting surplus water and increasing the intensity of the crop for rainwater conservation and lifesaving irrigation as much as possible, and reaping maximum benefits from the harvested water. Apart from increasing the availability of water in various ways, it also aims to arrest the losses associated with water use and achieve the highest input from each and every drop of stored water. The flagship program is the watershed management of the country used to increase the availability of water resources, which goal is to reduce drought, soil erosion, and flood, it optimizes the water, land, and vegetative uses, improve agriculture yield, and enhance the availability of fuel and feed on a progressive way [22] as given Figure 2.7.

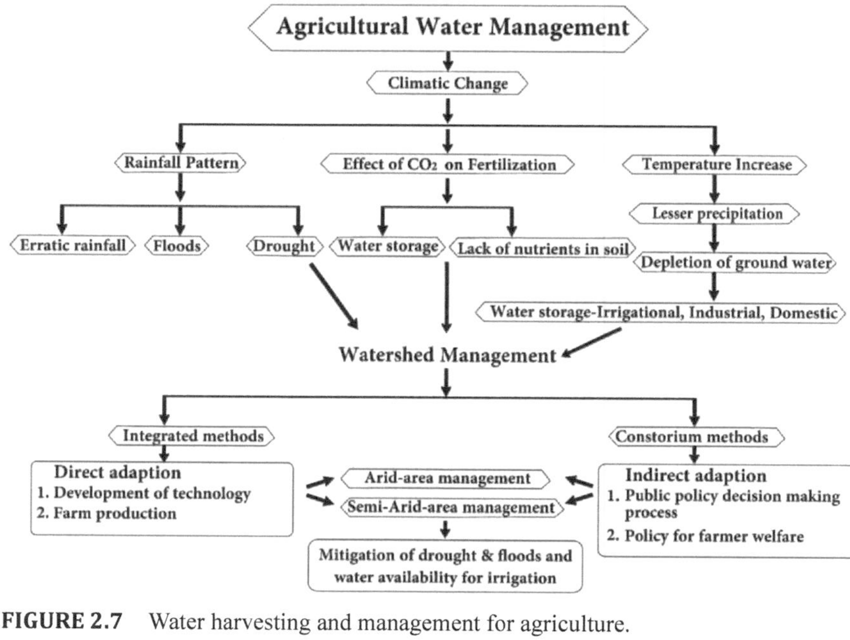


**FIGURE 2.7** Water harvesting and management for agriculture.

## 2.12 SUMMARY

Worldwide rainwater harvesting is the most promising method which is used to solve water scarcity issues. Depending on the dry conditions, micro-, and macro-water harvesting techniques are used for water storage in semi-arid and tropical regions. In agriculture, the production implementation of rainwater harvesting has a positive effect because it provides water to the crop growth stage, thus increasing production. It also advantageous in soil erosion and runoff velocity and, therefore, contribute in recharge of groundwater. Though, improper management, poor design, and lack of communication between engineers can lead to rainwater harvesting failure for farmers and government. The rainwater harvesting methods capacity to tolerate agricultural yield requirement of supporting other technologies, mainly information technology (IT). Management of soil and nutrient, in addition to the implementation, socioeconomic status of farmers, can be used to discover the victory of rainwater harvesting methods to increase local agricultural yield [24].

## ACKNOWLEDGMENTS

The authors are thankful to Dr. Anupam Srivastava for his expert guidance. The authors would like to thank Mr. Pramod Kulshreshtha and Mr. Sunil Kumar for designing such wonderful graphics of the present chapter. We would also like to express our sincere thanks to Mr. Gagan Kumar and Mr. Lalit Mohan Ji for their administrative support.

## KEYWORDS

- **information technology**
- **rainfed farming**
- **rainwater harvesting and management**
- **recommended management practices**
- **semi-arid areas**
- **water harvesting techniques**

## REFERENCES


1. Aggarwal, P., Vyas, S., Thornton, P., & Campbell, B., (2018). How much does climate change add to the challenge of feeding the planet this century? *Environmental Research Letters, 14*(4), 043001.
2. Aggarwal, P., Vyas, S., Thornton, P., Campbell, B. M., & Kropff, M., (2019). Importance of considering technology growth in impact assessments of climate change on agriculture. *Global Food Security, 23*, 41–48.
3. Anderson, J. R., (1998). Selected policy issues in international agricultural research: On striving for international public goods in an era of donor fatigue. *World Development, 26*(6), 1149–1162.
4. Barnett, B. J., Barrett, C. B., & Skees, J. R., (2008). Poverty traps and index-based risk transfer products. *World Development, 36*(10), 1766–1785.
5. Challinor, A. J., Watson, J., Lobell, D. B., Howden, S. M., Smith, D. R., & Chhetri, N., (2014). A meta-analysis of crop yield under climate change and adaptation. *Nature Climate Change, 4*, 287–291.
6. Comprehensive Assessment on Water Management in Agriculture. 2007; https://www.iwmi.cgiar.org/assessment/files_new/synthesis/Summary_SynthesisBook.pdf (accessed on April 29, 2021).
7. Dixon, J., Gulliver, A., & Gibbon, D., (2001). In: Hall, M., (ed.), *Farming Systems and Poverty Improving Farmers' Livelihoods in a Changing World* (pp. iii–412). FAO, World Bank, Washington DC.
8. Droogers, P., Seckler, D., & Makin, I., (2001). *Estimating the Potential of Rain-Fed Agriculture*. International Water Management Institute, 20. IWMI.
9. Duckham, A. N., & Masefield, G. B., (1970). *Farming Systems of the World*. Chatto & Windus, London.
10. FAO (Food and Agriculture Organization of the United Nations), (2006). *"Water for Food, Agriculture and Rural Livelihoods."* Chapter 7: In water: A shared responsibility. The United Nations World Water Development Report 2. United Nations Educational, Scientific and Cultural Organization (UNESCO), Paris.
11. *FAO (Food and Agriculture Organization of the United Nations): About Water Management*, (2020). http://www.fao.org/land-water/water/water-management/agriculture-water-management/en/ (accessed on 9 April 2021).
12. Gowda, C. L. L., Serraj, R., Srinivasan, G., Chauhan, Y. S., Reddy, B. V. S., Rai, K. N., & Upadhyaya, H. D., (2009). Opportunities for improving crop water productivity through genetic enhancement of dryland crops. *Rainfed Agriculture: Unlocking the Potential, 7*, 133–163.
13. Grigg, D. B., (1974). *The Agricultural Systems of the World: An Evolutionary Approach*. The Cambridge University Press, London.
14. Gupta, V. V., Rovira, A. D., & Roget, D. K., (2011). Principles and management of soil biological factors for sustainable rainfed farming systems. In: *Rainfed Farming Systems* (pp. 149–184).
15. Harrington, L., & Phil, T., (2011). Types of rainfed farming systems around the world. In: *Rainfed Farming Systems* (pp. 45–74). Springer, Dordrecht.
16. International Water Management Institute (IWMI), (2020). *Rainfed Summary*. https://www.iwmi.cgiar.org/issues/rainfed-agriculture/summary/ (accessed on 9 April 2021).

17. IPCC (Intergovernmental Panel on Climate Change), (2008). *Technical Paper on Climate Change and Water*. IPCC-XXVIII/Doc. 13. Geneva: IPCC Secretariat.
18. Irz, X., & Roe, T., (2000). Can the world feed itself? Some insights from growth theory. *Agrekon, 39*(3), 513–528.
19. Knox, J., Daccache, A., Hess, T., & Haro, D., (2016). Meta-analysis of climate impacts and uncertainty on crop yields in Europe. *Environmental Research Letters, 11*, 113004.
20. Lasage, R., & Verburg, P. H., (2015). Evaluation of small-scale water harvesting techniques for semi-arid environments. *Journal of Arid Environments, 118*, 48–57.
21. Lobell, D. B., & Asseng, S., (2017). Comparing estimates of climate change impacts from process-based and statistical crop models. *Environmental Research Letters, 12*, 015001.
22. Rao, C. S., Lal, R., Prasad, J. V., Gopinath, K. A., Singh, R., Jakkula, V. S., & Virmani, S. M., (2015). Potential and challenges of rainfed farming in India. In: *Advances in Agronomy, 133*, 113–181.
23. RBI, (2007). Reserve Bank of India (p. 11). College of Agricultural Banking Lecture Series, Cropping Patterns.
24. Reij, C., Mulder, P., & Begemann, L., (1988). *Water Harvesting for Agriculture*. World Bank Technical Paper Number 91. Washington, D.C.
25. Rosegrant, M. W., Cai, X., Cline, S. A., & Nakagawa, N., (2002). *The Role of Rainfed Agriculture in the Future of Global Food Production*, 581-2016-39482.
26. Rovira, A. D., & Ridge, E. H., (1983). *Soilborne Root Diseases in Wheat, Soils: An Australian View Point* (pp. 721–734).
27. Scheierling, S. M., Critchley, W. R. S., Wunder, S., & Hansen, J. W., (2012). *Improving Water Management in Rainfed Agriculture: Issues and Options in Water-Constrained Production Systems: Water Paper, Water Anchor*. The World Bank, Washington DC Water harvesting for crop production in Sub-Saharan Africa, 31.
28. Scherr, S. J., (2000). A downward spiral? Research evidence on the relationship between poverty and natural resource degradation. *Food Policy, 25*(4), 479–498.
29. Shahid, S. A., & Al-Shankiti, A., (2013). Sustainable food production in marginal lands- case of GDLA member countries. *International Soil and Water Conservation Research, 1*(1), 24–38.
30. Sharma, K. L., Mandal, B., Venkateswarlu, B., Abrol, V., & Sharma, P., (2012). Soil quality and productivity improvement under rainfed conditions-Indian perspectives. *Resource Management for Sustainable Agriculture*, 203–230.
31. Sharma, K. L., Mandal, U. K., Srinivas, K., Vittal, K. P. R., Mandal, B., Kusuma, G. J., & Ramesh, V., (2005). Long-term soil management effects on crop yields and soil quality in a dryland alfisol. *Soil Tillage Research, 83*(2), 246e259.
32. Sparks, D. L., (2012). *Advances in Agronomy*. Academic Press.
33. Thirtle, C., Beyers, L., Lin, L., Mckenzie-Hill, V., Irz, X., Wiggins, S., & Piesse, J., (2002). *The Impacts of Changes in Agricultural Productivity on the Incidence of Poverty in Developing Countries*. DFID Report No. 7946. Department for International Development (DFID), London, UK.
34. Velasco-Muñoz, J. F., Aznar-Sánchez, J. A., Batlles-delaFuente, A., & Fidelibus, M. D., (2019). Rainwater harvesting for agricultural irrigation: An analysis of global research. *Water, 11*(7), 1320.

35. Wani, S. P., Rockström, J., & Oweis, T. Y., (2009). *Rainfed Agriculture: Unlocking the Potential* (p. 7). CABI.
36. World Bank, (2005). *Agricultural Growth for the Poor: An Agenda for Development.* The International Bank for Reconstruction and Development. The World Bank, Washington, DC.
37. World Bank, (2006). *Reengaging in Agricultural Water Management: Challenges and Opportunities.* Directions in Development. Washington, DC: World Bank.
38. World Bank, (2007). *World Development Report 2008: Agriculture for Development.* Washington, DC: World Bank.
39. World Bank, (2008a). *Strategic Framework for the World Bank Group on Development and Climate Change*. Washington, DC: World Bank.
40. World Bank, (2008b). *Double Jeopardy: Responding to High Food and Fuel Prices.* G8 Hokkaido-Toyako Summit. Washington, DC: World Bank.
41. World Bank, (2008c). *Rising Food Prices: Policy Options and the World Bank Response*. Policy Note. Washington, DC: World Bank.
42. Zhao, C., Liu, B., Piao, S., Wang, X., Lobell, D. B., Huang, Y., Huang, M., et al., (2017). temperature increase reduces global yields of major crops in four independent estimates. *Proceedings of the National Academy of Sciences, 29, 114*(35), 9326–9331.

## CHAPTER 3

# Sustainable Irrigation Under Global Context: Surface, Subsurface, Surface Micro, Subsurface Micro, and Ultra Irrigation*

MEGH R. GOYAL


### ABSTRACT

Several strategies including conventional and organic farming approaches have been implemented worldwide to ensure healthy and sufficient food for one and all. Sustainable and efficient use of irrigation practices stands up to the utmost priority to meet the growing demand of agriculture produce. The present chapter deals with different irrigation methods both used on farms including organic farming. surface, sub-surface, surface-micro, subsurface-micro and ultra-irrigation practices which have been significantly implemented. Pros and cons associated with usage these irrigation practices have been briefly discussed in order to enhance their implementation and ultimately to benefit the society.


## 3.1 INTRODUCTION

The art of irrigation is very ancient and has been essential for the development and growth of some civilizations. The Code of Hammurabi, 6th King of the First Dynasty of Babylonia, indicates the importance of irrigated land. II Kings 3: 16–17 of the Holy Bible alludes to irrigation, in 2000 B.C. In the same year, the queen of Assyria diverted the Nile River to irrigate the Egyptian desert. The channels that they constructed during her kingdom still work. Irrigation is also mentioned in the old documents of Syria, Persia, India, China, Java, and Italy. The importance of irrigation in

*Modified and with permission from, Goyal, Megh R. Management of Drip/ Trickle or Micro Irrigation. Chapter 4: Irrigation systems; Oakville - ON: Apple Academic Press Inc.; 2013; pages 71–132.

our times has been defined accurately by N. D. Gulati: "*In many countries, irrigation is an old art, as much as the civilization, but for humanity, it is a science, the one to survive* [3]." Irrigation water is an important component in the hydrologic cycle (Figures 3.1).

| | The irrigation is applied to achieve the following objectives: |
|---|---|
|  | 1. To provide the necessary moisture for crop development.<br>2. To ensure a sufficient supply of water during droughts of short duration and unpredictable climate.<br>3. To dissolve soil salts.<br>4. It is a way to apply agrochemicals.<br>5. To improve the ambient conditions for vegetative growth.<br>6. To activate certain chemical agents.<br>7. To generate operational benefits. |

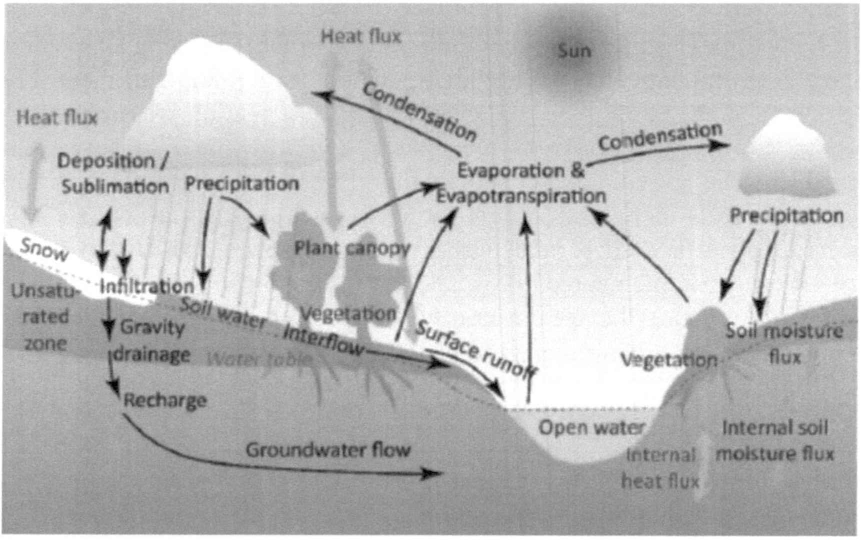


**FIGURE 3.1** Irrigation as a component in the hydrological cycle and water balance budget; modified from [10, p.73].

## 3.2 SELECTION OF IRRIGATION SYSTEMS

The irrigation can be accomplished by different methods: surface, subsurface, sprinkler, and drip [2]. With any of these means, it is necessary to select an irrigation system (Figure 3.2), before the design, the equipment specification and installation can proceed. In order to make a suitable selection of the system, one should carefully consider the capacities and limitations of all the potential alternatives. Tables 3.1 and 3.2 indicate the aspects that must be considered in the selection of an irrigation system. These aspects are listed as follows [9, 10]:

1. **The Crop and Related Factors:** Type, root depth, water consumption, development of diseases.
2. **Soil Characteristics:** Texture and structure, depth, and uniformity, infiltration rate, potential of erosion, salinity, and internal drainage, topography, and degree of irregularity.
3. **Water Supply:** Source, quantity available and reliability, quality, solids in suspension, chemical analysis.
4. Value and availability of land.
5. Limitations and obstructions of flooding.
6. **Phreatic Level:** Level of underground aquifer.
7. **Climatic Conditions:** Flood and furrow systems are usually associated with long irrigation cycles (many days between irrigations), so while the evaporation for the first few days after wetting may be high, the soil surface soon reaches "air dry" condition, and evaporating for the remaining days of the cycle is essentially zero—the average evaporation over the entire cycle is often low.
8. Availability and reliability of energy.
9. **Economic Considerations:** Investment of required capital, availability of credit and interest rate, durability of the equipment and annualized cost, inflation costs, crop yield and value.
10. Available technology.
11. **Social Considerations:** Political and legal issues, cooperation of the habitants, availability, and reliability of the manual labor, level of knowledge and specialization of the manual labor, expectations of the government and inhabitants, desirable level of automation, potential damage by vandalism, and health issues.

**TABLE 3.1** Conditions for the Selection of an Irrigation Method [10]

| Conditions | Irrigation Methods | | | |
|---|---|---|---|---|
| | **Flood** | **Furrow** | **Sprinkler** | **Drip/Micro** |
| Topography | Moderate to irregular | Moderate | Irregular | Irregular |
| Soil permeability | Good | Good | Excessive to good | Excessive to good |
| Potential for erosion | High | High | Low | Low |
| Crop characteristics | Sown by broadcast | Sown in rows | Crop value variable | Crop value variable |
| Water flow requirements | High | High | Moderate | Low |

**TABLE 3.2** Parameters for Comparison of Irrigation Methods [10]

| Parameters | Flood | Furrow | Sprinkler | Micro |
|---|---|---|---|---|
| 1. Evaporation loss | Low | Low | Medium | Minimum |
| 2. Wetting of the foliage | High | Medium | High | Minimum |
| 3. Water consumption by weeds | High | High | High | Minimum |
| 4. Surface drainage | High | High | Medium | Minimum |
| 5. Control of irrigation depth | Minimum | Minimum | Medium | High |
| 6. Crop yield per unit of applied water | Minimum | Minimum | Medium | High |
| 7. Uniformity in the crop yield | Little | Medium | Medium | High |
| 8. Soil aeration | Minimum | Little | Little | High |
| 9. Interference of other tasks by the irrigation method | Low | Low | High | Low |
| 10. Application of fertilizers and pesticides through the irrigation water | Minimum | Minimum | Moderate | High |
| 11. Operation and labor cost | Low | Low | Moderate | High |
| 12. Leveling of the land is required | High | High | Low | Minimum |
| 13. Automation of the system | Low | Low | High | High |
| 14. Energy requirements | Low | Low | High | High |
| 15. Quality of water | Minimum | Minimum | Moderate | High |
| 16. Use of filters | Minimum | Minimum | Moderate | High |
| 17. Control of diseases and pests | Minimum | Minimum | Moderate | High |

## *3.2.1 CAPACITIES AND LIMITATIONS [4]*

### *3.2.1.1 CROP, SOIL, AND TOPOGRAPHY*

1. **Surface Irrigation or Irrigation by Infiltration:** There are different methods of surface irrigation for almost every crop. In some regions, hay crops are irrigated by the furrow method. The flood irrigation method is adequate for forage crops after an adequate leveling. Row crops are irrigated by the furrow method.

   The irrigation system works better when the soil is uniform because of good infiltration of the water. The length of a furrow is limited to 100 meters in a heavy textured soil. However, it can go up to 400 meters in well-textured soils. Furrow irrigation is adaptable in all soil types; however, soils with fast or slow infiltration rates may need an excessive manual labor. Uniform and slopes of slight incline adapt better to the infiltration method of irrigation. The irregular topography and steep slopes increase the cost of leveling and reduce the length of a furrow. The deep plowing can affect soil productivity requiring a special fertilization. Steps, benches, or terraces may be required to control erosion in the case of high flow rates.
2. **Sprinkler Irrigation:** Almost every crop and soil can be irrigated with some type of sprinkler irrigation system, although the crop characteristics and crop height must be considered when the system is selected. Sometimes the sprinklers are used to germinate the seed and to cover soil. For this purpose, the light and frequent applications can be easily accomplished with some type of sprinkler system.

   Soils with infiltration rates less than 5 mm/h may require special considerations. Sprinklers can be used in soils with depths too shallow to permit effective surface irrigation. In general, the sprinklers can be used in any topography that can be cultivated. Generally, the leveling of the land is not required. The sprinkler system is designed (with some exceptions) so that it can apply to water at lower intensity than the infiltration rate of the soil in order that the quantity of water infiltrated at any point depends on the intensity of the application and time of application and not on the infiltration rate of the soil.

3. **Drip Irrigation:** Drip/micro or trickle irrigation is more convenient for vineyards, tree orchards and row crops. The principal limitation is the high initial cost of the system that can be very high for crops with very narrow planting distance. Forage crops cannot be irrigated economically with drip irrigation. Drip irrigation is adaptable for almost all soils. In very fine-textured soils, the intensity of water application can cause problems of aeration. In heavy soils, the lateral movement to the water is limited, thus more emitters per plant are needed to wet the desired area. With adequate design, use of pressure compensating drippers and pressure regulating valves, drip irrigation can be adapted to almost any topography. In some areas, drip irrigation is used successfully on steep slopes.

### *3.2.1.2 QUANTITY AND QUALITY OF WATER*

1. **Surface Irrigation:** It is important that the water applied for irrigation reaches the end of the field relatively quickly to obtain a uniform irrigation. The volume flow rate required varies from 15 to 300 lps. The furrow method is not well suited for leaching salts from the soil, since water cannot stay in the soil for a required time. However, flood irrigation with border strips is ideal for this situation.
2. **Sprinkler Irrigation:** It adapts particularly well to a situation with the high water table, since the sprinkler equipment can make the application with a determined volume of water. Sprinkler irrigation generally requires the application of a smaller quantity of water than flood irrigation.
3. **Drip Irrigation:** With this, the application of water is less intense during a period of time longer than in other methods of irrigation. The most economic design would utilize water flowing throughout the crop area during almost all the day, every day, during peak periods of water use. If the water is not available constantly, then it may be necessary to store water. Saline water can be used successfully, with special precautions. The salts tend to concentrate in the perimeter of the wet volume of soil. For longer time intervals between the irrigations, the movement of the water in the soil can be reversed, returning the salts to the root zone. When it rains after

a period of accumulation of salts, the normal irrigation must be continued until approximately 50 mm of rain have fallen to prevent the damage by salts. In the arid regions, where annual rain is insufficient (<300 to 400 mm) to leach the salts, the artificial leaching of salts may need more time, requiring the use of a complimentary sprinkler or surface irrigation method.

Sand (hydro cyclone) filters and screen filters are used to remove suspended solids in the irrigation water. A media filter can remove organic clogging agents. Chemical treatment of the water may be required to control biological activity, to adjust the pH or to avoid the chemical precipitation that can obstruct the emitters. Suitable design and periodic maintenance of the system for the treatment of the water are vital for the successful use of the drip irrigation. Drip irrigation can operate at low operating pressures and at low discharge rates. For more details, read Ref. [10].

#### *3.2.1.3 CROP YIELD*

1. **Surface Irrigation:** High potential crop yield is possible with surface irrigation, especially with the border strip method. The yields must be 80–90% in the design for the storage systems, except with soils having a very high infiltration capacity. Reasonable irrigation efficiency (IE) for border strip varies from 70 to 85% compared to 65–75% for furrow irrigation. One must take into account factors such as the runoff and drainage water characteristics.

   The engineer and the operator can control many of the factors that affect the IE. However, the potential uniformity of the water application in a surface irrigation is limited by the variation of the soil properties, mainly the infiltration capacity. Studies indicate that relatively uniform soils can have a uniformity of 80% in a single irrigation. Researchers have suggested that the calculations of uniformity of the surface irrigation based on the infiltration rate need to be reduced by 5%–10% due to variation of the soil properties.
2. **Sprinkler Irrigation:** The sprinkler irrigation system can give performance as tabulated below:

| | |
|---|---|
| Manual or portable | 65–75% |
| Lateral on wheels | 60–70% |
| Centre pivot | 75–90% |
| Linear move | 75–90% |
| Solid or permanent set | 75–80% |

3. **Drip Irrigation:** Properly designed and maintained drip irrigation is able to give high performance. The design efficiency can vary from 90 to 95%. With reasonable care and maintenance, the efficiencies of systems operation can range from 80 to 90%. Where obstruction is a problem, the emitter flow efficiency can be as low as 60%.

### *3.2.2 CONSIDERATIONS OF HAND LABOR AND ENERGY [7, 8]*

#### *3.2.2.1 SURFACE IRRIGATION*

The flood irrigation requires minimum use of manual labor compared to other surface methods. Furrow irrigation can be automated to a certain degree. To reduce the requirements for manual labor input, furrow irrigation requires expertise to obtain high performance. The need for this expertise can be reduced with the use of higher cost equipment. Putting siphons or tracks are useful to measure the desired flow rate. The surface irrigation method requires little or no energy, to distribute the water to the entire field. However, the energy is required for taking the water to the field, and for pumping the underground water. In some cases, these energy costs can be substantial, particularly with a low efficiency of water application. Some manual labor and energy for leveling and preparing the land are always needed.

#### *3.2.2.2 SPRINKLER IRRIGATION*

The requirement for manual labor varies depending on the degree of automation and mechanization for the equipment to be used. Manual systems require the least degree of ability, but more hours of labor. On the other end, the center pivot and linear move systems require considerable operational ability and fewer hours of labor. The energy consumption related to the requirements of operating pressure varies considerably among all methods of sprinkler irrigation. The traveling sprinkler system with progressive movement requires seven bars or more of pressure. Other systems can use of two to five bars, depending on the design of the sprinklers and nozzles.

#### *3.2.2.3 DRIP IRRIGATION*

Due to characteristics of low flow and relatively long irrigation set times, drip irrigation can be automated totally [10]. Therefore, manual labor is needed only for inspection and maintenance of the system and the initial installation. The manual labor requirements for maintenance are related to the sensitivity of the emitters to obstructions and the quality of the irrigation water. In a vineyard, for example, a worker can inspect and maintain approximately 20 ha per day.

Generally, the drip irrigation method requires less energy than other pressurized irrigation systems. The operating pressure ranges from 0.5 to 1.5 bars. The pressure of the system fluctuates from approximately 2 (small flat land systems) to 4 bars (steep slopes and uneven lands).

### ***3.2.3 ECONOMIC CONSIDERATIONS [5, 8, 9]***

#### *3.2.3.1 SURFACE IRRIGATION*

An important cost for surface irrigation is to level the land. The cost is directly related to the volume of earth that must be removed, the length and size of the border strips. The typical volumes of earth to be removed are in the order of 800 cubic meters per ha. For excessive costs, the design must be economical. The typical cost of the earth to be removed is $0.65 per $m^3$. The final, finished leveling with drag-scrapers controlled by laser after removing the earth will cost approximately $110 per ha. Finishing touches on previously leveled soil by laser will cost approximately $50.00 per ha.

Low pressure buried plastic pipes or concrete pipes for low flows, can approximately cost the double of the concrete ditches, and up to 5–10 times more for higher flow rates. A compact land dam of medium sizeable to store a water volume for 24 hours costs $250/ha for the small farm (<15 ha). For a larger farm, the price can be lowered up to $125/ha. A weir dam can cost two to five times more.

### 3.2.3.2 SPRINKLER IRRIGATION

Table 3.3 summarizes the factors that affect the cost of the sprinkler irrigation system. The capital cost depends on the type of the system and size of the area to be irrigated.

**TABLE 3.3** Cost of a Sprinkler Irrigation System [10]

| Type of System | Field Size | Capital Cost | Energy Use | Manual Labor | Maintenance Factor |
|---|---|---|---|---|---|
| | ha | $/ha | kvh/ha/mm | Hrs/ha/100 mm | ** |
| 1. Manual or portable | 65 | 175–250 | 0.86–2.05 | 1.65 | 0.020 |
| 2. Lateral on wheels | 65 | 740–1000 | 0.86–2.05 | 1.17 | 0.025 |
| 3. Linear move with progressive movement | 32 | 865–1100 | 3.42–4.79 | 0.68 | 0.060 |
| 4. Centre pivot (without system for corners) | 55 | 620–1000 | 0.86–2.23 | 0.09 | 0.050 |
| 5. Centre pivot (with system for corners) | 60 | 865–1100 | 0.96–2.33 | 0.09 | 0.060 |
| 6. Linear move or trench alignment | 130 | 1000–1200 | 0.86–2.23 | 0.19 | 0.060 |
| 7. Linear move or fed by hose | 130 | 1500–1850 | 1.20–2.57 | 0.19 | 0.060 |
| 8. Solid set | 65 | 2500–3000 | 0.86–2.05 | 0.97 | 0.020 |
| 9. Permanent | 65 | 2100–3000 | 0.86–2.05 | 0.09 | 0.010 |

**The maintenance cost is calculated by multiplying the capital cost by the maintenance factor.

The typical costs of investment (Table 3.3) assume that the water is available at the ground surface and near the field. The costs are necessary for the main pipe and the pumping unit. The energy costs vary widely from place to place. The energy requirements in Table 3.3 can be used to calculate the costs of adequate unitary energy. A pump efficiency of 75%

has been assumed. The units of energy are kilowatts per hour per ha per mm (gross) of applied water. The operational cost (manual labor) varies according to the type of system and the local cost for manual labor. Table 3.3 gives typical values per hour of manual labor that is required per ha per mm of applied water (gross).

It is difficult to predict the maintenance costs, but the numbers in Table 3.3 serve as a guide. The annual maintenance cost is calculated by multiplying the initial capital cost of the system by a maintenance factor.

#### *3.2.3.3 DRIP IRRIGATION*

The cost of the drip irrigation system can vary widely, depending on the crop characteristics and the type of drip laterals (disposable or permanent tubes). The cost of drip irrigation is about $2200/ha for vegetable crops with wider row spacing. For vineyards with narrow row spacing, the costs can be about $3400/ha. For vegetable crops with narrow row spacing (tomato, pepper, etc.), the cost can vary from $2900 to $4900/ha. For vegetable crops with disposable drip lines, the cost of replacing the disposable lateral lines can vary from $340 to $450/ha during the year.

These estimations are for a system of high quality and include pumps, filters, controls, the network of main lines and emitters. In situations where there is more than one basic pump, sufficient equipment for filtration and control, costs vary from 20 to 25% less of the calculated expenses. The typical expenses for operation and maintenance of a drip irrigation system vary widely depending on the local conditions and the IE. It is advisable to calculate the cost of maintenance and operation ($ by ha per year) as a fraction of the initial cost:

| Cost | Maintenance Factor |
|---|---|
| Manual labor | 0.0015 |
| Energy | 0.03–0.07 |
| Water | 0.04–0.06 |
| Maintenance | 0.01 |
| Taxes and Insurances | 0.02 |

*Depends on the performance of the system.

## 3.3 IRRIGATION WATER MANAGEMENT: PRINCIPLES AND TERMINOLOGIES

### 3.3.1 SOIL WATER BALANCE (FIGURE 3.2)

It is a process of keeping the record of the inputs, outputs, and water storage in the root zone based on the Law of Mass Conservation (or Continuity equation):

$$\Delta S = I + R - ET_c - P_d - RO + U$$

where; *storage (S)* is the maximum amount of water that soil can store within it. Total volume equals the water and soil particles with no air spaces left in the soil column. Drainage occurs under the force of gravity until the equilibrium achieves. When additional water level rises from the saturation point, then flooding occurs; *effective irrigation (I)* is the amount of irrigation that reaches down to the root zone. In an ideal scenario, it is equal to the net irrigation; *net irrigation* ($I_N$) is the quantity of irrigation, required to fulfill requirements of plant water; *total rainfall (R)* is the total amount of precipitation received in a specific area and time. It is also measured by the volume measured with a gauge; *net rainfall* ($R_N$) is the amount of rainfall which reaches a watercourse canal or the concentration point as a direct surface flow; *effective rainfall* ($R_E$) is the amount of rainfall which is incorporated and become the part of soil water reserves. From total daily rainfall < 5 mm in dry spell will not be taken as effective, because this quantity of rainfall will be evaporated from the surface before wetting the soil surface. At field capacity (FC) or above, the value of effective rainfall is zero. If the soil is below FC level, then $R_e$ would bring back the level of water in the soil to FC.; *up-flux from shallow groundwater table (U)* is amount of water which moves upward due to shallow groundwater table; *surface runoff (RO)* is the amount of water that fails to enter in the soil. It occurs when the irrigation or rainfall rate is greater than the infiltration rate of the soil (Figure 3.2).

Where; *crop evapotranspiration* ($ET_C$) is specific to the crop which is considered for irrigation and is quantified as [= ($ET_0$) x ($K_C$)]; *evaporation* is the removal of water from the surface due to energy source such as solar radiation or wind; *transpiration* is the removal of water from the plant surface, and it regulates the temperature of plant by producing the cooling effect; *evapotranspiration (ET)* is used for total water transferred from the root zone by evaporation and transpiration from the plants into the atmosphere; *reference evapotranspiration ($ET_O$)* is from the reference surface, which has sufficient amount of water and it is calculated as function of *weather variables*; *crop coefficient ($K_C$)* is an average ratio of ($ET_C$) to ($ET_O$). It varies with plant type, soil type, and weather:

$$ET_c = K_c \times ET_o$$

where; *deep percolation ($P_d$)* is the movement of water outside of the root zone. It can also be referred as leaching because it moves deeper into the soil.

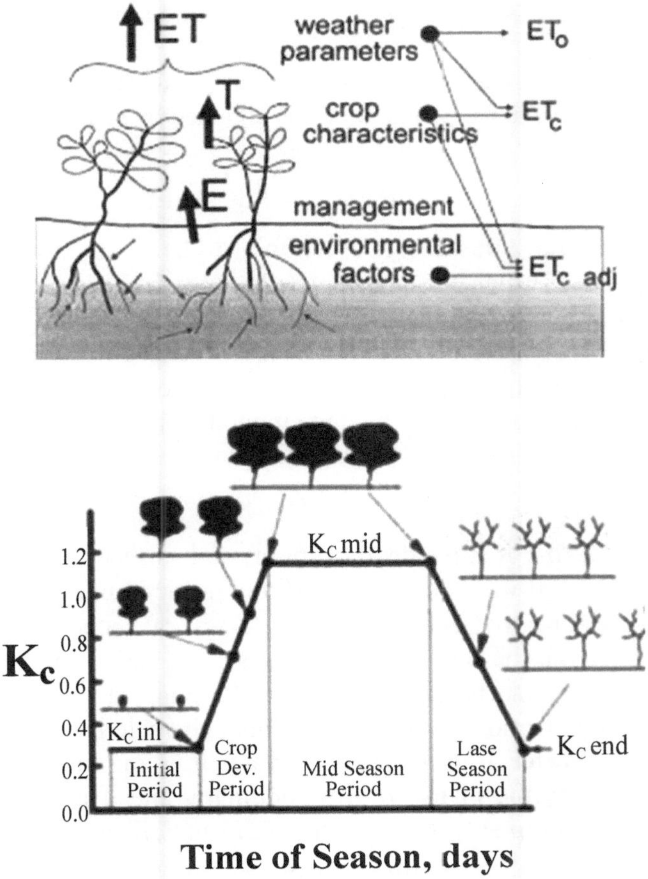


**FIGURE 3.2** Soil water balance; modified from [10, p.32 and 49].

### 3.3.2 TERMINOLOGIES RELATED TO IRRIGATION

1. **Infiltration Rate:** It is the speed of water that enters in the soil. It greatly varies with the type of soil, amount of water already in the soil and slope.
2. **Field Capacity (FC):** When a field is fully irrigated, some water retains within it without drainage due to gravity. When this water remains in water up to 48 hours after saturation, this level of water is called FC of soil. It varies with the type of soil.
3. **Permanent Wilting Point (PWP):** It is the least water amount that remains in the soil and is not available to the plants. Before attaining this point, permanent damage starts occurring in the plants.
4. **Plant Available Water (PAW):** It is the total amount of water which is available to the plants. Its value obtained by subtracting the value of FC and permanent wilting point:

$$PAW = FC - PWP$$

5. **Maximum Allowable Depletion (MAD):** This is the available portion of water which is depleted before the need of irrigation. It is considered as the initiating point.
6. **Readily Available Water (RAW):** It is the easily available portion of water to the plants when they exposed to minimal stress. Irrigation scheduling should be done immediately when the RAW level starts depleting. It can be calculated by:

$$RAW = MAD \times PAW$$

7. **Root Zone:** It is the depth from the soil surface to the deepest of the plant. It varies in agronomic and horticultural crops, while in former one root zone is only up to several inches which grows throughout the season to up to the several feet. Amount of water (no units) can be calculated by multiplying the root zone depth with the fraction of soil water-holding characteristics.
8. **Plant Water Requirement (PWR):** It is the amount of water needed by the plants to grow optimally and it is the difference between crop evapotranspiration (ET) and effective rainfall:

$$PWR = ET_c - R_E$$

9. **Irrigation Efficiency (IE):** It is responsible for the uniform and efficiency of irrigation system. When IE is low, more water in the form of irrigation will be needed to supply to the plants that can reach to its root zone. Generally, for furrow irrigation, IE ranges between 30 and 70%.
10. **Gross Irrigation ($I_G$):** It is the total amount of water which is needed so that water can reach down to the root zone. It increases as IE decreases.
11. **Gravimetric Water Content:** It is calculated by measuring the weight of the water compared to the weight of soil.
12. **Volumetric Water-Content:** It is water volume, compared with total volume represented as soil, water, and air. This calculation method gives soil water content in volume of water per volume of the soil column. Another way to determine it is by multiplying the gravimetric with bulk density.

$$Volumetric\ Water\ Content = Gravimetric\ water\ content \times Soil\ bulk\ density.$$

13. **Virtual Water:** It is calculated by the equation: $VWR\,[n, c, t] = cp\,[n, c, t] \times SWD\,[n, c]$, where: $VWR$ = virtual water requirement ($m^3$ $year^{-1}$) of the state $n$ for a crop $c$ in year $t$; $CP$ = crop production (ton $year^{-1}$) of the country $n$ in year $t$ of crop $c$; $SWD$ = specific water demand ($m^3$ $ton^{-1}$) of crop $c$ in state. Moreover, specific water demands are calculated by: $SWD\,[n, c] = CWR\,[n, c] / CY[n, c]$, where: $SWD$ = specific water demand; $GSWD$ = gross specific water demand; $n$ = country; $c$ = crop; $CWR$ = crop water requirement; and $CY$ = crop yield.
14. **Water Use Efficiency:** It is the crop yield per unit use of irrigation water.

## 3.4 SURFACE IRRIGATION (FIGURE 3.3)

In the case of the surface irrigation [2, 3], the water runs on the soil surface, providing necessary moisture to the plants for its development. The basic components are: Water source, supply line, control mechanism, dams or dikes of control, furrows of irrigation, system of drainage and system of reusability of the water (optional).

(a) (b)

**FIGURE 3.3** Surface irrigation: (a) flood irrigation; (b) gated pipe irrigation system; with permission from [10, p.80].

The supply lines can be PVC or flexible tubes (lay-flat); also, can be open channels of earth, asphalt, concrete, plastic, iron, or brick-lined. In addition to the pipes and open channels, other potentially useful structures are: Tunnels (to shorten to the channels for redelivery); hydraulic jumps (to reduce the energy of the water current without causing erosion and to reduce the speed of water, respectively), and siphons (to cross depressions of the land), etc. The control mechanisms may include: orifices, timer for mains (manual and automatic), Parshall flumes, garbage dumps, and control gates for channels. For the control, the flood gates are used (Figure 3.3).

### 3.4.1 *BORDER STRIP IRRIGATION*

In border irrigation, the water is applied to a leveled area surrounded by ridges. In level basin irrigation, each irrigated area is completely at level without slopes in any direction. The most common types of "border strip" irrigation use slope in the long direction of the border. It is not necessary that the edges are not rectangular or straight nor that the ridges are permanent. This technique is called "leveled flood or irrigation by border strip." The size of each border depends on the volume of available water, the land topography, the soil characteristics and the required leveling. The size fluctuates from a few square meters to 15 ha, approximately. An advantage of the border irrigation at leveled ground is that one needs to provide solely a specific volume of water. The border irrigation flood is more effective on uniform and accurately leveled grounds. It is possible to obtain high performance with low manual labor requirements.

### *3.4.2 FURROW OR BORDER STRIP IRRIGATION*

It is a traditional system and is more commonly used in agriculture. One adapts to extensive planting methods, and it is prone to diseases that are developed due to excess of soil moisture. It uses surface (rivers, lakes, pools, etc.), water rescurce, or a deep well. This method requires that the fields are prepared with gentle slopes so that the water runs slowly by gravity and arrives at the lowest part of the farm, where it is collected by open channels for elimination or is recycled for use.

#### *3.4.2.1 STRIP IRRIGATION WITH DIKES*

This type of flood irrigation is very popular. It is used mainly in narrow row crops like rice. It can be defined as the application of water between parallel strips. The strip between adjacent dikes does not have a slope in the transverse direction, but these have slopes in the direction of the irrigation. The strip irrigation method with dikes uses the soil accommodated in strips (made level through narrow measurement and with slopes throughout the wide measurement). These strips are surrounded by dikes in order to avoid the lateral flow of the water. The water is transported towards the superior end of the dike and advances downwards. High irrigation performance is rarely possible due to difficulty to balance the phases of advance and retraction in the application of water. The strip irrigation with dikes is most complicated among the all-surface irrigation methods. The design considerations include length and slope of each strip, volume of water per unit of width of strip, deficiency of soil moisture at the initiation of irrigation and infiltration capacity. However, due to wide variations in the field conditions, it is difficult for the irrigator to get high irrigation performance.

- **Advantages:**
  - It is easy to use;
  - It is of low operational control; and
  - The system is a traditional method of irrigation.
- **Disadvantages:**
  - Enough land for sowings is lost;
  - It is not possible to enter the field, once it has been irrigated;
  - Fertilizer application efficiency is low;

- Water losses by evaporation and runoff;
- May cause erosion problems; and
- Often the water cannot be reused, because it may be contaminated by fertilizers, etc.

### 3.4.2.2 *FURROW IRRIGATION OR IRRIGATION BY INFILTRATION*

The furrows are channels with slopes that are formed from the soil. The infiltration occurs vertically and laterally, and through the wetted perimeter of the furrow. The systems can be designed in a diversity of ways. The optimal length of each furrow is mainly determined according to the infiltration capacity and the size of the field. The infiltration capacities in the furrows can vary much, even when the soil is uniform, due to the crop practices. The infiltration capacity of a new furrow (or just cultivated) will be greater than the one of a furrows that has been watered before. The infiltration in furrows can be reduced due to compaction of rows (between the furrows) and due to compaction by farm vehicles and equipment. Due to the many controllable parameters of design and use, the furrow irrigation can be used in several situations within the limits of the uniformity and topography of the land. The uniformity and efficiency will depend on the soil characteristics and cultivation methods.

The furrow irrigation is designed so that the water runs throughout the desired field. In this system, the water under pressure arrives at the highest elevation of the field. It is distributed by channels or tubes, towards the fields where it will enter the furrows to flood the area. From the supply lines, the water enters the furrows by means of floodgates, siphons or by opening a furrow. Water is applied when the channel is opened. One may use flood gates to control the application of water to a particular field.

One can obtain high uniformity in the water application to the irrigated area by regulating the volume of water that is spilled into the furrow. For this purpose, one can use the pipes with lateral floodgates. The use of the pipes with lateral orifices has wider acceptance. The wing types of floodgates can facilitate the regulation of the water volume that arrives at each furrow. By means of these pipes, the exit of the water into the field can be automated. These floodgates can allow passage of water with flows

> 4 lpm. The galvanized iron or aluminum pipes with lateral flood gates are easy to install, simple to adjust and can be transferred quickly from one field to other. This system can be used for row crops: sugar cane, corn, fruit orchards, etc. The furrows are oriented directly toward the slope, but the cross-sectional slope is almost uniform (Figures 3.4–3.6).

- **Advantages:**
  - Low initial cost;
  - Less specialized manual labor; and
  - Easy operation, care, and maintenance.
- **Disadvantages:**
  - The application of fertilizers is manual;
  - Water loss by evaporation;
  - There are nutrient losses;
  - Planting area is reduced;
  - Land must be leveled. This increases initial cost and more time is lost between successive crops;
  - Low efficiency of water use;
  - Incidence of pests and diseases is high;
  - It requires irrigation duration and a large quantity of water; and
  - Losses by infiltration and runoff cannot be avoided.

### *3.4.2.2.1 Furrows Along the Contour Lines*

This system utilizes small channels with continuous slope and almost uniform by which the hilly areas are irrigated. The furrows follow the contours lines of the land. This system can be used in uneven lands, hilly areas, row crops and except for sandy soils. The furrows along the contours can be used to irrigate areas with pronounced slopes. The recommended slope must be between 1% and 2%. The system is efficient and acceptable if all the practices are followed. A proper method is necessary to avoid the overflow of the water in the furrows. The length of the furrows must be short to eliminate excess water that can destroy (dismantle) the furrows. The furrows along contours are used jointly with parallel terraces to provide protection against breakage of furrows. The superior and inferior drainage furrows must be protected.

### 3.4.2.2.2 Leveled Furrows

The leveled furrows with no slopes are formed by a furrow opener and are used to irrigate crops seeded on the furrows or on the sides of furrows. This method requires a fast supply of water. The leveled furrows adapt better to soils with a moderate to slow index abscrption and an index of retention capacity from medium to high. With the furrow irrigation, the best results are obtained in gentle and uniform slopes. The amount of water can be adjusted according to the variations of the furrows. High application efficiency can be obtained if it is designed properly and the surface runoff can be reduced. Unless high wind speed affects the water flow, the furrows can be doubled in length, since the water can be applied on both sides of the furrow. This procedure reduces the cost of construction and maintenance of the distribution system.

In places where the wind speed is greater than 24 or 32 Km/h, the application of the irrigation water becomes difficult if the wind direction is opposite to the watercourse. The capacity of the furrows must be sufficient to maintain the flow rate. These must be able to apply at least half of the volume of the irrigation for the crop use. For effective operation of the system, the field operations should not alter the topography, the shape, and the cross-sectional area of the furrows.

### 3.4.2.2.3 Sloping Furrows

These furrows consist of small channels with a uniform continuous slope that follow the direction of the irrigation. This method can be used for row crops, including vegetables. Furrows with slopes can be used in all soils except sandy soils, with a high degree of infiltration capacity and with a very little lateral distribution. This method must be used with extreme care in soils with a high concentration of soluble salts. Normally, most of the energy is needed for the preparation of the furrow. The flow to each furrow must be regulated carefully so that the distribution of the water is uniform, with the minimum runoff. The lands must be leveled. Necessary facilities should be provided to collect and dispose of the excess runoff. The method is not adaptable for shallow-rooted crops or low irrigation rates for the germination of the seed.

#### *3.4.2.2.4 Corrugated Furrows*

This system consists of a partial wrinkling of the soil surface. The irrigation water does not cover all the land, but is distributed in small channels or undulations at regular spaces. The water applied in the undulations infiltrates into the soil and extends laterally to irrigate the intermediate spaces between the furrows. Irrigation by undulations or corrugations adapts better to the drought areas and flatlands with slopes of l to 8%. It can be used on irregular slopes, but the undulations must have a continuous slope in the direction of the irrigation, and the cross-sectional slope must be so as to minimize the runoff. The undulations give best results in moderate to heavy textured soils. These are not suitable in soils with a high coefficient of absorption and in saline soils. The irrigation flow rates can vary from high to low, because the numbers of undulations that are watered at the same time, and can be adjusted to suit the flow available. The manual labor requirements are high. The irrigation flow rates must be regulated carefully so that the distribution of the water is uniform and the runoff is reduced to the minimum. This method does not adapt to areas with high rainfall during the irrigation season.

## 3.5 SPRINKLER IRRIGATION

In sprinkler irrigation, the water is applied on the soil surface in form of a rain [1, 2]. The spray pattern is obtained when the water at pressure is expelled through small orifices. The operative pressure is developed by an appropriate pumping unit. Use of this system expanded after World War II with the introduction of light and movable aluminum pipes. Sprinklers and fast couplers were developed to facilitate assembly and dismantling. Manually lateral movable and rotating sprinklers were then developed. Later to save energy and water, the fixed solid-set systems came into being. In plantations and fruit orchards, these consisted of lateral plastic pipes placed between or along the tree row and sprayers of low flow rates or mini sprinklers. Giant sprinklers mounted on small carts were also developed to facilitate to coverage of extensive areas.

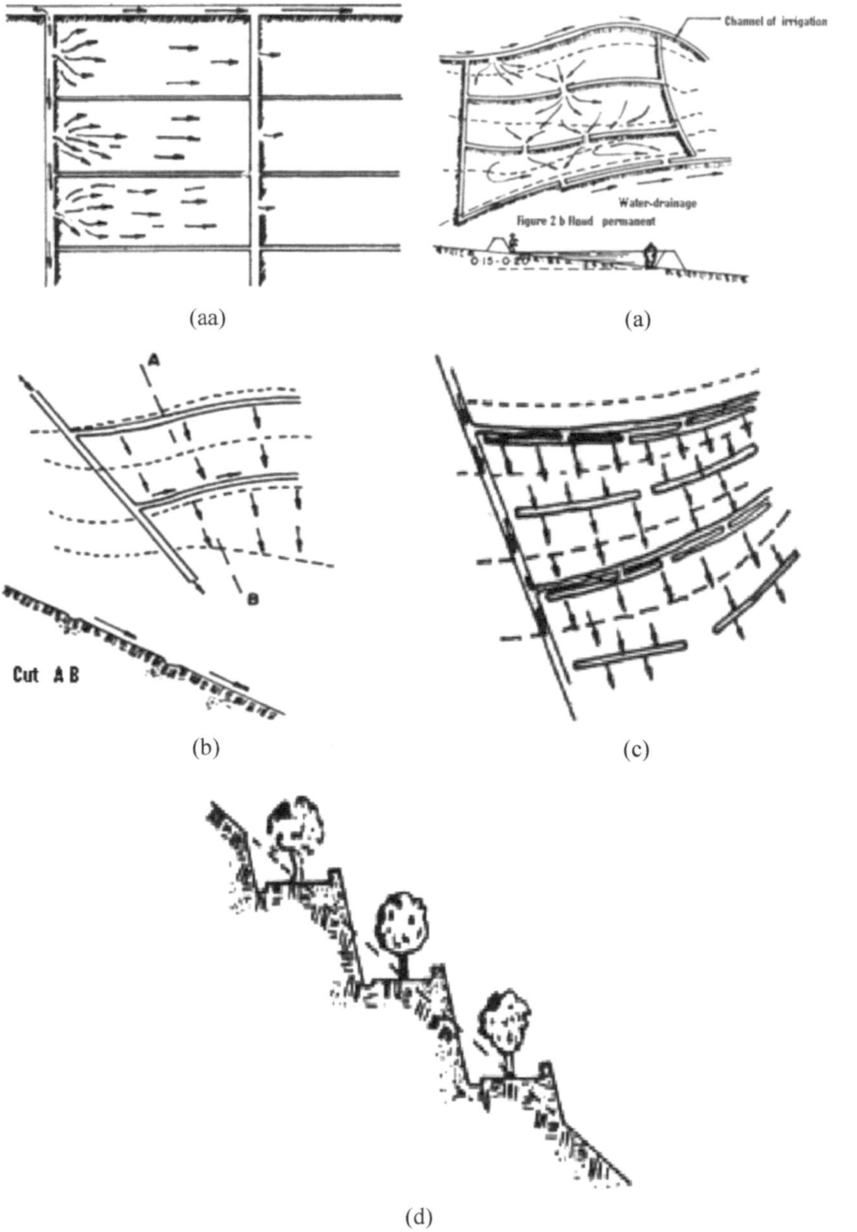


**FIGURE 3.4** Flood irrigation; with permission from [10, p.81]

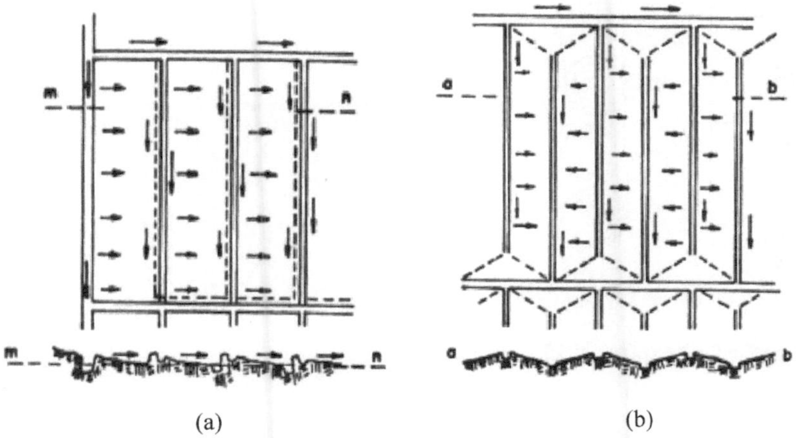


**FIGURE 3.5a** Flood irrigation; with permission from [10, p.81].

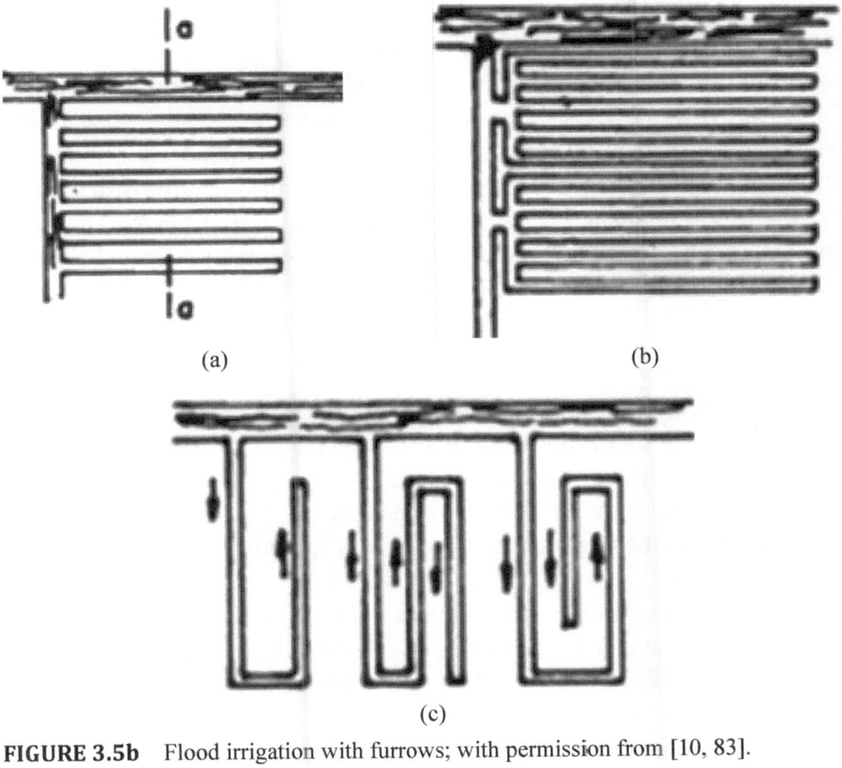


**FIGURE 3.5b** Flood irrigation with furrows; with permission from [10, 83].

(a) (b)

**FIGURE 3.6** Infiltration irrigation; with permission from [10, p. 84].

The simple automation is carried out by means of the use of volumetric metering valves to provide a known volume of water or timed valves to provide water for a known duration. The valves can work at a predetermined sequence. Then the automation systems included the use of solenoid valves to close and open according to an established schedule. Several fields can be watered by connecting the units from subfield to a "Master Central Control Unit." Although the sprinklers have increased in popularity, the surface irrigation continues to predominate. The basic components of a sprinkler system are: A water source, a pump to provide an operating pressure, one or more than one main/sub mains to distribute the water to all the field, sprinklers, and control valves, etc.

- **Advantages:**
  - Preparation of land is not required. It can be used on lands of rough topography and level lands;
  - It allows total use of the land;
  - It offers great flexibility in design;
  - It can be used in any type of soil including high permeability rates;
  - It is generally more efficient than the surface irrigation;
  - More water in comparison with the surface irrigation is saved;
  - One adjusts very well to additional irrigation;
  - It can be used for protection against frosts and the heat;
  - The use of portable pipe reduces the cost of the equipment;

- Fertilizers can be applied efficiently through the irrigation water;
- The root zone is developed better than with the surface irrigation;
- The cost for manual labor is lower than for the surface irrigation;
- The gaging of the water is easier with this system;
- The performance efficiency is high; and
- It is possible to make frequent applications at low volumes, when it is necessary.

➢ **Disadvantages:**

- Elevated initial cost;
- The evaporation loss is high and these losses can be reduced to a minimum by watering at night;
- Salty water can cause damage to the crop;
- Incidence of foliage diseases is high.
- Some crops can suffer by loss of flowers due to the impact of the water;
- The freezing conditions can affect the functioning of sprinklers;
- There is interference with the pollination; and
- The wind interferes so that distribution of water is affected.

### *3.5.1 TYPES OF SPRINKLER IRRIGATION SYSTEMS*

#### *3.5.1.1 MANUALLY OPERATED AND PORTABLE SYSTEM*

These systems include lateral pipe with sprinklers installed at regular intervals. The lateral pipe is, generally, of aluminum, with Sections of 6, 9, or 12 meters in length, and fast connections for pipes. The sprinkler is installed at the top of a riser (in the orchards, the height of a riser should go under the leaf canopy). The risers are connected to the lateral tube. The length of the tube is selected to correspond with the desired spacing of the sprinklers.

The lateral pipe with sprinklers is placed on a ground and it is used until the application has been terminated. Then, the lateral tube is dismantled and is positioned in the next section. This system has a low initial cost, but requires high manual labor. It can be used in almost all crops. However, it may be difficult to move the lateral pipes when the crop is mature.

### *3.5.1.2 LATERAL SYSTEM ON WHEEL (MOVABLE)*

This system is a variation of the manual system. The lateral pipe is mounted on wheels. The height of the wheels is chosen so that the axis exceeds the height of the crop for easy movement. A drive unit is commonly a motor-driven with gasoline and is located near the center of the lateral pipe. The system is moved from one place to another by wheels.

### *3.5.1.3 TRAVELING SPRINKLER SYSTEM WITH PROGRESSIVE MOVEMENTS*

This system uses a sprinkler (tube) gun of high pressure and high volume. The spray gun is mounted on a towable trailer. The water is supplied by means of a flexible hose or from an open ditch. The system can be used in one field for a desired time and then can be moved to the next field. The tow can be moved by means of a cable, or it is possible to be pulled ahead while the hose is coiled in a spool to the border of the field. These systems can be used in almost all crops. However, due to high intensities of application, these are not suitable for clay soils.

### *3.5.1.4 CENTRE PIVOT SYSTEM*

This system consists of a lateral pipe with simple sprinklers supported by a series of towers. The towers are impelled in such way so that the lateral pipe moves around the center, about the pivot point. The speed of the complete circular motion fluctuates from 12 hours to several days. The intensity of the application of water increases with the distance from the pivot in order to give a uniform amount of application. Because the center pivot irrigates a circular area, it leaves non-irrigated the corners of the field (unless additional special equipment is added to the system). Centre pivots are able to water almost all field crops.

### *3.5.1.5 LINEAR MOVE SYSTEM*

The linear move system is similar to the center pivot system. The line of the pipe extends in a perpendicular direction to the lateral one. The delivery of the water to the lateral is by a flexible hose or from an open ditch. The system irrigates the rectangular fields free of high obstructions.

#### *3.5.1.6 LOW ENERGY PRESSURE APPLICATION (LEPA)*

Low energy pressure applications (LEPA) are similar to the linear move systems of irrigation. The orifices in the lateral pipe and pipes can discharge at very low water pressure, exactly according to the soil moisture. Generally, LEPA systems are accompanied by soil surface management regimes such as minimum tillage or, most commonly, micro-dams to prevent high application rates from causing runoff. With suitable soil surface management systems, high infiltration rates are not an absolute requirement.

#### *3.5.1.7 SOLID-SET SYSTEMS*

In this system, sufficient lateral pipes are placed in the field and are not moved during the season. Solid-set systems utilize a network of aluminum tubes for irrigation. Enough lateral lines are used to cover all the area. The system reduces to a minimum the need for manual labor during the irrigation season.

### ***3.5.2 TYPES OF SPRINKLERS***

1. **Rotating Sprinklers, Impact Type:** These are commonly used over a wide range of pressure, discharge, spacing, and rate of application for different crops.
2. **Sprinklers, High Volume, or "Gum" Type:** These are rotating sprinklers with a discharge up to 60 $m^3/h$ at a pressure head of 60 meters. It can cover areas up to one ha simultaneously.
3. **Sprinklers with Low Flow Rate:** These sprinklers apply 120 to 350 lph at a pressure head of 15–25 meters. These are used mainly in irrigation of fruit trees.
4. **Mini Sprinklers:** Small sprinklers can apply 30–120 lph at a pressure head of 15–25 meters. These are used in vegetables and nurseries.

## 3.6 SUBSURFACE IRRIGATION [2–4, 8, 9]

In many areas, soil conditions and topography are favorable to irrigation using the water below the ground surface. These favorable conditions are: The existence of impermeable subsoil at a depth > 1.8 m; silt or silt-sandy permeable layer; uniform topography and moderate slopes. The irrigation may be applied by means of exposed drain ditches. The water table stays at a predetermined depth, normally from 30 to 40 cm, depending on the root characteristics of the crop.

The perforated subsurface pipes allow the infiltration through the soil. The pipes can be placed at a spacing of 45 cm and at a depth of 50 cm. These buried pipes can suffer damage by deep plowing. This method works, if the soil has a high horizontal and a low vertical permeability.

The open ditches are probably used on a greater scale. The feeding drain is excavated along the contours. The drain spacing must be sufficient to maintain and to regulate the level of the water. These are connected to a supply channel that runs downwards, following a predominant slope of the land.

### 3.6.1 ADAPTABILITY

Subirrigation is appropriate for uniformly textured soils with a good permeability so that the water is mobilized quickly, in a horizontal and vertical direction and to a recommended depth under the root zone. The topography must be uniform or almost level or very smooth with uniform slope. Subirrigation is adapted for vegetable and root crops, forage crops and gardens.

### 3.6.2 IMPORTANT CHARACTERISTICS

This method is used in soils with low capacity and when surface irrigation cannot be used and the cost of pressurized irrigation is excessive. The level of the water can be maintained at the optimal depth according to the crop requirements at different growth stages. The evaporation losses are reduced to a minimum. The irrigation does not allow the weed seeds to come to germinate.

### 3.6.3 LIMITATIONS

Water with a high concentration of salts cannot be used. The selection of crops is limited. The crops with deep root system (such as some citrus) are not generally suitable for subirrigation.

### 3.6.4 NATURAL SUBSURFACE IRRIGATION

When geological and topographical conditions are favorable natural subsurface irrigation is suitable for nearly flat topography with a deep surface layer and high lateral permeability. At a depth of two to seven meters from the soil surface, there is usually an impermeable rocky substratum. A constant water table at a particular depth is maintained. A set of parallel ditches can be arranged below the soil surface under the following conditions: For complementary irrigation in spring and summer in humid regions; a good drainage is needed in winter. The soils are sandy and very permeable. During excessive rainfall, water is removed by gravity or by pumping.

## 3.7 DRIP/TRICKLE OR MICRO-IRRIGATION

The application of water is by means of drippers that are located at desired spacing on a lateral line [10]. The emitted water moves due to an unsaturated soil. Thus, favorable conditions of soil moisture in the root zone are maintained. This causes an optimum development of the crop. For more details, read Ref. [10].

### 3.7.1 RECOMMENDATIONS: 9TH INTERNATIONAL MICRO IRRIGATION CONFERENCE (9IMIC) ON "MICRO IRRIGATION IN MODERN AGRICULTURE" ON 16–18 JANUARY, 2019 AT AURANGABAD, MAHARASHTRA, INDIA; ORGANIZED BY WWW.ICID.ORG

This congress is organized by ICID.org every 4–6 years. 9th International Micro Irrigation Conference was organized by ICID and INCSW (Indian National Committee on Surface Water). This forum generated opportunity for discussions and deliberated strategies with eminent stakeholders through seminars, exhibitions, field visits to Jain Farm and sessions to build public awareness for the use of micro-irrigation on large-scale and to get support to implement key strategies for conservation, preservation, inter-sectoral arrangements, advances in crop technology, precision engineering techniques, etc. It was an effort to create awareness, conserve, and use water resources in an integrated manner. 9th IMIC was a platform for the congregation of local, regional, national, and global ideas and opinion from decision-makers, politicians, researchers, and entrepreneurs in the field of Water Resources. The recommendations from the congress are listed below:

#### 3.7.1.1 IMPORTANCE OF THE MICRO-IRRIGATION DEVELOPMENT

- Water should be considered as a resource instead of a commodity to be distributed among the farmers. Investment in micro-irrigation can help in increasing GDP per unit of water deployed, as saved water can be used for economic activities that generate more GDP per unit of water.
- Micro-Irrigation builds resilience against climate change and to secure water for people, environment, and food security.
- Micro-irrigation needs to be developed as a viable agriculture business model for farmers with better productivity and economic returns in tangible and intangible forms. Ramthal (Karnataka), India's largest drip irrigation project, where the irrigated area has been almost doubled by adopting integrated micro-irrigation, is an excellent example of a technology-driven, community based, efficient, large scale irrigation system.
- In order to promote and make micro-irrigation an economically viable option, micro-irrigation industry needs to be considered as priority sector and brought under infrastructure industry status.
- All bore wells need to be integrated with micro-irrigation for efficient use of groundwater. In addition, integration of lift irrigation schemes with community drip irrigation should be considered.
- Adoption of on-farm micro-irrigation should be made an integral part of all irrigation projects as it results in efficient use of water and also results in low or no greenhouse gas (GHG) emissions and also prevents soil degradation.

- Adoption of micro irrigation need to be considered for all agriculture crops. All efforts should be made for making micro-irrigation mandatory for water intensive crops like oil palm, sugar cane, banana, papaya, citrus, coconut, etc.
- Responsive strategies are needed for propagating the micro-irrigation networks on a large scale. Apart from making necessary infrastructure available on a one-time basis, continuous, and appropriate responses to the needs at the farm level is required so as to achieve sustainability of the measures adopted.
- High initial investment, relatively higher maintenance cost, clogging of drip emitter, availability of water-soluble fertilizers and their compatibility and overdosing of fertilizers, etc., are few problems being faced in fertigation and chemigation. Further research work in this field can enhance effective usage of fertigation and can help Indian farmers to adopt micro-irrigation system effectively.

### 3.7.1.2 ECHNOLOGY, INNOVATIONS, AND PRODUCTS

- Hi-tech green-houses equipped with climate control, micro-sprinkler, subsurface drip irrigation (SDI) and mechanized farming techniques have the potential to increase the yield three to four times. Success of such prototype projects needs to be replicated on a large scale with appropriate hand-holding and capacity building of the farmers, so that it becomes a regular practice.
- In view of increasing emphasis on pressurized systems for irrigation: there is a need to consider both water and energy efficiencies simultaneously, and therefore the new concept: "more crop per drop per kilowatt" needs to be adopted. With any saving of drop of water, or a morsel of food or kW of energy, lives of more persons can be made better with same resources.
- Pressurized irrigation systems can be effectively bundled with solar pumps, which are best suited in areas with no or erratic power supply and for the farmers growing less water-intensive crops.
- More focus is required on the agronomic practices such as mulching along with drip irrigation for areas having water shortage, because this practice may provide better water efficiency and increased yield. Experiences need be propagated.

- Latest research carried out on different crops (viz. sugarcane, banana, horticulture, and vegetables) concludes that fertigation is the most effective and convenient means of maintaining optimum fertility level and water supply according to the specific requirement of crops under drip system. Field studies showed that drip irrigation along with fertigation increased the yield of the crop to a great extent.
- The pipe (pressurized) irrigation network (PIN) system requires less land as compared to conventional canal irrigation network (CIN) system. Its adoption can address the growing problems of land acquisition as well as for providing effective rehabilitation and resettlement. However, care has to be exercised for its sound case specific designs and selection of appropriate materials for construction for ensuring its sustainability. Custom-made software applications need to be developed for standardized and sound design of PIN network.
- It is assessed that micro-irrigation with reliable water source and Pre-Paid Meter and Smartcard could save around 50 to 90% of water in on-farm context.
- Sub-surface drip irrigation (SDI) system is a promising technique, especially in arid regions. SDI can reduce losses by evaporation and deep percolation significantly. SDI also promotes very good development of the root system and palm growth in arid regions. However, maintenance and clogging of drippers are a major concern.
- It is necessary to conduct the physical and chemical analysis of the water before designing a drip irrigation system and choosing an appropriate filtration system. It is important to analyze the sample for suspended solids, dissolved solids, and the acidity (pH), macro-organisms, and microorganisms. This aspect is particularly relevant while planning systems based on recycled water as well as for groundwater containing high salt content.
- Design of filters for providing sediment free water and their regular maintenance is necessary for longevity of the drip irrigation systems in the field.
- In future: ready irrigation projects, outlets with sufficient pressure should be provided to each farmer field instead of a group of farmers to promote adoption of micro-irrigation by every farmer. This will require changes in operation policies as well as delivery

systems in case of existing irrigation projects. These changes are worthwhile to factor in during planning of extension, renovation, and modernizing the existing irrigation distribution networks.

- While carrying out the viability of the project with micro-irrigation, project focus and evaluation should be based on life cycle analysis and not on initial investment cost only. The main criteria for design and evaluation should be a selection of material based on service life, energy consumption, recurring operation and maintenance expenses, impact on the environment, and overall sustainability.
- To overcome and manage water scarcity, there is need to consider and promote use of poor-quality water with due treatment. To ensure effectiveness and safe use of waste treated water, it is necessary to develop standards and guidelines for the safe use of wastewater in agriculture. Micro-irrigation is the most efficient way to use treated wastewater in agriculture.

#### *3.7.1.3 MICRO-IRRIGATION FUNDING AND SOCIAL INVOLVEMENT*

- Water pricing is one factor, which can act as a trigger for adaptation of large-scale micro-irrigation with the objective of water-saving and increasing productivity, and it should be increasingly factored in the irrigation policies. Resistance to payment of water charges can be softened with direct benefit transfer for changing the mindset of the consumers in the initial period.
- In addition to the horticulture and cash crops, the micro irrigation also needs to be adopted for cereals and pulses on a large scale, to broad base the water savings. There is a need for a review of funding policies for micro-irrigation for cereals and other crops.
- Micro irrigation fund needs to be created to facilitate States in mobilizing resources for expanding coverage of micro irrigation by taking up-special and innovative projects as well as for incentivizing micro irrigation beyond provisions under PMKSY-PDMC to encourage farmers to adopt micro-irrigation systems in a bigger way.
- There should be a gradual shift from grant-based to loan-based model and from Project-based to Program-based approach in micro-irrigation.

- Cost-effectiveness is the key in popularizing the MI as the initial higher cost of sub-surface drip irrigation gets offset by its long-term advantages in water-saving and hence sustainability. Low-cost drip lines have been useful in popularizing the MI. While ensuring cost-effectiveness, rapid tangible returns in terms of savings of labor and better outputs and economic returns should be provided for the success of the investments made.
- Studies indicate that energy subsidy in lift irrigated areas do not support water and energy savings. Shift of the subsidies from energy to water-saving technologies in lift irrigated areas will improve water and energy use efficiency.
- Small landholdings which are considered uneconomical for intensive farming need consolidation approach through local institutions as ecological units.
- It is necessary to make transition from field experiments and trials to routine practice in respect of technologies and devices developed. Strategies for provision of long-term handholding and ancillary services are required with dedicated manpower for translating the benefits to the ground and at the macro-management level.
- Social networks and neighborhood success stories can provide effective means of propagating the micro-irrigation techniques on a wide spread basis. Appropriate skill sets may also be deployed along with the technological and financial resources at the local level to ensure sustainable adoption.

#### *3.7.1.4 OPERATION AND MAINTENANCE (O&M) SERVICES AND CAPACITY DEVELOPMENT*

- The geometry of emitters leading to clogging should be handled during manufacturing. The roots blocking the emitters to be handled by change of design. The rate discharge affect emitters clogging, smaller the discharge rate more are the chances for clogging.
- Areas having higher hardness should have a different design of emitters and paths. There can be separate design of MI systems depending on the hardness of the source of water.
- Life cycle cost analysis of pipes of various materials needs to be carried out which includes construction cost, operation cost,

revenue loss, replacement cost and disposal cost. Findings of study on life cycle cost analysis on commercially available diameter may lead to ready reckoner for planning of micro irrigation and piped irrigation system.

- Ground water-based pressurized irrigation system run on the same pump or motor which lifts the water without any additional electricity.
- Imparting training, skill development, providing technical support through research institutes, NGOs, state, and central government institutes can encourage farmers to adopt fertigation and chemigation as part of micro-irrigation for increased field application efficiency and crop yield improvement.
- Safe and efficient utilization of sewage effluent is of great significance to ensure agricultural production and optimize balance between supply and demand. Reclaimed water use has special concerns, such as:
  - Great seasonal variability is observed in the quality of the treated water. However, it should remain in compliance with the reuse standards, especially in periods of high demand for irrigation water.
  - It is better to irrigate with treated wastewater because of the degraded quality of the groundwater.
  - Problems of clogging the drip irrigation network in the near future may occur due to the weakness of the filtration system.
  - With a view to health and safety, the use of wastewater should be avoided for crops that are eaten raw (salad, vegetables, etc.).
  - Creation of specifications for drip equipment for treated wastewater reuse.

## 3.8 SUBSURFACE DRIP IRRIGATION (SDI)

In this system, laterals with drippers are buried at about 45 cm depth. The fundamental intention is to avoid the costs of transportation, installation, and dismantling of the system at the end of a crop. When it is located permanently, it does not harm the crop and solve the problem of installation and annual or periodic movement of the laterals. A carefully installed system can last for about 10 years.

The components of the system are basically the same as those for the drip irrigation system: manual solenoid valves; pressure controllers; pressure gages; fertilizer tank; and hydro cyclone filters, sand filters and mesh filters.

The system can be automated by means of electronic sensors on the basis of soil moisture. The laterals are installed at a depth of 45 cm depth. The system is connected to a computerized central control station. Chemical agents can be applied through the system only if these are highly soluble. It is recommended to apply water at pressure, use fumigants and herbicides to prevent the penetration of roots into the tubes.

- **Advantages:**
  - High crop yields;
  - Conservation of water and energy;
  - Less manual labor;
  - Less damage to the lines and the system during the installation and dismantling process;
  - Life of irrigation lines and parts is increased, since the plastic pipes are not exposed to the solar light.

## 3.9 ULTRA-IRRIGATION OR XYLEM (TREE INJECTION) IRRIGATION

Xylem- or ultra-irrigation (Figure 3.7) is the direct application of water with necessary chemical agents into the xylem of the tree trunk using a series of injectors that depends on the age of the tree. Xylem irrigation is also called ultra-micro, high frequency, tension, tree injection, or chemo-therapy irrigation [10]. There is no difference in the concept these names represent.

The basic idea originated when various chemicals were injected into the internal circulatory system of tree. It is simple to inject water, fertilizers, micronutrients, growth promoters, growth inhibitors, pesticides, trace elements, gases, precursors of flavor/color and aroma, and in general any substance valuable for the improvement of fruit quality [10]. This system is in the experimentation stage and has not been evaluated commercially.

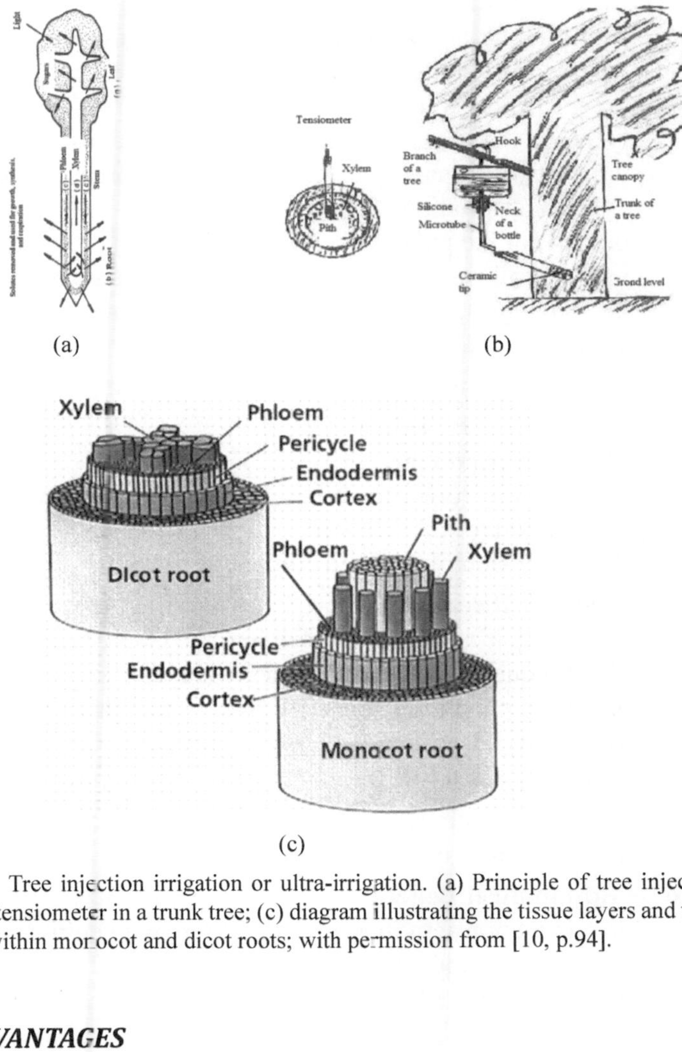


**FIGURE 3.7** Tree injection irrigation or ultra-irrigation. (a) Principle of tree injection irrigation; (b) tensiometer in a trunk tree; (c) diagram illustrating the tissue layers and their organization within monocot and dicot roots; with permission from [10, p.94].

### *3.9.1 ADVANTAGES*

1. **Efficient Use of Water:**
   i. No evaporation from the soil surface into the atmosphere;
   ii. No infiltration into the subsoil where roots are incapable of absorbing moisture;
   iii. No runoff;

iv. No wetting of foliage;
v. Inhibits non-beneficial consumptive use of water by weeds because terrain;
vi. is free of weeds;
vii. One can irrigate the entire field up to edges;
viii. Accurate quantity of irrigation water can be applied according to transpiration;
ix. rate of the plant;
x. Overall water application efficiency can go up to 99%;
xi. Savings up to 95% of water use can be achieved.

2. **Plant Response:**
   i. Crop growth characteristics can be manipulated;
   ii. Better fruit quality and uniformity of crop is expected.
3. **Root Environment:**
   i. Shallow root system;
   ii. Effective soil aeration;
   iii. Provision of required amount of nutrients.
4. **Pest and Diseases:**
   i. Pesticides can be injected into the plant system;
   ii. Frequency of sprays can be reduced;
   iii. Reduction in incidence of insects and diseases;
   iv. Reduced application rates of pesticides.
5. **Weed Growth:**
   i. It is a minimum;
   ii. No weeds in dry surface between trees.
6. **Agronomical Benefits:**
   i. Irrigation activities do not interfere with cultivation, spraying, picking, and handling;
   ii. Less inter-cultivation, soil crusting, and compaction problems;
   iii. No surface runoff;
   iv. No soil/water erosion due to irrigation;
   v. Fertigation and chemigation are possible thus savings in energy and quantity;
   vi. Other necessary chemicals can be applied along with irrigation water.
7. **Engineering and Economic Benefits:**
   i. Significant savings in energy;

ii. Cost is low compared to surface sprinkler and drip irrigation systems;
iii. Pipe sizes are significantly smaller compared to pipe sizes in other irrigation systems;
iv. Conveyance efficiency and water use efficiency can be increased up to 99%;
v. System can be installed in uneven terrains;
vi. It requires constant discharges at low pressures;
vii. Water and chemical use can be programmed with the crop response.

### 3.9.2 DISADVANTAGES

1. May not be applicable in vegetable crops as it is more convenient to inject into tree trunks.
2. May not be used in monocot species as the xylem is not as differentiated.
3. Introduction of new substances can cause toxic effects just as a man can overdose on drugs.
4. May cause fungus growth at the injection site.
5. Holes in tree trunk must be made to install injection tips thus causing physical injury to the plant.

### 3.9.3 OPERATIONAL PROBLEMS

1. No information is available on a number of injection sites, water application rates, dosages of various chemicals depending upon the age of the tree.
2. At what height and depth, should the injection points be located?
3. Injection tips can be easily clogged by gum, wax, and resins of tree.
4. Effective cleaning agent needs to be found to avoid clogging of tips.
5. Algae formation in the injector lines.
6. Laterals may contain air [from the tree] and thus obstructing the flow.
7. Leakage of water at the contact point between the injector tip and tree surface.

8. Excess pressure might loosen the sealing agent [silicon] and may throw out the injection tip.
9. Expert advice is needed to locate xylem.
10. Chemigation might disturb the osmotic and electrical internal equilibrium in the plant.
11. Screening of pesticides and chemicals suitable for xylem irrigation.
12. Salts in excess of 300–500 ppm may require desalinization of water.
13. A clean, pure or soil water is necessary.

### *3.9.4 PRINCIPLE OF OPERATION*

It is based on utilizing natural negative sap pressures within a plant to suction liquids and gases directly into the inner circulatory system, analogous to a human blood transfusion. The technique is accomplished by placing an injection tip [e.g., a ceramic implant] directly in the xylem layer, the negative pressure area. Liquid or gas is then made available to the implant through *a* plastic tubing at very little or no pressure. Fluids can then traverse in the plant in any direction. The roots of the plant continue to be nourished by the natural way, with sap, water, and nutrients. The roots still seek moisture and grow down using a stimulus called geotropism.

Plants give off water through a process called transpiration. The amount of water a plant "throws off" and the amount it needs are two different situations. A well-known "Hill Reaction" is:

$$6CO_2 + 6H_2O + H_2 = C_2H_{12}O_6 + 6O_2$$

Opposite of transpiration is called respiration. By careful measurement of the quantities of sugar synthesized in the leaves by unitary surface and time [10–15 mg of hexose/sq dm-h], it is readily calculable what stoichiometric quantity of water is required under the same conditions [e.g., 50–80 ml for a period of 8 hours considering a canopy surface from 8 to 10 m$^2$]. This quantity is very approximate to the quantity of water consumed by xylem irrigation during the same period under the same conditions. Primary water uptake occurs only during photosynthesis or daylight [12].

#### *3.9.4.1 SYSTEM MODIFICATION*

It is accomplished by simply placing the ceramic piece in the root zone of house or commercial indoor plants, nursery stock, or almost any plant too small to receive an implant in the trunk. The same efficient use of water and nutrients are applicable but some of the metabolic engineering techniques [Modulation of the plant metabolism with the aim of obtaining better fruits by injection of substances such as promoters of color, bouquet, flavor, aroma, metabolites, enzymes, or coenzymes] may not be effective. Seeds for greenhouse can be germinated and grown from an implant in the soil. The seed can actually be glued to the implant, then planted and grown through maturity.

Tree crops can be raised with other irrigation systems during the first 2 years, and then ceramic tips can be installed. Water usage of 1200 ml/day on older trees, 150 ml/day for grapes is enough. This is equivalent of approximately 186 lpm/acre of irrigation during 12 hours of photosynthesis or 134 liters/day/acre.

### ***3.9.5 DESCRIPTION OF ULTRA-IRRIGATION SYSTEM***

The system consists of a water resource, pump, chemigation system, filter, mainline, sub-main, laterals, and injection tips. The installation of injection tip should be done in the following manner:

1. Select the size of a ceramic tip.
2. Select the best location on the plant.
3. Bore a hole through the cambium layer approximately 1/4 diameter larger than the injector.
4. Use a sharp instrument to remove this plug of bark. It is important to bore past the phloem to cause leakage out of the plant.
5. Continue the bore into xylem [sapwood] portion to the same dimension as the length of the ceramic portion of the injector. The hole should allow a snug fit.
6. Use an inert sealing agent [silicone] for sealing the injector to the tree.
7. Hook water to be injected to the tip at a pressure of 1–8 ft of head. [Necessary pressure can be allowed by gravity, low pressure

pump]. Very minute quantities of chemical can be injected into the water stream using a plastic syringe [doctor's needle].

### *3.9.6 PRECAUTIONS FOR ULTRA-IRRIGATION SYSTEM*

- Sterilize the drill bit;
- Hole should allow a perfect fit;
- Use a good sealing agent;
- Water should be free from pathogens;
- Use a pesticide to avoid fungal growth;
- Injector site should be allowed to dry before starting irrigation;
- No leakage can be allowed between the tree and tip, as it will break the suction;
- High precaution is essential in determining dosage of the chemicals to avoid toxic hazards in the plant;
- Any injection holes, which cannot be used, should be left open. They heal with time.

## 3.10 HYDROPONICS [6]

The growth of plants without soil is known as hydroponics. From 1925 to 1935, extensive work was done to modify the nutrient culture in the nurseries. In 1930, W.F. Gericke at the University of California, defined hydroponics as a science to cultivate plants without soil use, but using inert materials such as sand and sawdust, among others.

## 3.11 SUMMARY

The need for additional food for the world's population has spurred the rapid development of irrigated land throughout the world. Vitally important in arid regions, irrigation is also an important improvement in many circumstances in humid regions. Unfortunately, often less than half the water applied is beneficial to the crop-irrigation water may be lost through runoff, which may also cause damaging soil erosion, deep percolation beyond that required for leaching to maintain a favorable salt balance. New irrigation systems, design, and selection techniques are continually being developed and examined in an effort to obtain the highest practically attainable efficiency of water application [10].

## KEYWORDS


- **canal irrigation network**
- **evapotranspiration**
- **greenhouse gas**
- **low energy pressure application**
- **permanent wilting point**
- **plant water requirement**


## REFERENCES


1. Bebr, R., (1954). *Agricultural Hydraulics* (pp. 212–220). S.A. Barcelona, Madrid: Salvat Publishing, Chapter 8.
2. Jensen, M. E., (1980). *Design and Operation of Farm Irrigation Systems* (p. 829). St. Joseph MI: Monograph #3 by American Society of Agricultural Engineers.
3. Israelson, W., & Hansen, V. E., (1965). *Principles and Applications of the Irrigation: S.A*. Barcelona-Madrid: Reverte Editorial.
4. On-Farm Irrigation Committee of Irrigation and Drainage Division, (1987). *Selection of Irrigation for Methods Agriculture* (p. 95). New York, NY: American Society of Civil Engineers.
5. Reed, A. D., (1980). *Irrigation Costs* (pp. 1–10). Leaflet 2875 October by Division of Agricultural Science, Berkley-CA: University of California.
6. Resh, H. M., (1985). *Hydroponics Food Production*. Santa Barbara, CA: Word Bridge Press.
7. Agriculture of the Americas Magazine (1983, July 1984, and February 1988).
8. Solomon, K. H., (1988). *Selection of an Irrigation System* (pp. 1–11). Publ. Núm. 880702-Institute of Agricultural Technology of California.
9. Turner, J. H., & Anderson, C. L., (1980). *Planning an Irrigation System* (p 120). Athens-GA: American for Association Vocational Instructional Materials.
10. Goyal, M. R., (2012). *Management of Drip/Trickle or Micro Irrigation* (p. 426). Oakville-ON, Canada: Apple Academic Press Inc.; ISBN: 9781926895123.

## CHAPTER 4

# An Insight into the Biological and Ecological Effects of Agrochemicals: A Global Scenario

ASHOK KUMAR MEHTA, RASHMI MITTAL, and RISHI VERMA

### ABSTRACT

To ensure food availability for a growing population and to uplift the social and financial status of agriculture, several novel technological advancements have been adopted. In recent times, agrochemicals have gained enormous popularity to serve the purpose. Previously, the impact of agrochemicals was evaluated on the basis of monetary benefits which were availed from the production without prioritizing its impact on the environment. Keeping in view the rapid deterioration of the environment due to excessive use of agrochemicals, different laws and policies were initiated to monitor and to control the application of agrochemicals globally. To ban the illegal use of agrochemicals, laws were strictly implemented, and new technologies have been adopted to pause its discharge into the surroundings. This chapter briefly described the impact of agrochemicals worldwide and the policies and strategies adopted to overcome the impact.

### 4.1 INTRODUCTION

To uplift the agriculture productivity, several agricultural inputs are currently under implementation. Agriculture inputs chiefly refer to the biological and chemical inputs along with farm machinery and technologically advanced equipment. Especially, chemical inputs stand for the application of chemical constituents in agriculture for production, harvesting, processing, and for other different processes. Chemical constituents include chemical fertilizers, pesticides, and plant growth regulators. Diverse agriculture management practices have been introduced into the frame regarding systemic use of chemicals or fertilizers [3, 27]. Benefits obtained from their usage are often evaluated on the scale of economic efficiency, which means a reduction in associated production costs and a significant increase in production yield [1, 5]. Impact over the environment due to their application is generally ignored. Rather, the enhanced produced have always been prioritized [33].

### 4.1.1 IMPACT OF CHEMICAL CONSTITUENTS ON AGRICULTURE

Chemical constituents are often called agrochemicals and are broadly classified into three categories:

- Chemical fertilizers;
- Pesticides; and
- Plant growth regulator.

#### 4.1.1.1 CHEMICAL FERTILIZERS

Chemical fertilizers are referred to as synthetic compounds possessing a high concentration of nutrients which are essential for plant growth and development [5, 21]. Chemical fertilizers contain synthetic form of macronutrients and micronutrients. Nitrogen, phosphorous, potassium, sulfur, and magnesium, which falls under the category of micronutrients are broadly used for making chemical fertilizers [6, 7]. Often chemical fertilizers are also distinguished on the basis of type and form of nutrients present in it (Figure 4.1):

- Nitrogenous fertilizers;
- Phosphate fertilizers;
- Potassic fertilizers.

#### 4.1.1.2 PESTICIDES

Pesticides are those chemical substances which are meant to control pests. Pesticides can be of both chemical and biological origin. Pesticide spraying can substantially reduce the economic loss due to pesticide attacks [8]. Most of the pesticides are used to serve as plant protection products (Table 4.1).

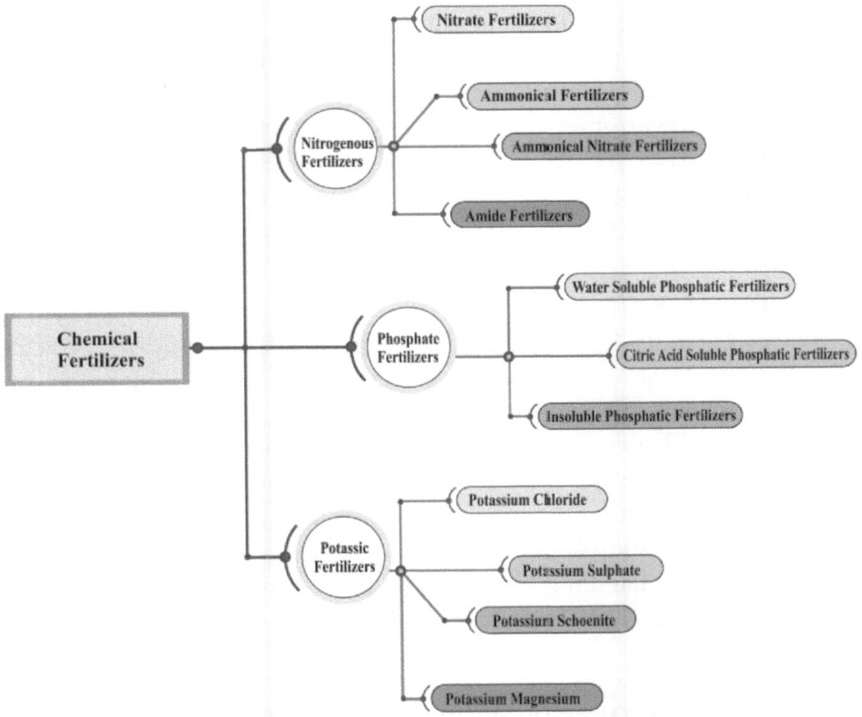


**FIGURE 4.1** Different types of chemical fertilizers on the basis of the form of nutrients present in it.

**TABLE 4.1** Classification of Pesticides on the basis of Target Organisms, Chemical Structure, and Their Physical State

| Target Organisms | Chemical Structure | Physical State |
|---|---|---|
| Herbicides | Organic | Solid |
| Insecticides | Inorganic | Liquid |
| Fungicides | – | Gaseous |
| Rodenticides | – | – |

#### *4.1.1.3 PLANT GROWTH REGULATORS*

Plant growth regulators are the third class of chemicals which are used in agriculture for growth promotion and better yield. There are several well-known plant growth regulators currently getting used globally, such as auxins, cytokinin, naphthalene acetic acid [9].

Application of pesticides and fertilizers over the crops helps in enhancing agricultural products and ensures timely supply of its products. However, many countries have announced an alarming situation due to excessive accumulation of pesticides and fertilizers in the food chain. Presence of agrochemical residues in air, soil, water, and its accumulation in adipose tissues of humans has pose a threat to our ecosystem [37].

- **Key Facts:**
  - Uncontrolled use of fertilizers and pesticides may induce acute or either chronic health effects on the basis of situation and quantity of exposure.
  - Rapid degradation of pesticides or fertilizers cannot be assured. Few countries have banned the list of highly toxic compounds for agricultural purposes, whereas they are still getting used in neighboring countries.
  - People handling these agrochemicals are more prone to health risk [9, 23].

### *4.1.2 WHO RESPONSE OVER CURRENT SITUATION?*

World Health Organization (WHO) along with 'Food and Agriculture Organization of United Nation (FAO)' have initiated several programs to assess the risk of pesticide exposure to humans either through direct exposure or through the residues available in food. WHO has expressed their concern over the harmful effect induced by inadequate use of agrochemicals on humans and the environment [12]. They stated that usage of pesticides to produce food, to feed both local population and for export purposes must follow approved agricultural practices irrespective of the economic status of a particular country. Strict guidelines need to be followed to limit the amount of pesticide used to the minimum needed to ensure crop protection [39].

#### 4.1.2.1 ASSESSMENT OF RISK FACTORS

At regular intervals, risk assessment of pesticide traces in food were carried out by Joint FAO/WHO. Risk assessment is based on the data submitted by national registration dealing with pesticides. Along with this, scientific studies published in National and International journals are also considered for carrying out the assessment. Based on the above-mentioned parameters of evaluation, safer limits for pesticide usage have been assigned. Maximum possible limit of pesticide in food intake have also been fixed to reduce the adverse health effects imposed by the agro-chemicals on human body [14, 30].

#### 4.1.2.2 CODEX ALIMENTARIUM COMMISSION (CAC)

CAC is the major body responsible for dealing with matters concerned with the implementation of programs initiated by FAO/WHO. CAC was established in November 1961 by FAO of the United States with the prime goal to protect consumer health and further ensures application of fair practices in international food trade [28]. Matter of concerns of CAC:

- Food labeling;
- Food additives;
- Contaminants;
- Pesticides and chemical residues in food;
- Risk assessment;
- Food hygiene;
- Methods of analysis and sampling.

### 4.1.3 ROLE OF AGROCHEMICALS IN AGRICULTURE

Agricultural production has profoundly enhanced since the $20^{th}$ century to deal the demographic growth. In the last century, a huge population explosion was observed with a rise from 1.5 to 6.1 billion from 1990 to 2000. From the presented figures it can be predicted that the world population may jump to 9.4–10 billion by 2050. Such a rapid increase in the world population rate would not have been possible in the absence of adequate amount of food supply needed to ensure sustainability of people and this could have been achieved only by the application of fertilizers [4]. But the agricultural practices adopted to meet such high food needs includes vast use of chemicals, which poses the ability to induce detrimental health effects both on humans and animals and simultaneously has adversely affected our ecosystem [23, 47] (Figure 4.2).

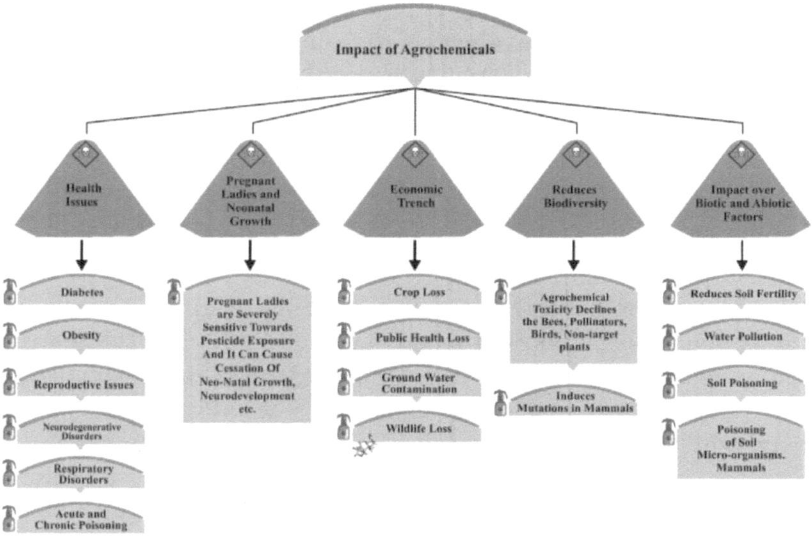


**FIGURE 4.2** Negative impact of agrochemicals over the environment and human life.

Several groups of agrochemicals named as organochlorides, growth regulators, organophosphates, pyrethroids, neonicotinoids, and a lot more have been developed so far. Sales of agrochemicals have shown an abrupt increase in the past few years with enhanced food demand ultimately stressing on the productivity [18].

Agriculture industrialization has brought a series of problem in front of Government Organizations including social, economic, and environmental impact. Current agriculture model has exaggerated the chemical burden on the local population. Concrete efforts are needed to adopt globally to deal with the current burden of hazardous pesticides. Several researchers revealed that pesticide ingestion has caused a 60% suicidal rate in rural areas belonging to China and in South East Asia. Till 2002, 877,000 people attempted suicide worldwide by ingesting pesticides. Apart from the death issues, there are several unwanted consequences of pesticide exposure, such as non-fatal self-harm, occupational poisoning, and death due to accidental exposure [11].

WHO/FAO revealed that 30% of pesticides available in developing countries did not meet the quality standards set up by the International authorities. Trade of illegal and forged pesticides have remained a major concern. Lack of coordination and collaboration was observed between South East Asian Countries and Pesticide Regulatory Authorities. Similarly, non-cooperative behavior was also observed between Government bodies and other stakeholders. All these issues have significantly contributed to death rate caused due to pesticide exposure [20]. Worldwide total suicide magnitude was computed in the WHO region and it was observed to be 877,000 (Table 4.2).

**TABLE 4.2** Total Suicidal Magnitude around the World Due to Agrochemical Poisoning

| WHO Region | Figure of Suicide Cases | Disability Adjusted Life Years (DALYs) |
|---|---|---|
| America | 63,000 | 1.0 |
| Africa | 34,000 | 0.2 |
| Southeast Asia | 246,000 | 1.7 |
| Eastern Mediterranean | 34,000 | 0.7 |
| Western Pacific | 333,000 | 2.6 |
| Europe | 164,000 | 2.3 |
| Worldwide | 877,000 | 1.4 |

➢ **Key Observations:**

- On an average, 3 million cases involved in pesticide poisoning are observed every year globally, which is possibly due to the increasing consumption of pesticides across the globe (Table 4.3 and Figure 4.3).
- Above mentioned mortality rate reflects around 900,000 people worldwide who are dying every year due to pesticide exposure.
- Intentional or either unintentional pesticide poisoning is considered as serious threat, especially in lower- and middle-income countries, including India, Sri Lanka, Vietnam, and China. Sales of agrochemical has seen an emergent growth in the past few years, which can be clearly predicted from their revenues (Table 4.4).

**TABLE 4.3** Pesticide Sales in 2016 Across the World

| Country | Quantity (Millions of Kilograms) |
|---|---|
| Spain | 78 |
| France | 73 |
| Italy | 60 |
| Germany | 47 |
| Romania | 11 |
| Portugal | 10 |
| Hungary | 10 |
| Belgium | 8 |
| Turkey | 50 |
| Czech Republic | 6 |
| Greece | 5 |
| Finland | 5 |
| Austria | 4.5 |
| Ireland | 4 |
| Denmark | 4 |
| Slovakia | 3 |
| Goetia | 3 |
| Latvia | 2 |
| Slovenia | 1 |

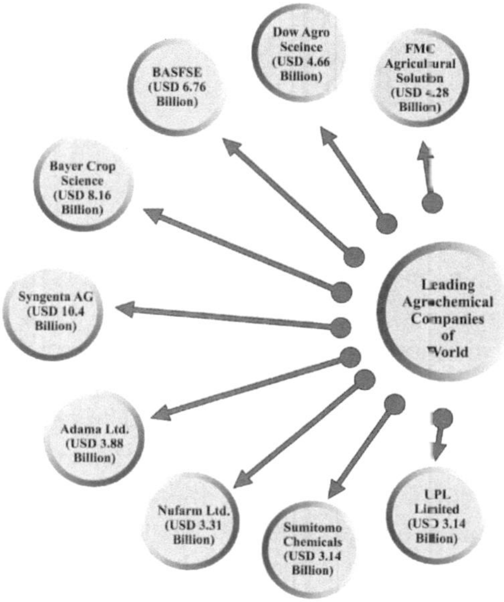


**FIGURE 4.3** Top most leading agrochemical companies around the world.

**TABLE 4.4** Expected Global Agrochemical Market Size in Coming Years in USD Billion Dollars

| Year | Market Size (USD Billions) |
|---|---|
| 2018 | 223.7 |
| 2019 | 234.2 |
| 2020 | 245.19 |
| 2021 | 256.7 |
| 2022 | 268.75 |
| 2023 | 281.37 |
| 2024 | 294.57 |
| 2025 | 308.4 |

- There exists an immense need to improve pesticide policies to decrease the morbidity and mortality rate caused because of pesticide poisoning [11, 26].
- Mental health of people is greatly affected due to pesticide poisoning in health care centers at both the National and International levels.
- To develop or strengthen the community programs to significantly reduce the risk of intentional or unintentional acute or chronic poisoning.
- To establish regional centers for prevention and management of pesticide poisoning to increase training, surveillance, and community action [22, 40].

## 4.2 STANDARD PARAMETERS SETUP BY USDA FOR UTILIZATION OF AGROCHEMICALS

Trend of pesticide utilization across the world have not remained the same and the possible reason behind this includes cost of chemicals (in which maximum of them are patented), man power costs and pests of particular climatic or geographical locations. WHO has estimated an average pesticide application rate per hectare of arable land which comes under the range of 6.5–60 kg/ha in Asia and a few of South American Countries. In Northern America and West Europe, herbicide usage has also bloomed in urban areas in the past few decades. While in Asia, herbicide usage remained low and contrasting in comparison to insecticides which is expected to be very high [2, 23].

Policymakers have admitted that unrestricted and un-systematic application of agrochemicals is an obstacle towards the sustainable development of agriculture. Several countries have designed policies to regulate the volume of usage and different types of agrochemicals applied [24].

- **1972 Federal Environmental Pesticide Control Act (FEPCA):** In the USA, FEPCA, and other amendments have recognized the hazardous impact of pesticide application on human health and on the ecosystem and have enforced compliances against the utilization of banned pesticides.
- **European Union (EU) Regulation:** In 2003, EU Legislation EC No. 2003/2003 have finalized the norms that, fertilizers involved in electrical conductivity should specifically follow the criteria regarding nutrient constituents, safety prospects and imposition of devastating effects over the environment due to the un-systematic use of agrochemicals. In 2015, an attempt named as 'Action to Achieve Zero Growth in Application of Fertilizers' was initiated by the Chinese Ministry with the prime goal of safeguarding and controlling the usage of agrochemicals [59].
- **Global Collaborative for Development of Pesticide for Public Health (GCDPP):** GCDPP was established for promoting safe and cautious usage of fertilizers in the interest of public health [18, 19]. Members of GCDPP includes:
    - National and Government Supported Agencies;
    - Research Institutes and Universities;
    - Regional and International Organizations.

It strengthens the collaboration with industry to make it:

- Cost-effective;
- Operationally acceptable;
- For identification of new technology;
- To promote the development of alternative compounds.

### *4.2.1 WHO PESTICIDE EVALUATION SCHEME (WHOPES)*

WHOPES was established in 1960 to promote and coordinate the evaluation and testing of pesticides for human health. The prime members of the scheme include Government officials, WHO collaborating Centers,

Research institutes and manufacturers of pesticides. WHOPES comprises of a four-stage evaluation and testing program [27, 60]:

- Assessment of safety;
- Determination of efficiency;
- Evaluation of operational feasibility;
- Development of specifications for quality control and international trade.

➢ **Major Goal of WHOPES:**
  - To expedite the hunt for less hazardous alternative pesticides, to search safe and cost-effective methods of application.
  - To enlarge and promote different policies, guidelines, and strategies for the specific and sensible application to benefit public health.
  - Proper monitoring of implemented strategies by Member States [13] (Figure 4.4).

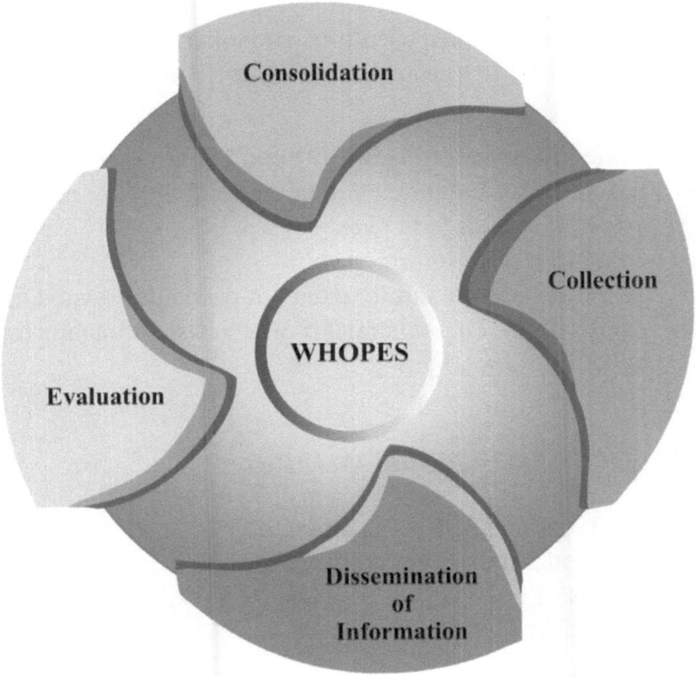


**FIGURE 4.4** Major goals of WHO Pesticide Evaluation Scheme.

- **Progress Made by WHOPES so far:**
  - Application of three types of pesticide equipment since 1997, testing, and evaluation of approximately 13 agricultural products.
  - 29 technical materials along with 53 formulations of public health pesticides were developed and revised on a regular basis.
  - Specific guidelines for vector-control equipment, insecticidal products and bacterial larvicides were developed.
  - FAO and WHO guidelines were further strengthened for the development of pesticide specifications [53].

### *4.2.2 POSSIBLE PREVENTIVE APPROACH ADOPTED BY POLICY MAKER*

- Building and maintenance of proper food system and infrastructure needed for it to manage risks associated with food safety covering entire food chain and emergency situations.
- Establishment of multi-sectorial collaborations covering public health, animal health, agriculture, and other sectors.
- Integration of food security into wide food programs and other policies.

- **Possible Initiatives Adopted by Food Handlers and Consumers:**
  To provide information regarding the food they are using:
  - To read label on food package;
  - To make a prior informed choice;
  - To get them more familiarize with pesticides related hazards;
  - Application of agrochemicals on the basis of parameters set up by USDA [30, 50].

### *4.2.3 PESTICIDE DATA PROGRAM (PDP)*

Pesticide Data Program (PDP) was initiated in 1991 as a fragment of the USDA Global Food Safety Initiative (GFSI). PDP in general is a National Pesticide Database Program [15, 31]. Data generated by PDP is prominently considered by:

- Federal Agencies;
- Academic Institutions;

- Food Producers;
- Food Processors;
- Chemical Manufacturers;
- Environmental Interest Groups;
- Food Safety Organizations.

With the aim to detect pesticide residues presence in food products which may potentially affect agricultural practices, its domestic and international trade (Figure 4.5). With the help of laboratory techniques, PDP is involved in testing both fresh and processed vegetables, grains, meat, fruit, dairy, poultry, drinking water and certain other commodities such as honey, fish, nuts, corn syrup to detect presence of any pesticide residues which possibly can be present in it. The prime focus of PDP is to provide the analysis of food consumed primarily by infants and children [32]. The need to include pesticides and other commodities in PDP is determined on the basis of EPA data (Figure 4.6).

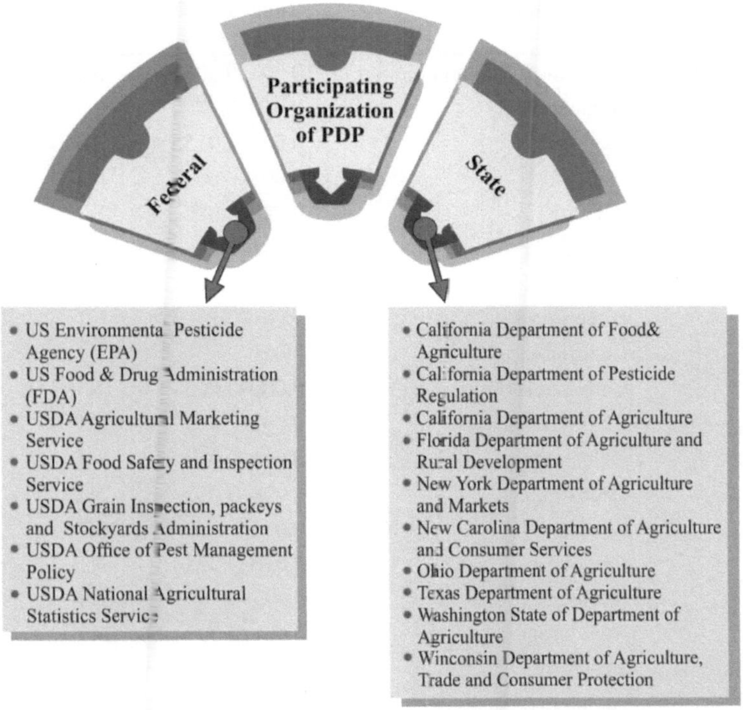


**FIGURE 4.5** Federal and state-level participating organizations of Pesticide Data Program.

### 4.2.4 PESTICIDE LIMIT IN FRUITS AND VEGETABLES

PDP have revealed residue availability in fruits and vegetables/commodity pairs in terms of:

- Limits of detection (LODs).
- U.S. Environmental Protection Agency tolerance for each pair.

In 2018, PDP evaluated 9,257 samples of fruits and vegetables, which includes 4949 fresh and 4308 processed products. The report also revealed tolerance violations to FDA. The below-mentioned table depicts the pesticide/commodity tolerance violation limit [33, 34] (Table 4.5).

Similarly, in 2018, PNP analyzed rice, wheat flour, and heavy cream for estimating available agrochemical residues in it whether they fall under the permissible limits in it (Figure 4.6).

**TABLE 4.5** EPA Tolerance Limit and LOD of Wide Category of Agrochemicals

| Category | Agrochemical | Range of LOD (ppm) | EPA Tolerance Level (ppm) |
|---|---|---|---|
| Insecticide | 2,4-Dimethylphenyl formamide | 0.003–0.005 | NT |
| | Abamectin | 0.050–0.10 | 0.01 |
| | Acephate | 0.005–0.015 | 0.02 |
| | Aldicarb | 0.001–0.0020 | NT |
| | Allethrin | 0.020–0.080 | NT |
| | Aspen | 0.005 | NT |
| | Bendiocarb | 0.001–0.003 | NT |
| | Bifenthrin | 0.001–0.10 | NT |
| | Cadusafos | 0.001 | NT |
| | Carbophenothion | 0.001–0.006 | NT |
| | Chlorpyrifos | 0.001–0.015 | NT |
| | Deltamethrin | 0.001–0.050 | NT |
| | Endosulfan | 0.008–0.030 | NT |
| | Famphur | 0.001 | NT |
| | Imipothrin | 0.010–0.095 | NT |
| | Neoprene | 0.005–0.10 | NT |
| | Novaluron | 0.001–0.005 | NT |
| Plant growth regulator | 2,6-DIPN | 0.003–0.005 | NT |
| | Paclobutrazol | 0.001–0.040 | NT |
| | Uniconazole | 0.001 | NT |

**TABLE 4.5** *(Continued)*

| Category | Agrochemical | Range of LOD (ppm) | EPA Tolerance Level (ppm) |
|---|---|---|---|
| Herbicide | Acetochlor | 0.001–0.050 | 0.05 |
| | Aclonifen | 0.001 | NT |
| | Alcohol | 0.002–0.020 | NT |
| | Ametrine | 0.001–0.005 | NT |
| | Amicarbazone | 0.005 | NT |
| | Anilofos | 0.001 | NT |
| | Asylum | 0.001 | NT |
| | Beflubutamid | 0.001 | NT |
| | Chlorsulfuron | 0.001 | NT |
| | Dichlobenil | 0.001 | NT |
| | Ethnoformesate | 0.050 | NT |
| | Hexazinone | 0.001–0.005 | NT |
| | Pencil | 0.005 | NT |
| | Mesotrione | 0.020–0.050 | NT |
| | Metolachlor | 0.001–0.005 | NT |
| | Oryzalin | 0.020–0.20 | NT |
| | Oxyfluorfen | 0.001–0.040 | NT |
| | Simazine | 0.001–0.005 | NT |
| | Sulfosulfuron | 0.001 | NT |
| | Tri-Allate | 0.001–0.005 | NT |
| Fungicide | Ametoctradin | 0.001–3.4 | 50.0 |
| | Benalaxyl | 0.005 | NT |
| | Benthiavalicarb isopropyl | 0.001–0.010 | NT |
| | Carboxy | 0.003–0.0025 | NT |
| | Dimethomorph | 0.01–0.020 | NT |
| | Fenarimol | 0.01–0.015 | NT |
| | Fenthion | 0.001–0.015 | NT |
| | Imazali | 0.001–0.005 | NT |
| | Iprodione | 0.005–0.075 | NT |
| | Mepanipyrin | 0.001 | NT |
| | Penconazole | 0.001–0.010 | NT |
| | Tebuconazole | 0.001–0.010 | NT |
| | Vinclozolin | 0.001–0.005 | NT |
| Plant activator | Acibenzolar S Methyl | 0.004–0.024 | NT |

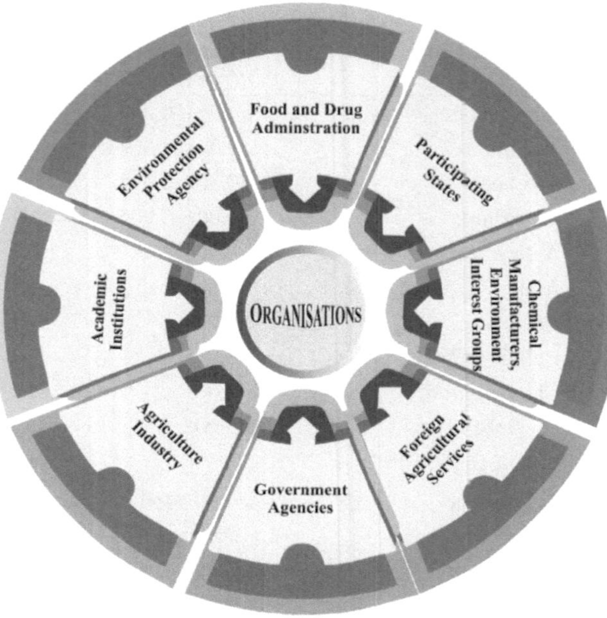


**FIGURE 4.6** PDP policy and planning contributors.

## 4.2.5 PESTICIDE LEGISLATIONS

1. **Insecticidal Act of 1910:** This particular act has banned the manufacturing, sale or delivery of the pesticides which are either adulterated or misbranded. It protects farmers from fake or duplicate products by disrupting the supply chain of misleading products.
2. **Federal Food, Drug, and Cosmetic Act of 1938 (FFDCA):** It facilitates safe tolerance limit to be set for residues belonging to unavoidable poisonous agrochemicals available in fresh or processed food [14].
3. **Federal Insecticide, Fungicide, and Rodenticide Act of 1947 (FIFRA):** It ensures the registration of pesticide before commencement of its soil. Under this act, a product needs to be properly labeled along with its content specifications. It is mandatory to specify the poisonous substances on the label present in agrochemical.
4. **Miller Amendment to FFDCA of 1954:** It amended the FFDCA with the focus to mention pesticide tolerance limit for food and

feed. Risks and benefits associated with the usage of pesticides have also been considered while setting up the tolerance standards.

5. **Food Additives Amendment to FFDCA of 1958:** FFDCA was amended to authorize the presence of food additives. Pesticide residues present in processed food are called food additives. When the pesticide residues were applied to raw agricultural commodity, they are considered as processed products and were decided to be not called food additives if its concentration falls under the permissible limits [58].
6. **Federal Environmental Pest Control Act (EPCA) of 1972:** Certain amendments have been made in FIFRA to increase the authority towards pesticide regulation. Registration of pesticide is permissible only if it does not cause any unreasonable adverse effects to the environment including humans.
7. **Federal Pesticide Act of 1978:** The present act has reviewed the previously registered policies for pesticide registration.
8. **FIFRA Amendments of 1988:** Act has accelerated the pesticide registration process, but all the active ingredients of the pesticides are needed to be registered with the authority. According to the act, additional financial support will also be provided, and the annual maintenance fees which are generally imposed for the registration is also levied [54].
9. **Food Quality Protection Act of 1996 (FQPA):** FIFRA and FFDCA have been amended to regulate the safety standards to ensure the complete avoidance of health hazards. Pesticide residues have no more remained to be Delaney Clause of FDCA. Fresh as well as processed food which might contain pesticides but it should fall under the 'carcinogen tolerance limit' which is considered to be safe. EPA needs to revise the pesticide tolerance limit at regular intervals.
10. **Pesticide Registration Improvement Act of 2003:** FIFRA were amended to cover long lists of pesticides, to provide free service for registration actions involved in the pollution prevention, anti-microbial activity, and bio-pesticides [26, 36, 40].

### *4.2.6 RULES AND REGULATIONS IMPOSED BY USDA*

- **Pesticide Record-Keeping (PRK):** Along with the 1990 farm bill, private applicators are needed to keep a record of consumption of restricted pesticides. But the PRP operation abruptly stopped in September 2013 due to the non-availability of funding sources. But no standard Federal Form is required for maintaining application of 'restricted use pesticide (RUP)' [3, 44]. There are nine major elements which are needed to be maintained within 14 days of RUP, which areas in subsections (Figure 4.7).

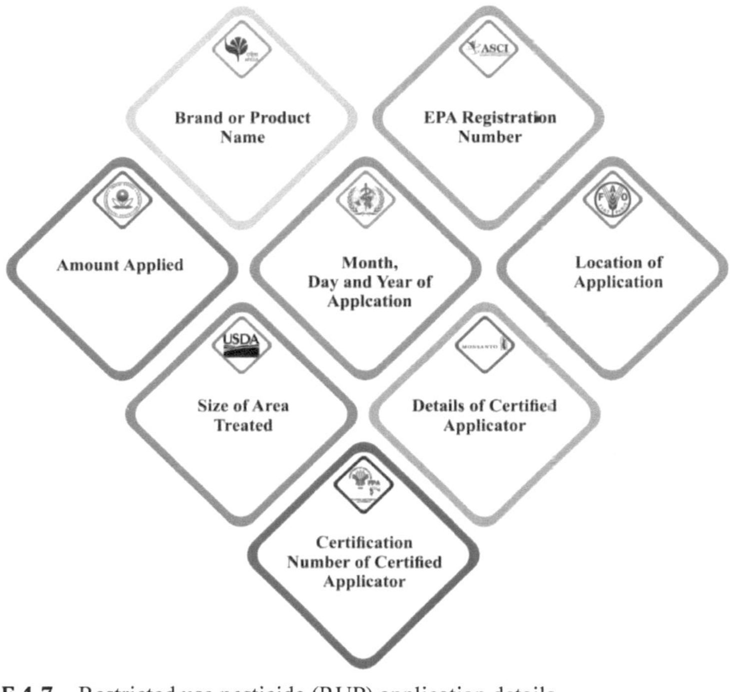


**FIGURE 4.7** Restricted use pesticide (RUP) application details.

### *4.2.6.1 MATERIAL SAFETY DATA SHEET (MSDS)*

MSDS contains entire information regarding potential hazardous impacts, including health reactivity, environmental effect, and how to deal with these chemical substances while working with them. MSDS is the preliminary standpoint that needs to be fulfilled for initiating health and safety programs. It also includes information regarding its usage, storage, handling, and emergency precautionary measures. MSDS comprises of detailed information even more than the labeled ones. It also reflects the hazardous impacts associated with the usage of chemicals. At the international platform, many countries have made it mandatory to accompany the MSDS details along with chemical products [24, 25]. The details available on MSDS involves (Figure 4.8).

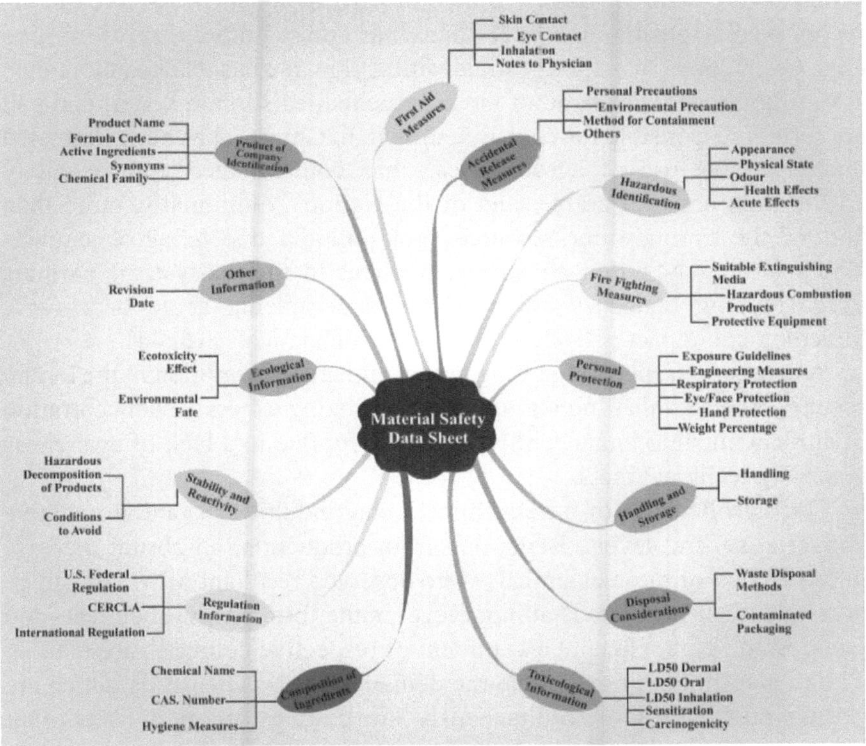


**FIGURE 4.8** Details need to be available on the material safety data sheet.

## 4.3 AN ACCOUNT OF NON-STANDARDIZED CONSUMPTION OF FERTILIZERS AND PESTICIDES

Research carried out at both the national and international levels revealed that the uncontrolled usage of inorganic fertilizers is directly linked to massive accumulation of its residues such as cadmium (Cd), mercury (Hg), arsenic (As) in agricultural soil. Surveys conducted by the U.S. Geological

Survey unit revealed that in 51 paramount river basins and aquifer, 97% of time samples were detected with a massive amount of pesticides in it. In Japan, pesticide residues were detected in air, which further contributes to air pollution [44, 57].

Condition is observed to be more severe in the case of developing countries and also made a significant contribution in causing pollution. Whereas in China, the usage volume of agrochemicals is observed at the record peak. Consumption of fertilizer has crossed more than 59 million tons. On the other hand, the pesticide utilization rate has reached more than 1.8 million tons. Discharge of these agrochemicals in air, water, and soil are further responsible for contaminating it. In China, it has been estimated that excessive use of agrochemicals has contaminated approximately 150 million acres of arable land of the Nation. Additionally, more than half of the groundwater resources got polluted due to agrochemicals which ultimately posed an emergent threat to the ecosystem. Farmers of developing countries are currently experiencing acute or chronic toxicological impact due to exposure to agrochemicals [16, 44].

A number of reports regarding agrochemical consumption in the Indian scenario revealed that Indian farmers are utilizing excessive concentration of agrochemical to enhance their productivity due to a lack of awareness regarding technical facts.

Uncontrolled use of agrochemicals may exhibit substantial negative impact at several levels. Beginning from production, an abrupt increase in production of these chemicals were observed resultant of which a huge investment has been made at input level in the form of raw chemicals and fuels. At the global level, Government of respective countries are making a huge investment to overcome the demand for agrochemicals which are getting produced and simultaneously are getting imported from other countries [38, 51].

At the secondary level, these agrochemicals are further responsible for contaminating and inducing air, soil, and water pollution. Moreover, industries engaged in the production of agrochemicals are responsible for discharging industrial effluents enriched with heavy metals and toxic chemicals into the environment. They are disrupting and disturbing the ecosystem, contaminating groundwater, several acute and chronic disorders such as cancer, diabetes, respiratory, and digestive disorders. Evidences emerged out revealed that pesticide poisoning has deeply affected South America and Africa's economy. A high suicidal rate in tobacco-growing regions of Brazil was observed, which is depicted to be due to the wider availability and use of pesticides. In the Southern rural area of Trinidad, 80% of suicide cases were found to be due to pesticide poisoning. Both fatal suicidal cases, which accounts for 55% of suicidal cases, involved excessive pesticide exposure in Suriname [12, 48].

Data obtained from American countries, especially from Zimbabwe, revealed that poisoning due to organophosphate exposure has led to three-quarters of hospital admission due to suicidal behavior [46] (Figure 4.9).

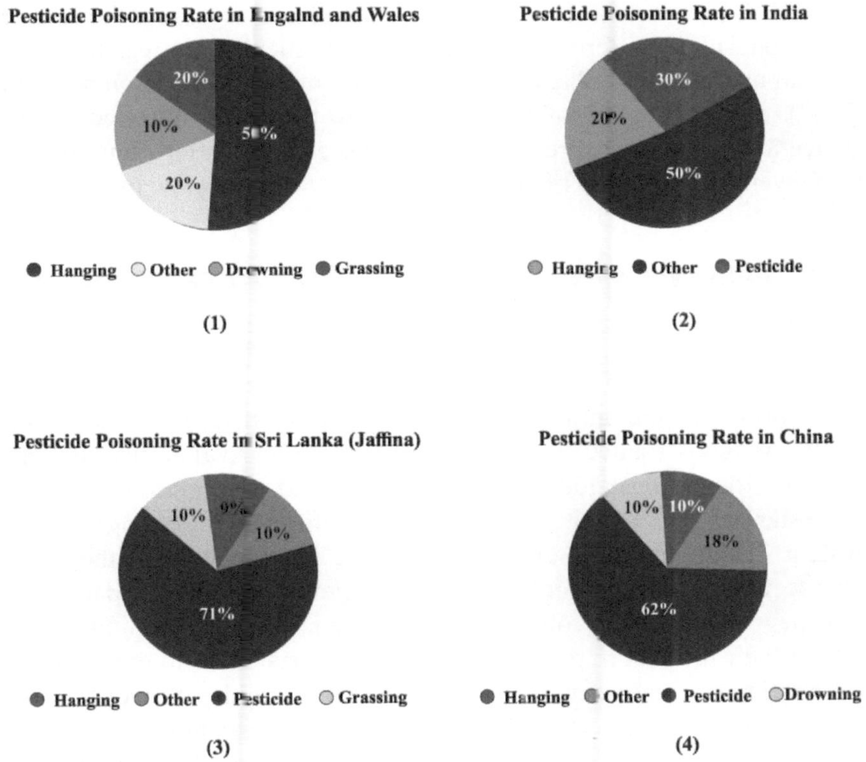


**FIGURE 4.9** Pesticide poisoning rate worldwide.

### 4.3.1 IMBALANCE IN CONSUMPTION OF AGROCHEMICALS

Data regarding exact consumption of worldwide is slightly controversial. Indian ranks 2nd worldwide after China in consumption of agrochemicals.

According to fertilizer association of India, production, and consumption of fertilizers have increased from 201.6 and 65.6 tons to 41427.8 and 25949.9 tons, respectively since 1951–1952. A lot of imbalance in consumption of agrochemicals is also observed along with an abrupt increase in its production and consumption in India. In States like Punjab and Haryana, the ratio of NPK usage is higher than 31.4:8.0:1 and 27.7:6.1:1, respectively. Excessive usage of agrochemicals has forced Government to impose a ban on the overuse of certain excessive toxic or hazardous chemical constituents [6, 41].

In U.S, ban over its 11 States have been placed regarding the sale or use of phosphorous fertilizers. Around 288 pesticides are registered for rational use alone in India. As per the reports of pesticide action network (PAN), out of 288, use of 85 pesticides are banned in other countries. Hence it can be clearly predicted that norms and regulations imposed by the Government overuse of agrochemicals vary country wise [43, 55].

### *4.3.2 PRESENCE OF NON-GENUINE AND ILLEGAL PESTICIDES IN MARKET*

Globally, presence of illegal pesticides in markets have always raised a question over the authorized bodies regarding strict implementation of the protocol framed by Government officials. Availability of non-genuine, illegal pesticides have greatly affected several countries and a substantial effect of it was observed against the Indian agriculture system. Technical grade agrochemicals have been imported without any license from Central Boards and Registration Committee. Apart from this, new issues have evolved in which counterfeits are facilitating the sale of their product in the name of 'Bio-products' so as to escape from registration processes. In India alone illegal pesticides of worth value 3200 Cr are currently available in market which accounts for 25% by value of current 30% by volume of domestic pesticide industry. The current system is expected to grow by 20% approximately with every passing year [15, 17, 36].

Non-genuine/illegal pesticides have induced substantial negative effect on the environment. Environmental degradation, impact over soil fertility, targeting non-target organisms are resultant of all these circumstances.

#### *4.3.2.1 INAPPROPRIATE UTILIZATION OF AGROCHEMICALS*

Pesticide concentration either consumed directly or come in contact with pests are extremely smaller in comparison to the amount of application. It has been estimated that only 0.1% of applied agrochemicals falls on targeted pests which reflects that 99.9% of pesticide moved into environment which potentially may adversely affect flora and fauna and can further contaminate air, water, and soil. Several reasons are found to be affecting the target missing of pesticide, but the most prominent among them is their mode of application and the extent to which person engaged is spraying pesticide whether, has been trained properly or not.

Agrochemicals poses a great threat to our ecosystem because of their fat solubility rate and bioaccumulation in the non-targeted organisms [23]. Potential hazardous impact exhibited by agrochemicals over the ecosystem (Figure 4.10):

- Weakens the aquatic life by lowering the sunlight exposure;
- May induce toxicity in humans, aquatic organisms, and wildlife;
- Leads to eutrophication;
- Induces mutagenic and carcinogenic effects when consumed above the tolerance level [34, 36, 51].

## 4.4 CURRENT CHALLENGES AND ACTION PLAN FOR REHABILITATION OF SITUATION

Pesticide expenditures are directly associated with pesticide use. In the past half-century, U.S pesticide expenditure has observed steady growth. Pesticide expenditure increased from 2.3 to 12 billion dollars from 1960 to 2008. Despite over the continuous arguments over the potential hazards imposed by pesticides over both humans and the ecosystem. These escalated concerns raised recently due to the lost trust over the agricultural and industrial methods carried out for production and on the concerned authority involved in framing rules and regulations to control the situation [49, 52]. Therefore, various uncertainties also need to be evaluated regarding:

- Pesticide safety;
- Scientific data;
- Policy guidelines;

- Professional judgment must be considered while evaluating that whether the pesticide can be utilized in a beneficial manner within the permissible limits or not.

The possibility to overcome the risk-on environment because of pesticide exposure is very less. Furthermore, the producers believed that lowering the risk is associated with either increased input or reduced output as a resultant of the substitution of pesticide inputs. Therefore, policies are needed to be implemented for reducing the possible negative effect related to the utilization of agrochemicals which has put pressure on agricultural commodity regarding its safety aspects [29].

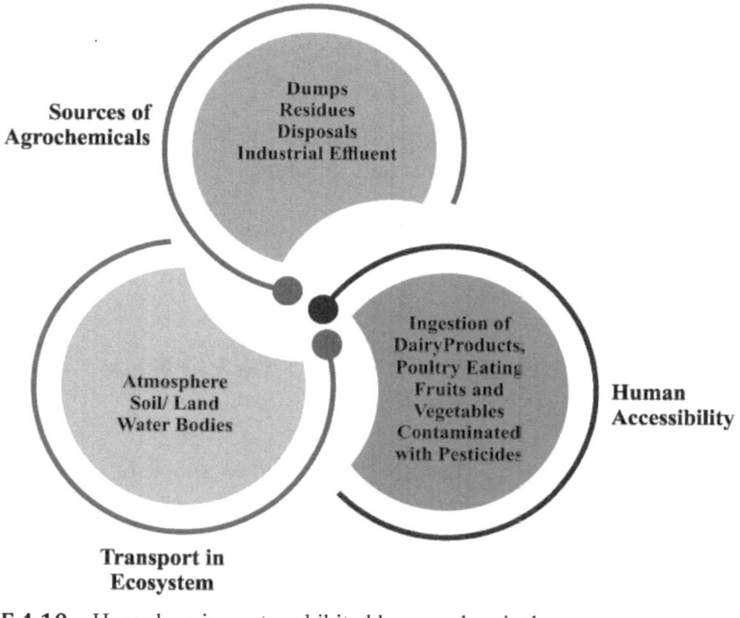


**FIGURE 4.10** Hazardous impacts exhibited by agrochemicals on ecosystem.

### 4.4.1 COST FUNCTION-BASED PREDICTION MODEL

The present model was proposed with the goal to focus on the substantive cost which could be possibly imposed on the present sector with the aim to decrease the risk on environment due to excessive pesticide exposure. Present cost is directly linked with enlarged demand of active pesticides [56].

Concerns regarding the adverse impact of pesticide application on health and environment has led the EU to adopt a 'Thematic Strategy on Sustainable Use of Pesticides.' Scientists belonging to the Agriculture community have initiated an alternative crop management system to reduce the hazardous effect of farming practices on the environment and human health. To be specific, integrated crop management (ICP) practices includes guidelines which are needed to be followed by farmer union to implement the actions needed to produce safe agriculture products along with the prior concern over the environment. Along with this, ICM promotes the usage of complementary methods for pest management with the aim to decline the animal pest or population of weeds below the economic injury level to reduce the negative impact of agrochemicals over the ecosystem [10, 22, 30].

Due to the increasing concerns regarding pesticide usage, ICM allowed pesticides to be used only through the IPM program. Few criteria have already been framed concerning the pesticide selection. Certain set of protocols are needed to be followed regarding the pesticide application over the crops. Analysis of pesticide residues have been prioritized in IPM. Pesticide which are authorized to be used in IPM should be:

- Biologically effective;
- User friendly;
- Environmentally compatible;
- Economical;
- Profitable.

Specific set up of instructions which are needed to be followed during pesticide spraying involves:

- Application of pesticides in permissible limits either in presence of disease-causing organisms or as a precautionary measure.
- Optimize the pesticide to ensure economic saving by adjusting the pesticide dose according to population density of pest.
- Minimize the dependence on pesticides by making certain alterations in the cultivation system, thereby lowering the pesticide risk [13, 34].

All the above-mentioned preventive measures indicated that the IPM introduction could potentially contribute towards the reduction of pesticide usage and can significantly reduce the negative effect of pesticide on human health and our ecosystem without harming the crop productivity level and any possible loss on the crop due to pest attack [35].

Along with above mentioned precaution measures, chemical crop protection has also got altered a lot in the past few years, not only in terms of technological advancements in active ingredients but also in the evaluation of behavior and impact of chemicals over the environment. Availability and degradation of these chemicals in crops plants and the potential side effects associated with their usage will also be evaluated under this initiative. This is further ascribed as the technological advancement in several disciplines such as molecular biology, chemistry, and has certainly improved the method of searching for the development of new agrochemicals with their novel mode of action with improved version of safety profile. These novel agrochemicals along with their effective and appropriate application procedure may be considered safe and reliable thus making the chemical crop protection method as the best suited technological advancement in agriculture sector and it is predicted to play a vital role in agribusiness despite of swift emergence of biotechnological advancements [30, 42, 52].

## 4.5 SUMMARY

Agricultural inputs have played a pivotal role in uplifting its status globally. The application of agrochemicals over the crops has ensured timely availability of products to feed the growing population worldwide. To deal with demographic growth, a substantial increase in agrochemical consumption has raised several questions in front of WHO, USDA, FAO, etc. Reflective analysis of current paradigm revealed that these agrochemicals are posing a great threat to the human population and our environment. Accumulation of its residues in air, water, soil, and adipose tissues of humans has led to the emergence of several acute and chronic disorders. WHO and FAO initiated several programs for proper management of their usage. They have clearly stated that at both the national and international levels, it is mandatory to follow the guidelines imposed by USDA, irrespective of economic status of particular country. Codex Alimentarium Commission was brought into power to ensure implementation of fair practices in international food trade. Health issues, economic trench, reduction in biodiversity, and negative impact over biotic and abiotic factors have still remained to be the major issues of concern. WHO has reported 877,000 suicide cases due to pesticide poisoning and 1.4 of disability-adjusted life year. On an average mortality rate of 900,000 is getting observed every year on account of pesticide poisoning. To mark a regulatory check, many countries have framed efficient policies to monitor the volume of usage and type of agrochemical being applied. In this initiative, WHOPES played a prominent role by initiating the hunt for safe, cost-effective, and less hazardous alternative pesticides. Discrete guidelines for analysis of agricultural products and vector control equipment were commenced. PDP was recorded to be the supportive strategy of the USDA global food safety program to analyze the presence of pesticide residues in food products and have set up an EPA tolerance limit and LOD. More than 15 pesticide legislation have been initiated for proper monitoring of the channel. Material safety data sheets (MSDS) have further brought transparency between the manufacturer and end consumer regarding the hazardous impact associated with the usage of agrochemicals, if any. Many more legislative approaches, monitoring, and regulatory parameters are needed to be brought into practice to meet the current challenges linked with agrochemical consumption and rehabilitation of the situation enforced due to it.

## ACKNOWLEDGMENTS

We would like to thank Dr. Bhaskar Joshi for supporting us throughout the work. The authors are grateful to Mr. Ajeet Chauhan for graphical support. The authors are highly indebted to Dr. Rishi Arya and Ms. Mamta Chaudhary for their administrative support.

## KEYWORDS

- **Agrochemicals**
- **Codex Alimentarium Commission**
- **Federal Environmental Pesticide Control Act**
- **Food and Agriculture Organization**
- **integrated crop management**
- **legislations**
- **pesticide action network**
- **WHO Pesticide Evaluation Scheme**
- **World Health Organization**

## REFERENCES


1. Adaime, M. B., Botega, M. P., Prestes, O. D., & Zanella, R., (2014). Pesticides and the environment: Insertion of the theme in school through an interdisciplinary approach. *Science and Natura, 36*(2), 250–257.
2. Aerts, M. J., Nesheim, O. N., & Fishel, F. M., (1998). *Pesticide Record Keeping*. University of Florida Cooperative Extension Service, Institute of Food and Agriculture Sciences, EDIS.
3. Alister, C., & Kogan, M., (2006). ERI: Environmental risk index. A simple proposal to select agrochemicals for agricultural use. *Crop protection, 25*(3), 202–211.
4. Alvarez, A., Saez, J. M., Costa, J. S. D., Colin, V. L., Fuentes, M. S., Cuozzo, S. A., & Amoroso, M. J., (2017). Actinobacteria: Current research and perspectives for bioremediation of pesticides and heavy metals. *Chemosphere, 166*, 41–62.
5. Baker, B. P., Benbrook, C. M., & Benbrook, K. L., (2002). Pesticide residues in conventional, integrated pest management (IPM)-grown and organic foods: Insights from three US data sets. *Food Additives and Contaminants, 19*(5), 427–446.
6. Barooah, A. K., (2011). Present status of use of agrochemicals in tea industry of Eastern India and future directions. *Science and Culture, 77*(9, 10), 385–390.
7. Bernstein, J. A., (2002). Material safety data sheets: Are they reliable in identifying human hazards? *Journal of Allergy and Clinical Immunology, 110*(1), 35–38.
8. Bijman, J., & Tait, J., (2002). Public policies influencing innovation in the agrochemical, biotechnology, and seed industries. *Science and Public Policy, 29*(4), 245–251.
9. Bongiovanni, R., & Lowenberg-DeBoer, J., (2004). Precision agriculture and sustainability. *Precision Agriculture, 5*(4), 359–387.
10. Burth, U., Gutsche, V., Freier, B., & Roßberg, D., (2003). *Defining the 'Necessary Minimum' of Pesticide Use*. Electronic citation. Available at: https://citeseerx.ist.psu.edu/viewdoc/download?doi=10.1.1.398.1362&rep=rep1&type=pdf (accessed on 28 April 2021).
11. Carey, A. E., & Kutz, F. W., (1985). Trends in ambient concentrations of agrochemicals in humans and the environment of the United States. *Environmental Monitoring and Assessment, 5*(2), 155–163.
12. Casida, J. E., (2009). Pest toxicology: The primary mechanisms of pesticide action. *Chemical Research in Toxicology, 22*(4), 609–619.
13. Chester, G., Sabapathy, N. N., & Woollen, B. H., (1992). Exposure and health assessment during application of lambda-cyhalothrin for malaria vector control in Pakistan. *Bulletin of the World Health Organization, 70*(5), 615.
14. Cornell Agrochemical Database, (2004). http://pmep.cce.cornell.edu/ (accessed on 29 April 2021).
15. Cossée, O., Hermes, R., & Mezhoud, S., (2006). *Report of the Second Mission*. http://citeseerx.ist.psu.edu/viewdoc/download?doi=10.1.1.365.6894&rep=rep1&type=pdf (accessed on 29 April 2021).
16. Dosman, J. A., & Cockcroft, D. W., (1989). *Principles of Health and Safety in Agriculture*. CRC Press.
17. European Food Safety Authority, (2016). The 2014 European Union report on pesticide residues in food. *EFSA Journal, 14*(10), e04611.

18. Fernandez-Cornejo, J., Jans, S., & Smith, M., (1998). Issues in the economics of pesticide use in agriculture: A review of the empirical evidence. *Applied Economic Perspectives and Policy, 20*(2), 462–488.
19. Fernandez-Cornejo, J., Nehring, R. F., Osteen, C., Wechsler, S., Martin, A., & Vialou, A., (2014). Pesticide use in US agriculture: 21 selected crops, 1960–2008. *USDA-ERS Economic Information Bulletin*, (124).
20. Fresenburg, B., & May, M., (2017). *USDA Pesticide Record-Keeping Requirements for Certified Private Applicators of Federally Restricted-Use Pesticides*. https://extension.missouri.edu/publications/mp692 (accessed on 29 April 2021).
21. Garbarino, J. R., Snyder-Conn, E., Leiker, T. J., & Hoffman, G. L., (2002). Contaminants in Arctic snow collected over northwest Alaskan sea ice. *Water, Air, and Soil Pollution, 139*(1–4), 183–214.
22. Gartiser, S., Lüskow, H., & Groß, R., (2012). *Thematic Strategy on Sustainable Use of Plant Protection Products-Prospects and Requirements for Transferring Proposals for Plant Protection Products to Biocides*. Environmental Research of the Federal Ministry of the environment, nature conservation and nuclear safety.
23. Gilliom, R. J., (2007). *Pesticides in US Streams and Groundwater. https://pubs.acs.org/doi/ pdf/10.1021/es072531u* (accessed on 29 April 2021).
24. Good, A. G., & Beatty, P. H., (2011). Fertilizing nature: A tragedy of excess in the commons. *PLoS Biology, 9*(8).
25. Greenberg, M. I., Cone, D. C., & Roberts, J. R., (1996). Material safety data sheet: A useful resource for the emergency physician. *Annals of Emergency Medicine, 27*(3), 347–352.
26. Gunnell, D., & Eddleston, M., (2003). *Suicide by Intentional Ingestion of Pesticides: A Continuing Tragedy in Developing Countries*. https://academic.oup.com/ije/ article/32/6/902/775135?login=true (accessed on 29 April 2021).
27. Herath, G., (1998). Agrochemical use and the environment in Australia. *International Journal of Social Economics*.
28. Herrman, J. L., (1989). Round table on the European registration of agrochemical compounds-actual and perspective status: Joint FAO/WHO meeting on pesticide residues. *Food Additives and Contaminants, 6*(S1), S103–S104.
29. Hester, R. E., & Harrison, R. M., (2017). *Agricultural Chemicals and the Environment: Issues and Potential Solutions* (Vol. 43). Royal Society of Chemistry.
30. Hillocks, R. J., (2012). Farming with fewer pesticides: EU pesticide review and resulting challenges for UK agriculture. *Crop Protection, 31*(1), 85–93.
31. Kott, P. S., & Carr, D. A., (1997). Developing an estimation strategy for a pesticide data program. *Journal of Official Statistics-Stockholm, 13*, 367–384.
32. Kuchler, F., Chandran, R., & Ralston, K., (1996). The linkage between pesticide uses and pesticide residues. *American Journal of Alternative Agriculture, 11*(4), 161–167.
33. Kuiper, H. A., (1996). The role of toxicology in the evaluation of new agrochemicals. *Journal of Environmental Science and Health Part B, 31*(3), 353–363.
34. Kumar, S., & Singh, A., (2014). Biopesticides for integrated crop management: Environmental and regulatory aspects. *J. Biofertil Biopestici, 5*, e121.
35. Leake, A., (2000). The development of integrated crop management in agricultural crops: Comparisons with conventional methods. *Pest Management Science: Formerly Pesticide Science, 56*(11), 950–953.

36. Lim, C. J., (2014). *Food City*. Routledge.
37. Little, C. H. A., & Savidge, R. A., (1987). The role of plant growth regulators in forest tree cambial growth. In: *Hormonal Control of Tree Growth* (pp. 137–169). Springer, Dordrecht.
38. Nickell, L. G., (1982). *Plant Growth Regulators. Agricultural Uses*. Springer-Verlag.
39. NSW EPA, (2013). *What Are Pesticides and How Do They Work?* Available from: http://www.epa.nsw.gov.au/pesticides/pestwhatrhow.htm (accessed on 9 April 2021).
40. Osteen, C. D., & Szmedra, P. I., (1989). *Agricultural Pesticide Use Trends and Policy Issues*. US Department of Agriculture, Economic Research Service.
41. Pearce, G. R., (1998). *Agrochemical Pollution Risks Associated with Irrigation in Developing Countries: A Guide. https://www.gov.uk/research-for-development-outputs/agrochemical-pollution-risks-associated-with-irrigation-in-developing-countries* (accessed on 29 April 2021).
42. Pimentel, D., & Burgess, M., (2014). Environmental and economic costs of the application of pesticides primarily in the United States. In: *Integrated Pest Management* (pp. 47–71). Springer, Dordrecht.
43. Rasul, G., & Thapa, G. B., (2003). Sustainability analysis of ecological and conventional agricultural systems in Bangladesh. *World Development, 31*(10), 1721–1741.
44. Rivera, X. C. S., Bacenetti, J., Fusi, A., & Niero, M., (2017). The influence of fertilizer and pesticide emissions model on life cycle assessment of agricultural products: The case of Danish and Italian barley. *Science of the Total Environment, 592*, 745–757.
45. Russell, M. H., Layton, R. J., & Tillotson, P. M., (1994). The use of pesticide leaching models in a regulatory setting: An industrial perspective. *Journal of Environmental Science and Health Part A, 29*(6), 1105–1116.
46. Savci, S., (2012). An agricultural pollutant: Chemical fertilizer. *International Journal of Environmental Science and Development, 3*(1), 73.
47. Singh, K. N., & Merchant, K., (2012). The agrochemical industry. In: *Handbook of Industrial Chemistry and Biotechnology* (pp. 643–698). Springer, Boston, MA.
48. Smith, L. A., & Thomson, S. J., (2003). United states department of agriculture-agricultural research service research in application technology for pest management. *Pest Management Science: Formerly Pesticide Science, 59*(6, 7), 699–707.
49. Tirado, R., Englande, A. J., Promakasikorn, L., & Novotny, V., (2008). *Use of Agrochemicals in Thailand and its Consequences for the Environment.* Greenpeace Research Laboratories Technical. Bangkok, Thailand.
50. U.S. Department of Agriculture (USDA), Economic Research Service (ERS), (1984). *Inputs Outlook and Situation Report*. IOS-6.
51. United States Department of Agriculture, (2018). *Pesticide Data Program*. USDA.
52. Van, D. W. H. M., (1996). Assessing the impact of pesticides on the environment. *Agriculture, Ecosystems and Environment, 60*(2, 3), 81–96.
53. Watts, M., (2016). *Highly Hazardous Pesticides in the Pacific.* National Toxics Network, New Zealand.
54. World Health Organization, (1990). *Public health Impact of Pesticides Used in Agriculture*. World Health Organization.
55. World Health Organization, (2001). *Report of the 4th WHOPES Working Group Meeting: WHO/HQ, Geneva, 4–5 December 2000: Review of: IR3535; KBR3023;(RS)-Methoprene 20% EC, Pyriproxyfen 0.5% GR; and Lambda-Cyhalothrin 2.5% CS* (No. WHO/CDS/CPE/WHOPES/2001.2). Geneva: World Health Organization.

56. World Health Organization, (2006). *The Impact of Pesticides on Health: Preventing Intentional and Unintentional Deaths from Pesticide Poisoning*. Geneva: WHO.
57. World Health Organization, (2008). *Report of the 6th Meeting of the Global Collaboration for Development of Pesticides for Public Health (GCDPP)* (pp. 24, 25). WHO/HQ, Geneva (No. WHO/HTM/NTD/WHOPES/GCDPP/2008.1). Geneva: World Health Organization.
58. World Health Organization, (2017). *Determination of Equivalence for Public Health Pesticides and Pesticide Products, Report of a WHO Consultation, Geneva, Switzerland* (No. WHO/HTM/NTD/WHOPES/2017.1). World Health Organization.
59. Yan, D., Wang, D., & Yang, L., (2007). Long-term effect of chemical fertilizer, straw, and manure on labile organic matter fractions in a paddy soil. *Biology and Fertility of Soils, 44*(1), 93–101.
60. Yearbook, F. S., (2013). *World Food and Agriculture*. Food and Agriculture Organization of the United Nations, Rome.

# CHAPTER 5

# Organic Farming: A Promising Approach for Sustainable Agriculture

ACHARYA BALKRISHNA, PRIYANKA CHAUDHARY, RITIKA JOSHI, and VEDPRIYA ARYA


## ABSTRACT

Organic farming is a traditional method of agriculture that mainly keeps the crops growing in the soil alive and provides healthy food crops. Organic farming uses natural fertilizers and good agricultural techniques resulting in the production of sustainable crops without harming the quality of production. It receives considerable support from various policies due to its contribution to environmental development along with providing amenities such as biodiversity and cultural landscapes. This chapter deals with the evaluation of global policies, global importance, and opportunities in organic farming with suggestions for better functioning as well as economic growth. Farmers are continuously fighting to improve their crop yield with tremendous profit. This can only be obtained by the proper marketing channels and by providing a certificate with good governance.


## 5.1 INTRODUCTION

Organic farming is a production system that avoids the use of synthetic fertilizers, growth regulators, pesticides, and livestock feed additives. The objectives of environmental and socio-economic sustainability are the basics of organic farming. The key features include protecting the soil's long-term fertility while maintaining levels of organic matter, soil biological activity, effective recycling of organic materials (including livestock wastes and weed, diseases, and pest control and crop residues, etc.,) and nitrogen self-sufficiency uses (biological nitrogen fixation and legumes) [37]. Increasing the pollution level is the most common problem in the earth which is due to the lifestyle of people which causes deterioration of the environment. The use of modern agricultural methods such as synthetic fertilizers, pesticides, etc., contributes to environmental pollution to maximize crop production. These methodologies initially disturb soil nutrient balance due to which soil fertility decreases. Inorganic fertilizers are the principal factor for pollution in the field of agriculture. Nowadays, organic farming is a promising and efficient agricultural technique used for crop production in which eco-friendly, plant, and animal-based indigenous organic resources are used [8, 27]. They help in the increment of microbial activity and soil quality.

In soil-based ecosystems, the microbiota in soil plays an essential role such as erosion control, nutrient cycling, pest, and disease regulation. Several studies have reported the positive effects of organic farming on organic quality and soil health as well as microbial community traits.

However, no systematic quantification of whether organic farming systems include larger and more active soil microbial communities than traditional farming systems has been determined [19, 33]. Organically managed soil become biologically active and produce powerful crops which can resist diseases, drought, and insect attack [20, 24]. Worldwide development of organic farming has gone through three steps, emergence, expansion, and growth [15].

The production difference is not so outstanding, mainly in developed countries where organic farming has traditions and the society can support these farm systems by technological innovations. In contrast, the yield difference between the two systems is significant in developing countries where the production system is not fast and farmers have limited access to natural resources, and their purchasing capacity is relatively low. Organic farming increases the production of nutritious and safe food. The demands for organic food are increasing significantly look the consumers for organic foods that are considered healthy and safe. Therefore, organic food possibly attests to food safety from farm to plate. The organic agriculture practice is environmentally safe as compared to conventional farming. Organic farming retains soil to recover and maintains the environment reliable thereby stimulates the health of consumers. Also, the organic products market is now the fastest-growing market in the entire world, including India. Organic farming promotes the financial development of a nation through the health of a nation's consumers, the green health of a nation, and income generation [29]. Presently, India is the world's largest organic producer, and from this point of view, we can conclude that encouraging organic farming in India can lead to the creation of an ecologically, nutritionally, and economically healthy nation shortly soon as given in Figure 5.1.

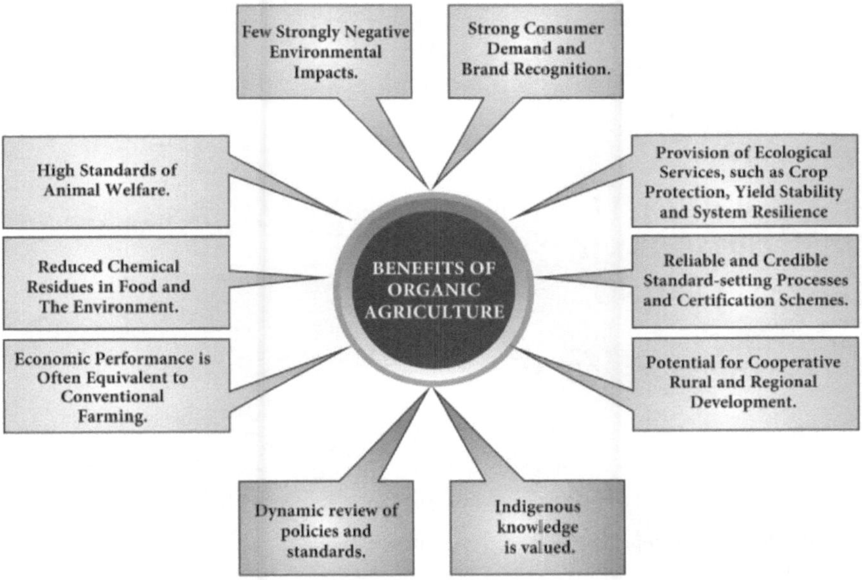


**FIGURE 5.1** Benefits of organic farming.

### *5.1.1 CONCEPTS OF ORGANIC FARMING*

According to Chandrashekar [5], major concepts of organic farming are:

- Do more and more viable work under a closed system and attract local resources;
- Maintain fertility of fertile soil;
- Reduce the utilization of fossil fuels in agricultural methods;
- Pollution prevention due to agricultural application;
- Give livestock an environment that meets their biological needs for livelihood;
- Make it viable for agricultural work to earn a living and to develop abilities for survival.

Organic farming is necessitating and wide-ranging for the development of economic, social, and ecologically safe food production systems. The four basic ethical principles of organic farming suggested by the International Federation of Organic Agriculture Movements (IFOAM) which are shown in Figure 5.2.

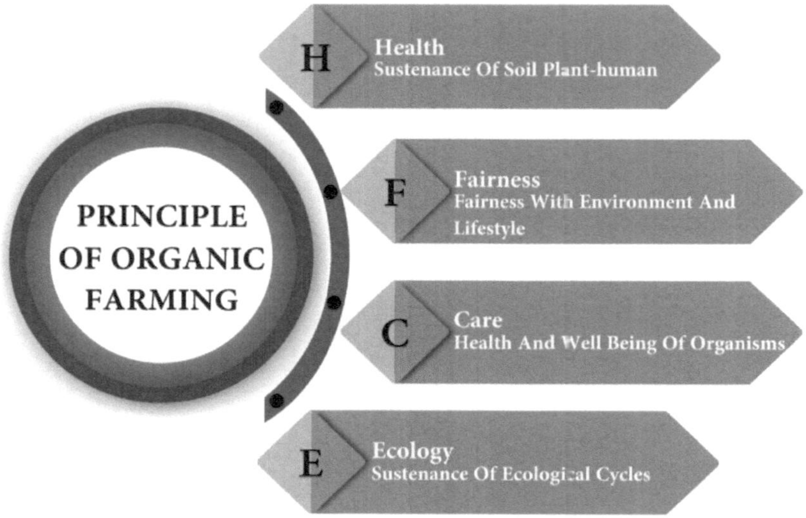


**FIGURE 5.2** Ethical principles of organic farming.

### *5.1.2 KEY POINT OF ORGANIC FARMING*

The key bases of organic farming are as follows [25]:

- Organic materials threshold standards;
- Authentic methods;
- Technology allowance;
- Effective and possible marketing network.

### *5.1.3 THE KEY POINT OF ORGANIC FARMING*

- Long term prevention of soil fertility by encouraging soil biological processes, maintaining organic materials, and careful mechanical intervention.

- Insoluble substances in plant nutrients are indirectly provided to soil microbes.
- Uses of biological nitrogen fixation and legumes to obtain strong recycling of organic matters including livestock manures and crop residues for Nitrogen self-sufficiency.
- Diseases, pest control, and weed rely mainly on crop rotations, organic manuring, natural hunters, resistant varieties, diversity, limited thermal, biological, physical, and chemical intervention.
- Focusing on your animal welfare issues about concerning comprehensive management of livestock, nutrition, breeding, and husbandry, evolutionary adaptation, evolutionary adaptation, health, and housing.
- The main focus on the impact of farming methods on the wider interaction of environment and wildlife.

## 5.2 CONCEPTUALIZATION OF CURRENT APPROACH FROM GLOBAL PROSPECT

The worldwide 20th investigation of certified organic agriculture was achieved by the Research Institute of Organic Agriculture (FiBL) with the collaborative effort of different teams from all over the world, and results were published by IFOAM-Organic International [11]. It is established in 1972 and is headquartered in Bonn, Germany. It is a membership-based organism working on true agricultural sustainability globally as given in Table 5.1.

**TABLE 5.1** Global Survey on Organic Agriculture in 2017

| Sr. No. | Regions | Share Countries That Provide Data (%) | Countries with Data on Organic Agriculture | Countries per Regions |
|---|---|---|---|---|
| 1. | World | 79% | 181 | 230 |
| 2. | Asia | 84% | 41 | 49 |
| 3. | Africa | 79% | 44 | 56 |
| 4. | Europe | 96% | 47 | 49 |
| 5. | Latin America and Caribbean | 72% | 33 | 46 |
| 6. | North America | 60% | 3 | 5 |
| 7. | Oceania | 52% | 13 | 25 |

Globally, many interventions and regulations in the form of government plans are designed for improving the culture and practices of organic agriculture. Most of the countries focused on organic food production statics for organic farming [2]. Based on IFOAM and FiBL reports, the trend of organic farming and high-rate growth was quite volatile. It has been assessed that organic farming reached its highest market value of 97 billion USD [8]. This document has been prepared from 181 countries data given in 2017 is given in Table 5.1. At present, USA has 1st position in world organic market with net worth 40 billion euro's [17] followed by Germany (10 billion Euros), China (7.9 billion Euros), and France (7.6 billion Euros) [26, 34]. However, the fastest organic market growth rate, i.e., 18% was recorded in France [8]. These statistical-based market analyzes are important in the policy preparation to understand the worldwide economic impact.

India has the highest number of organic farmers (835,000) in the world with 3rd rank in wild crop production and 9th in organic agriculture area, i.e., 1.49 ha. During the financial year 2015–2016, India export 1.35 million metric tons of certified organic food products worth INR 1,937 crores across the globe and hold 1st rank in the organic cotton export sector. The Indian organic market is making constant progress with 25% CAGR compared to 16% global growth rate. Even though the impressive presentation in terms of transport, local expenditure of organic products is still in an immature state with a market contribution of just under 1% [27] given in Table 5.2.

## 5.3 CERTIFICATION STANDARDS TO VALIDATE ORGANIC PRODUCE AROUND THE GLOBE

In several countries, the standard for organic product production and marketing are validated by government organizations. In Japan, European Union (EU) and the USA have a wide range of organic laws, terms, and conditions for the certification of its producers as well as retailer at both national and international platforms. Countries without biological laws and governmental guidelines are known where certification for organic farming is regulated by non-commercial and private organizations. In 1992, EU countries received organic legislation with the enforcement of the EU-Eco-Regulation at national level. At international platform, an

**TABLE 5.2** Present Status and Global Policy in the Organic Farming Sector

| Sr. No. | Locations | Policies and Legislations | Current Status of Organic Market |
|---|---|---|---|
| 1. | Asia | In 1994, China established OFDC for certification of organic products. CHAC (China) was established in 2002. China established CNOPS in 2005.<br>The "Paramparagat Krishi Vikas Yojana" cluster program by the Indian authority brings about 500,000 acres under organic farming. India launches organic farming scheme value chain based in the North East regions. Scheme implemented from 2014–2015, which combine development of horticulture and National Mission for Sustainable Agriculture. | The total arable land is 6.1 million has hectare (ha) (0.4% of the total arable land).<br>The organic agriculture sector recorded a 25% increment during 2016–2017.<br>After India, China leads the most organic cultivated area.<br>Estimates of the organic agriculture market in Asia at 9.6 billion euros. |
| 2. | Australia | Development of private certification organizations in the 1980s.<br>These organizations came under the AQIS in 1990.<br>Establishment of BFA (Bio nutrient Food Association) and NASAA (National Association for Sustainable Agriculture, Australia Limited) to promote biological practices. In 1992, adopting national standard which was later reviewed in 1998. | The Australian organic market is the net worth of $ 2.4 billion.<br>The year 2012 has seen an 88% growth rate.<br>12% of Australians are now buyers of organic products. |
| 3. | Denmark | In 1987, Organic Agriculture Act was adopted by the government.<br>In 1992, the government began providing support for biological research. | World's leading country in organic farming.<br>It holds 8.4% share in the global market. |

**TABLE 5.2** *(Continued)*

| Sr. No. | Locations | Policies and Legislations | Current Status of Organic Market |
|---|---|---|---|
| | | In 1995, an overview of permanent subsidy of the organic farming sector and education program for farmers in organic fields was started.<br><br>In 2004, Permanent organic expenditures and flat conversion replaced the enduring subsidy. | |
| 4. | Europe | European Union Rural Development Program and National Action Plan has launched a scheme. The goal is to increase the organic land amount. | In 2017, the organic farmland reached 12.6 million has.<br><br>In 2016, 250,000 organic product producers were reported in the European Union. Europe has become the 2nd highest user of organic products (34.4 billion euros), and in 2018, 3.3 million tons of organic products were imported. |
| 5. | Germany | The German government has implemented of Organic Farming Act and the formation of BOLW to recover organic farming. From 2001, organic produces were labeled by Uniform EcoLabel. | 8% increase in organic cultivated land and from 2018, 5.5% growth increased. |
| 6. | North America | In 1990, Organic Food Production Act was passed by the United States to regulate organic food production and processing, which was introduced by the USDA labeling and National Biological Program. | 7% agricultural land presence of which 2.2 million has is used for organic farming. In 2017, FiBL and IFOAM received an estimated net worth of $48.7 billion in the organic market. |

equivalent discussion was conducted and signed several agreements to coordinate certification between countries for the international trade of organic farm products. The chief certificated organizations are Organic Crop Improvement Association (OCIA) [22], IFOAM [14] and Ecocert [10] (Figure 5.3).

**FIGURE 5.3** Procedure of organic certification in India.

OCIA is a member-owned and non-profiting international agriculture organization established in 1985 in Quebec and it is headquarter is in Lincoln, Nebraska. It is the largest, oldest, and most trusted organization that created by farmer group on the basis of "crop improvement association" model for providing certificate for organic agriculture trade in the world. FOAM was established in 1972 and it is headquarter is located in Bonn, Germany. It serves with the guideline based on private and public standard bodies can made more specific organic standards organization. In 1991, Ecocert (organic certification organization) was established in France and continually supports to stakeholders for implementing and promoting sustainable methods through consulting, training, and providing certification [10, 14, 22]. There is no formal contract for the exportation of organic products at the global level, often agencies from the importing countries can certified their traders. In 2006, the EU under the NPOP process likened with India's organic certification by USDA's NPOP evaluation and approval (Table 5.3).

### *5.3.1 DEVELOPING REGION UNDER CERTIFIED ORGANIC AGRICULTURE*

- Agricultural land up to 35 million ha is achieved systematically by about 1.4 million makers.
- Europe, Latin America, and Oceania have 8.2, 8.1, and 21.1 million ha, the largest areas of managed organic agricultural land, while Falkland Island (36.9%), Liechtenstein (29.8%), and Austria (15.9%) have organically managed agrarian land.
- 340'000 organic farm producers are located in India, followed by Uganda (180'000) and Mexico (130'000).
- As per the report of 2007, organic agricultural land regions are in a growing phase in the world.
- More than 26% of land under biological management system was documented in with high growth rate in Argentina, Latin America, more than half million ha in Europe, and 0.4 million in Asia.
- About one-third of the world's arable land: 12 million has been found in developing countries. The major part of this domain is in Latin America, with Asia and Africa ranked second and third. The countries with the largest area under biological management are Argentina, China, and Brazil.

**TABLE 5.3** Organic Farming Certification around the World

| Sr. No. | Countries | Organization |
|---|---|---|
| 1. | Australia | In 2006, 7 (Administration of Quality Supervision, Inspection, and Quarantine) AQIS approval to get Organic Produce Certificates. In 2004, 2345 certified bodies got approval and the highest number of organic product certified by Australia and it becomes the nation's biggest organic farmers' group. |
| 2. | Canada | National organic standard guidelines were released by the government in the legislation process. Private sector organizations provide certification. In Quebec, provincial legislation is provided by the government for the monitoring of organic certification. |
| 3. | China | A and AA are two standards of Green Food Development Centre in China; whereas the former standard license some use of inorganic farming. |
| 4. | India | The Ministry of Commerce is the governing agencies for organic certification for transportation under Agricultural Processed Foods Export Development Authority (APEDA). For documentation, there is no original standard of organic products. Till today, 11 authorized and 13 under processed certification agencies under the National Program for Organic Production (NPOP). |
| 5. | Japan | In 2001 law in April, the Japanese Agricultural Standard (JAS) was completely applied. In 2005 Ministry of Agriculture, re-establish all JAS certifications. |
| 6. | Sweden | KRAV is the private label of the Swedish organic market. KRAV requirements must be followed by those other than those requested by the European Union-Organic. |
| 7. | United Kingdom | Organic certification has been used by different organizations, the largest of which are organic farmers, soil association, and growers. All certified agencies are subjected to the rules of the UK Register of Organic Food Standards (UKROFS). |
| 8. | Unites States | The National Organic Program (NOP), was passed in October of 2002 as a federal law. It prohibits the use of "organic" for certified organic growers. The state certification is regulated, personal, and non-commercial organization that has been accepted by the US Department of Agriculture (USDA). |

- There are 31 million has of organic wild gathering area and land for beekeeping. Most of this country is in developing countries as opposed to agricultural land, two-thirds of which is in developed countries. Also, 0.43, 0.01, and 0.32 in organic areas include aquaculture, forest, and non-agricultural land.

- About two-thirds of the arable land under biological management is pasture (22 million ha). Of the cropped area (cultivable land and permanent crops), 8.2 million ha (10.4% since 2007), representing a quarter of organic arable land [36].

### 5.3.2 ORGANIC CERTIFICATION AGENCIES

#### 5.3.2.1 AFRICA

At present, East African Organic Product Standards (EAOPS), Asia Regional Organic Standard (AROS), EU organic standards, and AROS are the major organic standards in Africa. The EAOPS represents the 1st multi-country biological standards in Africa that synchronize presenting standards of organic farming and practices in the five states of Africa. IFOAM, in association with African Organic Agriculture Movements, develop an organic agriculture training manual to provide best farming practices to farmers and related workers. Since not all African countries have national biological standards, there is insufficient regulation of organic products in the African continent [30].

#### 5.3.2.2 EUROPE

In France, Ecocert (organic certification organization) was established in 1991, which is continuously working to support organic agriculture stakeholders in the implementation and promotion methods by providing consulting, training, and certificate. Now, Ecocert has extended its efforts to many other sectors [10].

#### 5.3.2.3 AMERICA

On 15 May 1862, Abraham Lincoln established the Department of Agriculture as "People's Department" in America under the chairmanship of 1st agricultural commissioner Isaac Newton. The law authorizing the USDA NOP was the Organic Foods Production Act of 1990. Several USDA organizations are working on the development of the organic sector. Whether people are already certified organic, are considering changing all or part of the operation or functioning with organic producers, they all are part of it [32].

### *5.3.2.4 AUSTRALIA*

Several organizations of Australia, such as the National Association for Sustainable Agriculture, Australia (NASAA) and biological farmers of Australia (BFA), developed organic standards for organic farming and their trading, which were utilized in the in local market [35]. When OPEC was renamed as Organic Industry Export Consultative Committee (OIECC), the membership included not only representatives of AQIS and Certificates (BDRI, BFA, NASAA, OHGA, OFC, SFQ), but also the Commonwealth and Agriculture of the two states Department (Victoria) (NSW); The Organic Federation of Australia (OFA; National Organic Peak Body) and the Organic Produce Program of the Rural Industries Research and Development Corporation (RIRDC), is a body that researches organic agriculture in Australia [35].

### *5.3.2.5 INDIA*

Since 2001, organic farming has been encouraging by the 3rd party certification under NPOP through the government. The national program includes accreditation programs for standards for organic production, certification bodies, promotion of organic farming, etc., and 32 certificated agencies have been issued till now by NPOP [1]. Agricultural And Processed Food Products Export Development Authority (APEDA) was founded in December 1985 by the Parliament of India under the Agricultural and Processed Food Products Export Development Authority. It works under the supervision of the Ministry of Commerce and Industry, Government of India. The main focus of APEDA is to promote and develop new export platform agricultural products of scheduled products such as fruits and vegetables [31].

At present, the Ministry of Agriculture and Farmers Welfare, Government of India implemented the National Agricultural Project (NPOF), Department of Agriculture, Cooperation, and Farmers Welfare under participatory Guarantee System for India (PGS-India). The PGS is locally suitable guarantee highlights the involvement of stakeholders with farmers and users and works outside the frame of third-party certification by IFOAM [14]. NPOF is meant to promote organic farming in the country through technical capacity building of stakeholders by the development of human resources, technology transfer, and production of quality organic and organic inputs. It serves as the nodal quality control laboratory for analysis of bio-fertilizers as per the requirement of Fertilizer Control Order, 1985 (FCO). It is also responsible for the development, procurement, and efficacy evaluation of organisms, bio-fertilizer strips, and maternal cultures supply to production units, to maintain the National and Regional Culture Collection Bank of Bio-fertilizer and biocontrol [31].

#### 5.3.2.5.1 Organic Certification Procedure in India

In India, the organic certification processes (Figure 5.3) follow a set of standard guidelines which is laid down by NPOP [31]. These guidelines are as follows:

- Land must be converted for organic farming.
- Farm inputs must be in natural form.
- No genetically modification or technological radiation in inputs is allowed.
- The integrity of all physical, biological, and mechanical procedures must be maintained.
- Contamination is not allowed in agricultural or farmlands.
- Permanent methods should be applied in the farm.

## 5.4 KEY RESTRICTIONS IN THE IMPLICATION OF CONCEPT

The agricultural sector serves as a pollutant in several environmental issues such as biodiversity loss, soil degradation, climate change, soil infertility, and water pollution. It is generally believed that organic farming methods have little impact on the environment than conventional farming, which is the only reason governments subsidize the organic sector [24].

It should be stated that some of the studies are environmental effective are not still completely defined. Limited data is available from developing countries, so current data may not be representative worldwide. Furthermore, the available investigations mostly compare traditional and organic methods unless controlled for surprising factors, so that the observed differences cannot be interpreted as a result of the definitive organic parameters. Ultimately, it should be emphasized that large disparities in environmental impacts exist in both conventional and organic systems. Eco-friendly farming methods (e.g., long process crop rotations, nutrients management, and semi-natural landscape) are especially promoted by organic standards, but they are used by conventional farmers [12], as shown in Figure 5.4.

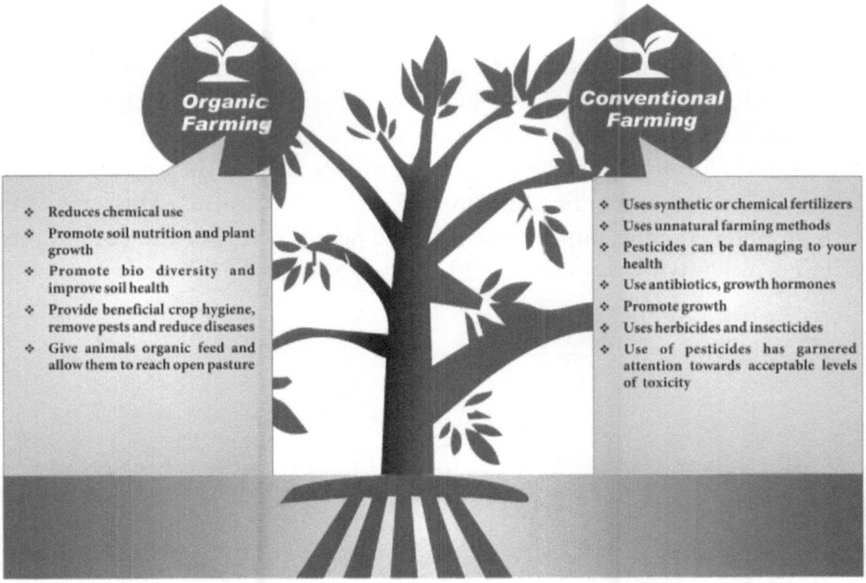


**FIGURE 5.4** Benefits of organic farming over conventional farming approach.

### 5.4.1 SOCIAL AND ECONOMIC IMPACT OF ORGANIC FARMERS

Agricultural growth can only benefit from organic farming methods when income derived from organic products is economically viable for farmers, at least higher than conventional farming. Organic farming is 22–35% more cost-effective than conventional agriculture on average. While organic production is quite low, farmers in certified organic markets get a good price for the products, which is a 30% price premium at the agricultural level [27]. Crop production by organic farming is tough and costly because it requires more labor, higher demand, higher cost of fertilizer, cost of crop harvesting, certification, and licensing, and better conditions for livestock. Therefore, the ability of organic movement to provide a comprehensive solution to global food and fiber production problems is limited under current circumstances. Although there are some well-recognized benefits of organic farming systems, widespread adoption of organic agriculture as the only method of farming is not possible in the medium to short term, even in the most progressive countries of Europe. Demonstrating large-scale conversion to organic agriculture at realistic rates (10–20%) shows that such conversion is unlikely to cause adjustment problems in the agricultural sector, for example, significantly lower yields as a result [13]. American society is recovering higher prices for organic products; hence it becomes somewhat inaccessible. Along with this organic farming provide benefit to farmers such as they do not handle synthetic and chemical fertilizers which are very harmful. This results in less absenteeism and improvement in the health status of the community at large, as shown in Figure 5.5 [23].

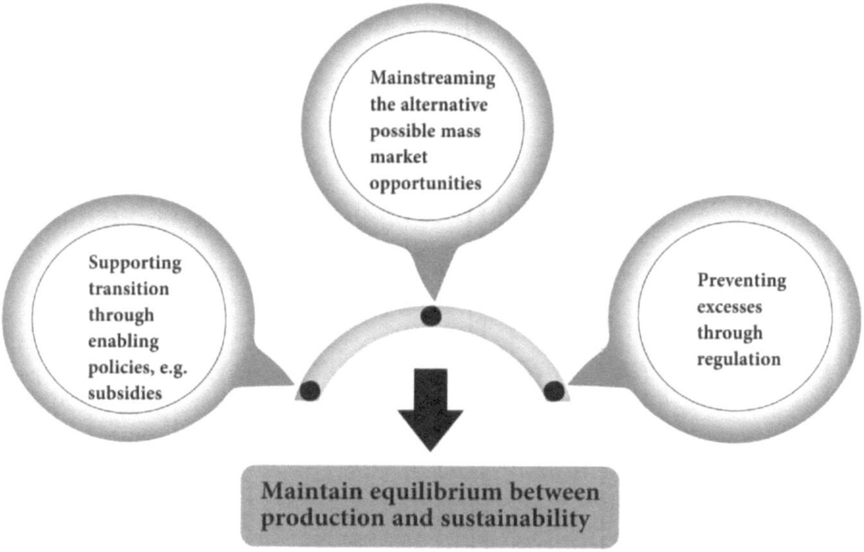


**FIGURE 5.5** The range of feedback mechanisms is likely to evolve to maintain equilibrium between production and sustainability.

## 5.5 OPPORTUNITIES FOR ORGANIC FARMING BEHIND THE FRAME

Organic farming is an ideal agricultural system in which economically divergent products are cultivated. The morality of the organic farming is based on the proper documentation and knowledge like use resources sparingly, nature is inherently valuable, treat livestock well and use the least harmful method. Food security is depending on personal connections of honesty and trust between agribusinesses, farmers, purchasers, suppliers, and agricultural stores chain. Organic agriculture has emerged as one of the known alternative farming systems established in reply to the drawbacks of mainstream agriculture [34], as shown in Figure 5.5.

Organic farming sector is growing day by day in the world. During the last 3 years, the total area of organic cultivation land is increases rapidly as compared to 1985. In 2017, 69.8 million ha organically managed land was recorded across the world, which represents 20% (11.7 million ha) increment as compared to 2016. In 2017, 35.65 million ha organic land was identified in Australia while India holds 8$^{th}$ position in the world with 1.78 million ha organic agriculture land [34]. In India, the growth of organic farming is very slow and only 41,000 ha organic land was recorded which is only 0.03% of the total agricultural land. The major issue in the progress of organic farming is the gaps in the government policies related to promotion of organic farming, inappropriate marketing system, lack of awareness, lack of financial support, lack of biomass, high input costs of farming, inadequate agricultural infrastructure, lack of quality manure, inefficiency of export demand and low yield [4]. However, the production rate of organic products was 14,000 tons, out of which 85% of products exported in over the world in 2002 [6].

The Millennium Ecosystem Assessment report list of several general strategies to promote more sustainable land use that are very applicable to organic agriculture. These strategies include creating appropriate governance and marketing structures, overcoming social and behavioral barriers, encouraging investment in development. Promoting appropriate technologies, knowledge, and skill and also suggested some specific policy options [21]:

1. Removing production subsidies has economic, social, and environmental impacts.
2. Investment in and dissemination of agricultural science and technology maintaining the necessary growth of the food supply

without the harmful trade-off from excessive use of water, nutrients, or pesticides.

3. Use of response policies that recognize the role of women in the production and use of food and which are designed to empower women and ensure the use and control of resources necessary for food security.
4. Application of a mix of regulatory and incentive-based, and market-based mechanisms to reduce the overuse of nutrients [21].

### *5.5.1 ORGANIC FARMING AND BIODIVERSITY*

The intensive farming methods interrupt the surrounding environment and their diversity. Organic as well as low-input farming methods can cope with these undesirable effects by maintaining biodiversity, controlling weeds and insects [28]. Letourneau and Bothwell [18], presented confirmation that improved biodiversity in organic farms results in increased pest control. Additionally, organic farming can help to reverse habitat species decline in areas, where usually conventional farming is applied [16]. The constructive consequences on biodiversity are the finest benefits of organic farming that is mostly indicates in contrast to conventional agricultural system. In 30 years, a meta-analysis of data from organic farming areas showed a nearly 30% increase in species productivity, with study groups varying in size and effect size for crop studies. A recent study suggested that the activity of different organisms in organic managed regions is not the same [29]. Doring and Kromp [9], initiate that in the organic farms, some predators are generally bigger than the population thicknesses as compared with conventional ones. Yet, the general opinion is that the loss of pests on many crops is usually not much in the case of a well-prepared organic field.

Other researcher recommended that the favorable effects of organic farming on species can be observed in actively managed agricultural lands except small-scale situations that contain isolated, non-crop biotopes. Research on organic farming system shows that the biodiversity is in increasing phase which is due the positive effect of organic farming techniques. Normally, biodiversity in organic farming land is 10–30% higher than conventional agricultural field [37] Though, calculation between the diversity and crop production in agricultural land of organic and conventional farms recommended that intensive production systems unfortunately linked with the biodiversity conservation and reduction crop yield. The main goal of organic farming is to identify the balance between sustainable productivity and conservation of agro-ecosystems [12], as shown in Figure 5.6.

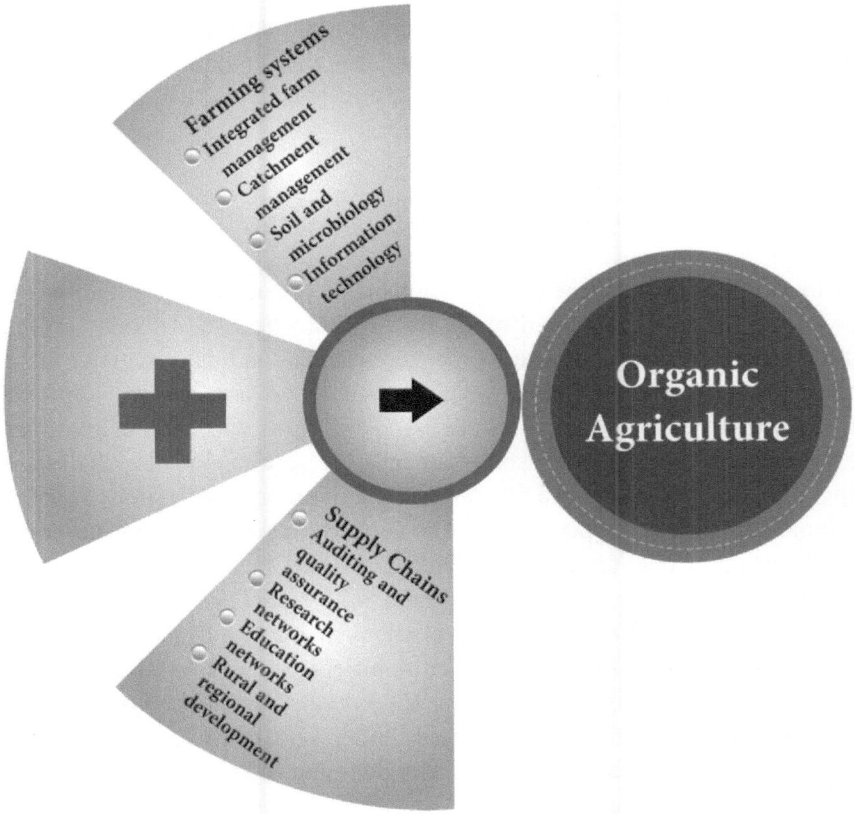


**FIGURE 5.6** Technologies and conceptual approaches relevant to organic agriculture.

### *5.5.2 ORGANIC FARMING AND SOIL FERTILITY*

The organic farming system supports human civilization with retaining soil structure, fertility, and fulfill the need of food for the growing human population [3]. The principal objective of organic farming is to raise farm production, food security, and constructive effect on three issues, namely crop rotation, minimum soil disturbance by mechanical means and permanent soil cover. In agroecosystem, the positive effects of conversation or sustainable agriculture have been widely seen in the improvement of water efficiency, prevention of soil erosion, organic carbon loss, mitigation of greenhouse gas (GHG) and nutrient cycling. Generally, "organic zone" of soil contains a high amount of organic matter, which can be due to the activity of organic fertilizers and activity of microorganisms in the decomposition of organic residues. To better understand these processes, investigators should work on the soil analysis, total content of humus, microbiota, and tillage processing and its physical and biochemical properties.

### *5.5.3 CHALLENGES IN THE ORGANIC FARMING*

With the rise in global health consciousness, organic food is determined to knock on every door and make its way into healthy kitchens around the world. People around the world use organic food as a hygiene factor rather than a product of their own. Organic food is a holistic approach in the Indian environment that begins at the farm and ends at the thorn of the consumer. The main stakeholder is the source, and the challenges encountered during organic farming can be overcome with smart strategy, scientific planning, responsible public activity, and government support.

A series of challenges highlighted are listed in Figure 5.8. Some of these challenges are conflict with each other (such as global cohesion versus local adaptation), and some challenges are also opportunities such as dynamic review of policies and stand orders. Specific local agricultural requirements may also create pressure to revise standards. For example, phosphorus (P) fertilizers permitted under certification standards are unable to supply sufficient P is naturally deficient soils in parts of Southern Australia [13]. Consequently, the ban on citrate-soluble superphosphate has been questioned. The question of whether organic food is better for human health than traditional foods is central in the minds of organic consumers. Yet research suggests that no definite nutritional difference has been found. The reviews in several countries all concluded that there is no evidence for any direct health benefits associated with the intake of organic foods. These reports also concluded that there were no health risks associated with organic food. Yields in organic agriculture may be equal to or better than traditional agriculture, although often they are not, simply due to insufficient plant-available nutrients, weed infection, crop rot or non-cash stages in inexperienced management.

Yield performance is very location and management specific and many of the underlying drivers of yield (like soil carbon, weed seed bank) have long-term responses. Some researchers have also highlighted alternative agricultural ecological criteria such as the value of flexibility and sustainability. Although organic agriculture causes fewer pesticides in food, people, and the environment, it is the former when claiming that organic agriculture is completely environmentally sustainable. In particular, some soil nutrients have a negative budget in some organic cropping systems, leading to depletion of the soil reserves of that nutrient, as shown in Figure 5.7.

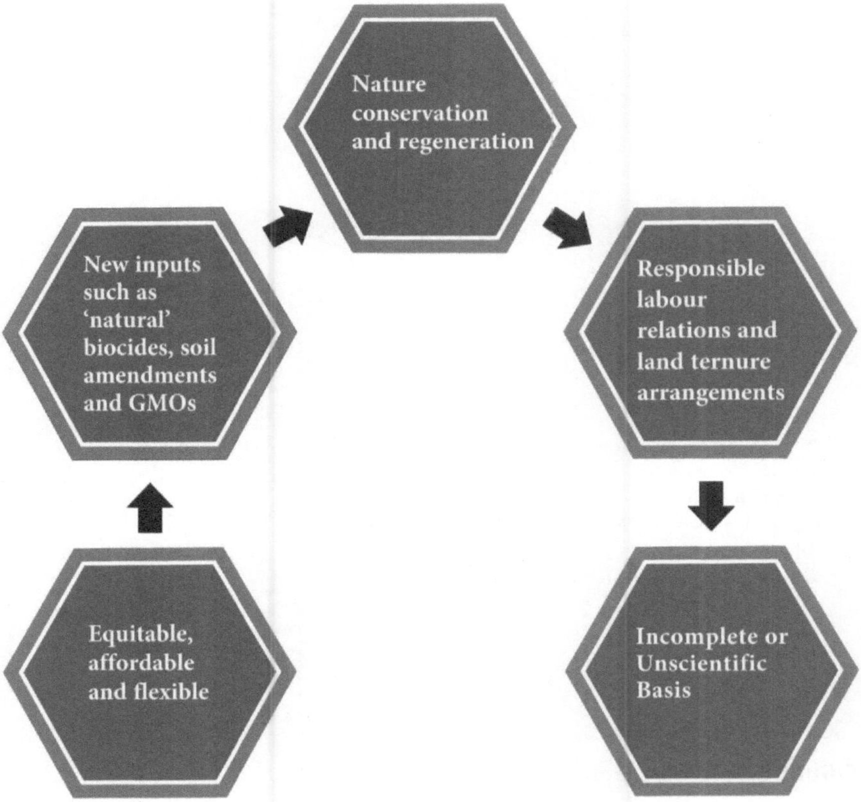


**FIGURE 5.7** Challenges for organic agriculture.

### 5.5.4 CHALLENGES IN THE ORGANIC SECTOR

Manufacturers of organic products are constantly struggling to optimize the scale of their operations while maintaining profitability. This is mainly due to gaps in the regulatory framework for organic products in India. In addition to the procedural challenges related to certification and quality assurance, the rising cost of inputs and the long-term conversion period from traditional to organic agriculture are some of the major challenges faced by producers, most of them small or marginal farmers. Some of the major challenges facing the biological sector today can be divided into three main heads (Figure 5.8) [27].

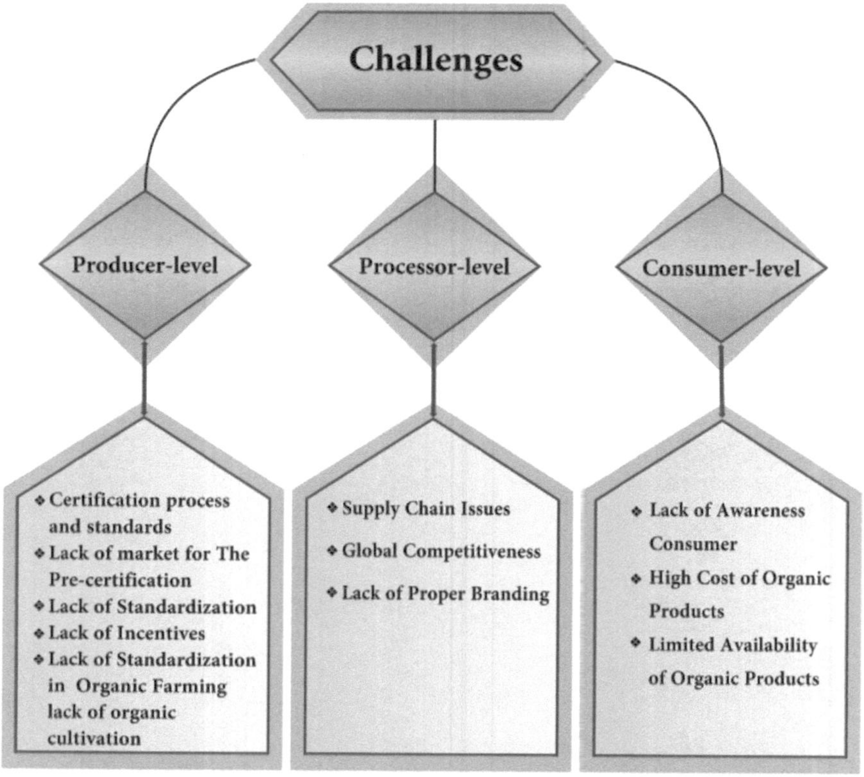


**FIGURE 5.8** Challenges for organic sector.

## 5.6 SUMMARY

This chapter emphasized the importance and extreme need for organic farming in the modern agricultural sector. It is a very ancient and suitable agricultural method. Many legislations and policies have been formulated around the world in the development of organic farming, but at the regional level, there is still a need for proper implementation of policies in some parts of the world. Organic farming is increasing the attraction of people all over the world. With its effectiveness, the biological sector has to deal with constant challenges. Therefore, time is needed to implement and implement organic sector policies with the involvement of stakeholders of society which ultimately improves organic farming. It has been decided that proper adherence, grassroots implementation, impact analysis, and public awareness are the keys to the success of every policy. An improved policy leads to sustainable development and helps in achieving the goal of a sustainable environment.

## ACKNOWLEDGMENTS

The authors are highly thankful to Dr. Rajesh Mishra for his extreme intelligence. We are also pleased to thank Mr. Pramod Kulshreshtha and Mr. Sunil Kumar for their graphical support. The authors would also like to thank Mr. Kapil Kumar for backend support.

## KEYWORDS

- **Asia Regional Organic Standard**
- **Certification Standards**
- **European Union**
- **National Agricultural Project**
- **Organic Farming**
- **Sustainable Agriculture**

## REFERENCES


1. *Agriculture and Processed Food Products Export Development Authority (APEDA)*, (2015). Ministry of Commerce and Industry, Government of India. https://www.indiafilings.com/learn/organic-farming-certification-in-India (accessed on 9 April 2021).
2. Assocham, E. Y., (2018). *The Indian Organic Market: A New Paradigm in Agriculture*. https://www.indiaretailing.com/2018/03/21/food//food-grocery/organic-packaged-foodmarket-to-cross-871-million-by-2021-assocham-ey-report/ (accessed on 28 April 2021).
3. Badgley, C., Moghtader, J., Quintero, E., Zakem, E., Chappell, M. J., Aviles-Vazquez, K., et al., (2007). Organic agriculture and the global food supply. *Renewable Agriculture and Food System, 22*(2), 86–108.
4. Bhardwaj, M., & Dhiman, M., (2019). Growth and performance of organic farming in India: What could be the future prospects? *Journal of Current Science, 20*, 1–8.
5. Chandrashekar, H. M., (2010). Changing scenario of organic farming in India: An overview. *International NGO Journal, 5*, 34–39.
6. Chopra, A., Rao, N. C., Gupta, N., & Vashisth, S., (2013). Come sunshine or rain; organic foods always on tract: A futuristic perspective. *International Journal of Nutrition, Pharmacology Neurological Diseases, 3*, 202–205.
7. Crowder, D. W., & Reganold, J. P., (2015). Financial competitiveness of organic agriculture on a global scale. *Proceedings of the National Academy of Sciences, 112*(24), 7611–7616.
8. Dhiman, V., (2020). Organic farming for sustainable environment: Review of existed policies and suggestions for improvement. *International Journal of Research and Review, 7*(2), 22–31.
9. Doring, T. F., & Kromp, B., (2003). Which carabid species bene € fit from organic agriculture? A Review of comparative studies in winter cereals from Germany and Switzerland. *Agriculture Ecosystem and Environment, 98*, 153–161.
10. ECOCERT, (2020). *Act of Sustainable World*. https://www.ecocert.com/en/certification (accessed on 9 April 2021).
11. FiBL and IFOAM, (2019). *The World of Organic Agriculture: Static and Emerging Trends*. https://ciaorganico.net/documypublic/486_2020-organic-world-2019.pdf (accessed on 9 April 2021).
12. Gabriel, D., Sait, S. M., Kunin, W. E., & Benton, T. G., (2013). Food production vs. biodiversity: Comparing organic and conventional agriculture. *Journal of Applied Ecology, 50*(2), 355–364.
13. Graham, R. F., Wortman, S. E., & Pittelkow, C. M., (2017). Comparison of organic and integrated nutrient management strategies for reducing soil $N_2O$ emissions. *Sustainability, 9*(4), 510.
14. International Federation of Organic Agriculture Movements (IFOAM), (1998). *The IFOAM Basic Standards for Organic Production and Processing*. General assembly, Argentina, IFOAM, Germany. Organic Food Production Act of 1990 (U.S.C) s. 2103.
15. Joachim, S., (2006). Review of history and recent development of organic farming worldwide. *Agricultural Sciences in China, 5*(3), 169–178.

16. Kehinde, T., & Samways, M. J., (2012). Endemic pollinator response to organic vs. conventional farming and landscape context in the Cape floristic region biodiversity hotspot. *Agriculture Ecosystem Environment, 146*,162–167.
17. Lernoud, J., & Willer, H., (2017). *Key Results from the FiBL Survey on Organic Agriculture Worldwide 2017: Key Data, Crops, Regions*. https://orgprints.org/33355/5/lernoud-willer-2019-global-stats.pdf (accessed on 9 April 2021).
18. Letourneau, D. K., & Bothwell, S. G., (2008). Comparison of organic and conventional farms: Challenging ecologists to make biodiversity functional. *Frontiers in Ecology and the Environment, 6*(8), 430–438.
19. Lori, M., Symnaczik, S., Mäder, P., De Deyn, G., & Gattinger, A., (2017). Organic farming enhances soil microbial abundance and activity: A meta-analysis and meta-regression. *PLoS One, 12*(7), e0180442.
20. Makadia, J. J., & Patel, K. S., (2015). Prospects, status and marketing of organic products in India: A review. *Agriculture Review, 36*(1), 73–76.
21. Millennium Ecosystem Assessment, (2005). *Millennium Ecosystem Assessment Synthesis Report*. Island Press, Washington DC. https://www.millenniumassessment.org/en/index.html (accessed on 9 April 2021).
22. *OCIA: About OCIA*, (2020). https://ocia.org/ (accessed on 9 April 2021).
23. Ranch & Farm, (2020). *About Organic Farming*. https://bsranchandfarm.com/is-organic-farming-cheaper-than-conventional-farming/ (accessed on 9 April 2021).
24. Röös, E., Mie, A., Wivstad, M., Salomon, E., Johansson, B., Gunnarsson, S., Wallenbeck, A., et al., (2018). Risks and opportunities of increasing yields in organic farming: A review. *Agronomy for Sustainable Development, 38*(2), 14.
25. Roychowdhury, R., Gawwad, M. R. A., Banerjee, U., Bishnu, S., & Tah, J., (2013). Status, trends, and prospects of organic farming in India: A review. *Journal of Plant Biology Research, 2*(2), 38–48.
26. Schaack, D., (2019). *The Organic Market in Germany-Highlights 2018*. Nürnberg. https://orgprints.org/23178/3/schaack-2019-gernany.pdf (accessed on 9 April 2021).
27. Singh, R., Jat, N., Ravisankar, N., Kumar, S., Ram, T., & Yadav, R., (2019). *Present Status and Future Prospects of Organic Farming in India* (pp. 1–25).
28. Tsvetkov, I., Atanassov, A., Vlahova, M., Carlier, L., Christov, N., Lefort, F., Rusanov, K., et al., (2018). Plant organic farming research-current status and opportunities for future development. *Biotechnology and Biotechnological Equipment, 32*(2), 241–260.
29. Tuck, S. L., Winqvist, C., Mota, F., Ahnström, J., Turnbull, L. A., & Bengtsson, J., (2014). Land-use intensity and the effects of organic farming on biodiversity: A hierarchical meta-analysis. *Journal of Applied Ecology, 51*(3), 746–755.
30. Tung, O. J. L., (2018). African organic product standards for the African continent? prospects and limitations. *Potchefstroom Electronic Law Journal/Potchefstroomse Elektroniese Regsblad, 21*(1), 1–38.
31. Tyagi, O. S., & Dhingra, P., (2018). *The Indian Organic Market a New Paradigm in Agriculture*. https://www.agroworldindia.com/souvernir/AgroWorld-2018-souvenir.pdf (accessed on 9 April 2021).
32. USDN, (2020). *Agriculture Marketing Service*. About Certification Agencies. https://www.agmrc.org/food/farmers-markets (accessed on 9 April 2021).

33. Verbruggen, E., Röling, W. F., Gamper, H. A., Kowalchuk, G. A., Verhoef, H. A., & Van, D. H. M. G., (2010). Positive effects of organic farming on below-ground mutualists: Large-scale comparison of mycorrhizal fungal communities in agricultural soils. *New Phytologist, 186*(4), 968–979.
34. Willer, H., & Lernoud, J., (2019). *The World of Organic Agriculture. Statistics and Emerging Trends*. Research Institute of Organic Agriculture (FiBL), Frick and IFOAM-Organics International, Bonn.
35. Wynen, E., (2007). Standards and compliance systems for organic and bio-dynamic agriculture in Australia: Past, present, and future. *Journal of Organic Systems, 2*(2), 38–55.
36. Yadav, A. K., (2010). Certification and inspection systems in organic farming in India. *Training*. https://cuts-cart.org/pdf/Useful_Information-Certification_and_Inspection_Systems_A_K_Gupta.pdf (accessed on 28 April 2021).
37. Yadav, S. K., Babu, S., Yadav, M. K., Singh, K., Yadav, G. S., & Pal, S., (2013). A review of organic farming for sustainable agriculture in Northern India. *International Journal of Agronomy*, 8.

# PART II

# Food Security: A Realm of Reality

# CHAPTER 6

# Post-harvest Management Practice (PHMP): A Systematic Program to Attain Food Security

DEVIKA SHARMA, HIMANI MALHOTRA, and RASHMI MITTAL

## ABSTRACT

Food wastage is of alarming concern at the international level. This issue is needed to be tackled threadbare in order to save the people from starvation. It cannot be done single-handedly unless problems regarding food security, fixing wastage loopholes in value-added supply chain are corroborated and fixed appropriately. Food losses and wastage are the results of improper management of food supply systems. The first priority of FAO and global developmental banks is to limit the post-harvest losses (PHL) to fill the gap significantly between the production and availability of food. Several strategies and policies are needed to be initiated both at National and International level to facilitate the idea and thereby to ensure the availability of food for a growing population. This chapter provides a brief overview of post-harvest management (PHM) practices followed globally along with the technologically equipped intervention which have been initiated to facilitate the idea and the possible strategies which can be adopted worldwide to attain the maximum benefit of the concept.

## 6.1 INTRODUCTION

Food is not merely a traded item but an essential thing without which no life is possible on the Earth. Utmost care is required to be adopted for stopping food grain wastages in any manner. To prevent food grain wastage post-harvest management (PHM) plays a major role. During the post-harvest period, both the parties viz producers and supervisors do pay attention in preserving the quality of the food grain and meager loss. The most important challenges are to determine that how to reduce food wastage in the world and place a leak-proof strategy when the world is over flooding with population which is disproportionately increasing.

It is also imperative, if the population growth is not checked in time, there would be a dearth of shortage of food, and in that case the world would need to increase its production of food by 70%. It is also feared that the world population would reach at the alarming stage roughly 9 billion by 2050 [16]. It is widely evident that an increase in urban population, change in lifestyle and modification in diet intake patterns in middle class, constant emerging economy together with a change in climatic conditions are drastically putting additional pressure on the natural resources. Drying of freshwater resources and changes in biodiversity, loss of fertile land due to excessive use of hazardous chemicals/fertilizers, etc., is compounding adverse effects on the soil health. Therefore, there is a necessity for a sincere and integrated approach by the global efforts to ensure food sufficiency and consumption, but it cannot be implemented unless food security measures are strengthened. There is an urgent need to revise re-look in the agriculture arena effectively to resolve the issues associated with post-harvest losses (PHL).

### 6.1.1 OBJECTIVES OF POST-HARVEST HANDLING

The most prominent objective of post-harvest handling is to ensure that products remain cool, reduce the chances of moisture loss, maintain nutritional components, retard the chemical changes, and avoid physical deterioration to prevent spoilage. Sanitation and proper upkeep are also important factors, to reduce the impact of pathogens which could accompany with fresh produce. Different crops exhibited varied ranges of temperature and humidity for storage. So far, post-harvest handling is concerned; it has been seen that ripening can be delayed by preventing tissue respiration, whereby storage of perishable commodities can be prolonged. This technique enthused scientists to discover and to fully explore the concept and the fundamental principles and mechanisms involved behind it, with regard to post-harvesting. Most often, food grain is lost in the initial and end stages. PHL occur before the start of the process, which of course is due to weeds, hailstorm, insects, rusts, and dust. Harvest losses occur prior and at the completion of the harvesting period. The chain of losses includes on-farm, transportation, storage, and processing. There is significant loss during the entire grain harvesting process and while carrying out the agricultural marketing chains.

### 6.1.2 FOOD WASTAGE

Indeed, it is of great concern that 1/3rd food grain produced 1.7 billion tons is wasted these days. It is worth noticing that in developing countries, food wastage was prominently observed during the post-harvest period due to insufficient infrastructure. It is estimated that by 2050 food wastage would be cut by half, which could fulfill the food requirement of one-quarter of the shortage [19].

### 6.1.3 POST-HARVEST MANAGEMENT (PHM)

The wastage of food crop begins at the onset of harvesting. The sole idea of PHM is that of slowing decaying process and provides better quality to the consumers. The main concept of post-harvest technology is put in place for treating agricultural produce after harvest to ensure its protection, processing, packaging, distribution along with proper marketing, and uniform utilization to guarantee its nutritional parameters. This technology must be developed with an aim to fulfill the needs of society around. A study conducted regarding the loss of fruits and vegetables revealed that we are wasting fruits and vegetables worth USD 2.55 per annum in India. India ranks at 102 positions out of 117 countries, according to Global Hunger Index [13].

### 6.1.4 GLOBAL SCENARIO

There exists a substantial need to increase global food production by 60% up to 2050 so as to fulfill the needs of the rapidly growing world population. The financial cost of food wastage is substantial and is calculated which was observed to be around USD 1 trillion every year. Food wastage also causes serious environmental impacts [12]. In large quantities cultivated globally and consumed are horticulture and tomato (*Solanum lycopersicon* L.) [1]. On account of weight, tomatoes rank second to potatoes in worldwide production. At the global level, nutritional constituents, calorie percentage, digestion ability, and acceptability of the food products are affected the most. These losses are often occurred more commonly in developed countries [23]. It is alarming that in 2010, approximately 133 billion pounds which accounts for approximately 31% of available food was observed to be wasted in the USA only at the retail outlets. It is estimated that 19, 20, and 44% losses of cereals, tuber crops, fruits, and vegetables occur every year, respectively.

'*Languages spoken by empty stomachs are heard by the heart.*' The thought that some lives matter slightly lesser in comparison to others is the root cause of all the wrongs with the world. Hunger has been chasing India for a long period of time. Despite having tremendous development in the agricultural sector and country's granaries are overflowing, thousands of people starve.

United Nations' Sustainable Development goal is to secure food security and zero hunger. Various Government bodies and International organizations took steps to achieve the target, their efforts remain an elusive attempt as more than 790 million people are facing consequences of hunger. Several factors contribute to these problems, one of which is too little in controlling PHL. Data available from 2009 the UN's Food and Agriculture Organization (FAO), revealed that approximately 32% of food produced globally was observed to be either getting lost or wasted.

According to World Bank (1999) reported 7–10%, a major portion of food grain has lost in PHO and 4–5% of losses alone occurred in India during marketing and in distribution outlets [25]. Estimates further highlighted that 12–16 MMT of grains is wasted every year, which would have sufficed the hunger of nearly 1/3$^{rd}$ of India's poor population. Despite criticism, the authentic data on PHL remained a challenge. Vast research has been carried out to assess the loss in major economies such as India, China, and Brazil. It is required to be highlighted that food insecurity is a vital issue which otherwise is understood subconsciously but consciously ignored to a large extent. It may not be out of place to point out that food security emerges when all the consumers have the availability of food at all times and have easy access to nutritiously rich food sufficiently. If done with judiciously, there is no doubt that dietary needs and food requirements could be met for a healthy life for one and all (Figure 6.1).

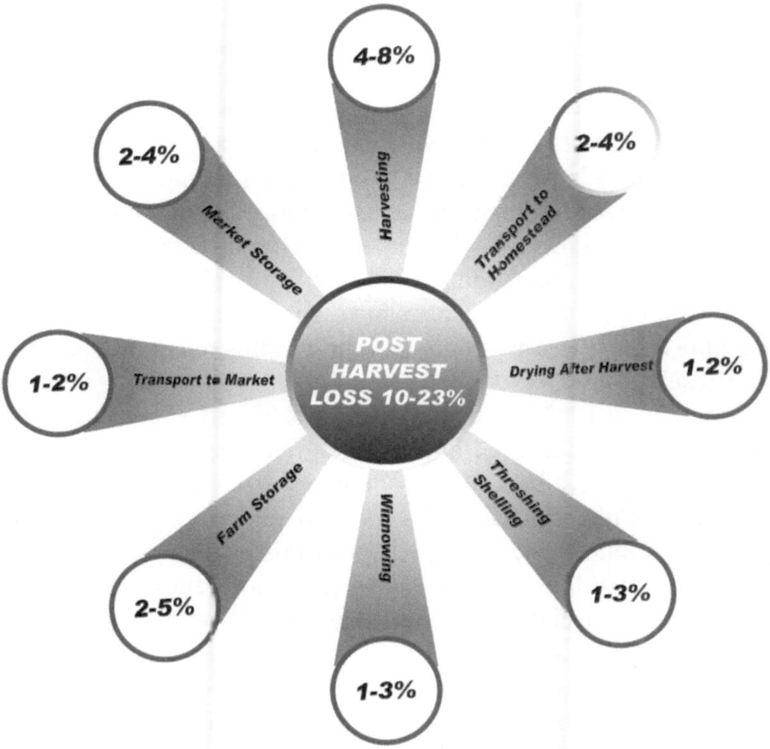


**FIGURE 6.1** Cumulative percentage of post-harvest loss at different steps.

## 6.2 PRACTICES FOLLOWED AT GLOBAL LEVEL: CURRENT SCENARIO

Each country, whether developed or developing, struggles to provide the required ratio of quantity of food to its inhabitants It is a very sorrowful state of affairs that despite the farmer's struggles day and night for producing varieties of crops and vegetables, no proper and adequate storing facilities exist whereby the good amount of food grain and perishable items such as vegetables and fruits spoiled and tore the pockets of farmers, who otherwise are the breadwinners of the world as a whole. A considerable amount of food grain says about 30% to 45% (30–45%) is drained in post-harvest handling, which is causing adverse effects on the nutrition quality and food security. This section is aimed at educating farmers, and agricultural products, technologists, employees working processing sector and elsewhere on PHM. It can be said with certainty that the state of the agriculture economy on account of PHL observed in both developing and as well as in developed countries is following the same pattern. It is of great concern that food grain is produced by small-scale farmers out of their personal and private landholdings where the good share is kept behind for personnel consumption, and the left-over quantity is ferried to the markets for sale. Under the circumstances, the small-scale farmers cannot hire heavy-duty machinery or implements. Furthermore, weather conditions from one plateau to another also play a major role. Though the researchers are doing their best in this regard but the outcome of their results cannot be applied throughout the pan-world, continent, or countries alike in 100%, thus preventing wastage. The scientific research mostly remained focused on improving agricultural production while no research or steps have been taken for the safety of PHL. India must invest in post-harvest research technologies with regard to mitigate losses of agricultural produce. The main precept of this chapter is to put forth the importance of PHM/technology and to create awareness for a fool proof PHM program, especially to farmers, agriculturalists, technologists, policymakers, administrators, and industrialists.

It would be helpful in reducing losses in agriculture production in developed and developing countries, increase in annual revenue improvement in economic health of farmers while enhancing lucrative nature of agriculture in their respective countries. There is a dire necessity for adopting various agroecosystem functions.

As per estimates, 16–36% post-harvest loss occurs due to mechanical, microbial, and physiological techniques to fruit crops. Perishable horticultural commodities need utmost attention. Lack of well-planned strategies does hamper promotion for processing and value addition. The value addition chain for processing, which plays a vital role, is considered to be a central tool in the improvement of food safety and strengthening the Nation's food security (Figure 6.2).

### *6.2.1 AFRICAN CONTINENT*

Foodgrains of approximately worth the US $4 billion per year is lost on account of PHL in sub-Saharan Africa (SSA) [1]. A large number of the populations is undernourished in Nigeria even after surplus production and large imports of rice each year. In Nigeria, the Rice Loss is 24.9% of the total food grain damage in the Supply chain. Total foodgrain loss has been accounted to Nigerian Naira (NGN) is 56.7 billion approximately each year [5]. As per the American National Academy of Science, wheat loss in Sudan and Zimbabwe were 6–19% [18, 21]. In Uganda, Tanzania, and Malawi, the PHL of agricultural produce in maize have been estimated at 1.4–5.9% at the farm level. Insects and pests are considered to be majorly responsible for grain losses in storage. In this list, corn constitutes 36%, which is a necessary element of the diet in SSA.

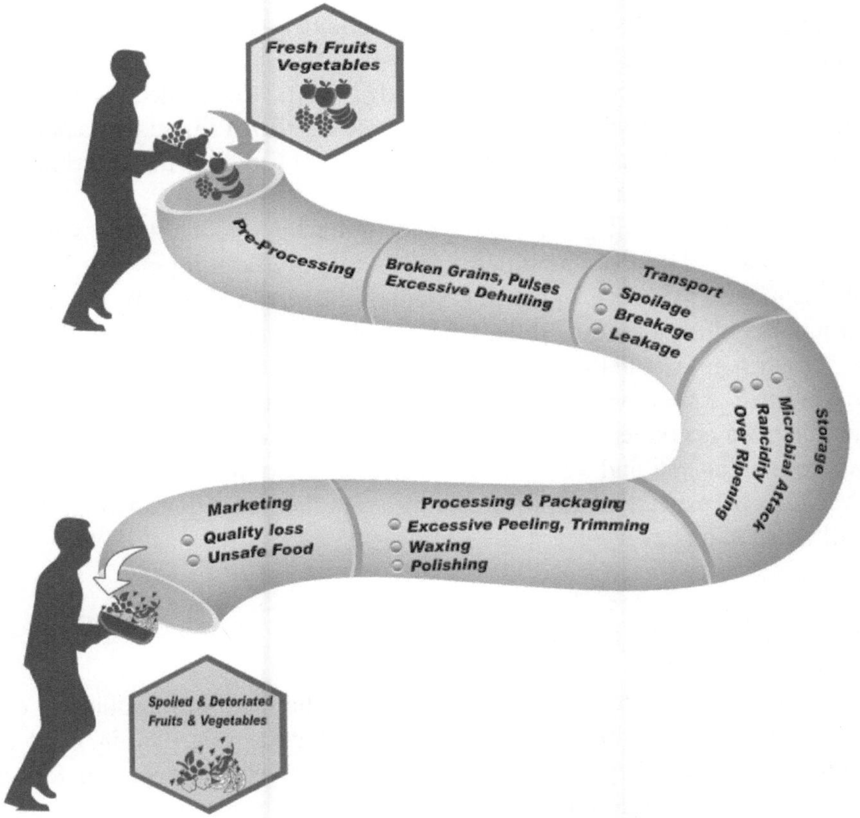


**FIGURE 6.2** Distribution chain between producer and consumer and deterioration of quality in between at different steps of post-harvesting procedures.

Corn is invariably stored locally within the space ceiling and the roof above the cooking places to allow hot air passage to guard against insects. In Western Africa, legume seeds and other cereals are usually kept at home or at the farm sites either in jute or in polypropylene bags, on erected platforms in conical shaped infrastructure or either in baskets as well. In some countries such as Eastern and Southern Africa, grain is kept in small sacks, wooden containers, pits, drums mixed with cow dung ashes covered with mud slabs. The loss of food grain does start from the very beginning till PHL. In the countries like Kenya and Malawi "Nkokwe" warehouse structures are used for storing grain [5]. In Ghana, around 50% of maize is lost due to insect attacks. 80–90% loss is done by pests and insects during the seed storage. Pest identified as *Callosobruchus maculates* (F) is the main cause of losses which exclusively spoiled 24% seeds in the storage in Nigeria [2, 18, 21].

Also, in Cameroon, 12–44% of maize seeds are spoiled by various insects viz mites, mold mite, and flour mite [5]. Such insects' attacks are more visible in African countries where, PHL of food cereals is estimated at 25% of the total crop harvested.

Some crops (fruits, vegetables, and root crops) which are less hardy than cereals, PHL could attain up to 50%. In dairy farming in East Africa, the economic losses are of worth US$90 million/year. Similarly, in Kenya, around 95 million liters of milk, whose costs is around US$22.4 million is lost each year. All the losses are either due to poor handling of the foodstuff or because of non-availability of proper infrastructure. Cumulative losses in Tanzania accounts to be around 59.5 million liters each year, which is approximately 16 and 25% of total dairy production during the dry and wet season, respectively. In Uganda, milk products of around US$23/year millions are lost every year [3].

### 6.2.2 AMERICAN CONTINENT

The Survey conducted in Latin America and the Caribbean countries on PHL revealed that the losses were estimated from 1.4–30% in various regions. The majority of losses in the regions were observed in farms ranging in size from small to medium due to the absence of adequate harvesting, drying, and storage techniques and lack of information about them. Due to deficiency of appropriate storage facilities and high humidity of Guatemala region caused 40–45% of loss of maize. The insect infection was the main cause of food grain loss [5]. In general, the United States of America (USA) listed the loss of 30% worth US$48.3 billion (€32.5 billion). About 26% of losses occur from the farms to the markets, whereas 1% loss is done in supermarkets. Overall, losses amount to around US$90 billion-US$100 billion a year. Research conducted recently indicated that the changes in climatic conditions and rising of global warmth will reduce the yield of corn in the region while bringing the commodity.

### 6.2.3 ASIAN CONTINENT

It is a well-known and visualized fact that farming never remained a commercial activity all over the world. Farming was being done by small sections of the society for their own consumption that is why they go on changing the places and tend to roam wherever suitable climate and fertile land is found. It is when the people settled at particular places and stopped roaming, they changed their way of life and start cultivation of food products for their consumption and to sell the surplus.

However, small-scale farmers remained dominated on agricultural systems but due to limited access to the resources owing to high initial costs in managing pre- and post-harvest technologies 20–44% fruit and vegetable stock which is perishable suffers damage [21]. Owing to ineffective handling techniques for transporting their products roughly 25%–40% high-value crops such as rice, wheat, and corn, perish in Asia. Bangladesh being the 4$^{th}$ largest producer of rice, still face a shortage of food grain and compelled to import rice to the tune of 1 million tons each year. The estimated loss of 10.74–11.7% occurs alone in the rice value chain in Bangladesh only [1, 5].

In India, rice loss is assessed at 3.5% per annum. Stringent efforts are needed to reduce the grain losses both at pre- and post-harvest operations [2, 5]. Data compiled on PHL observed in the case of rice value chains on the basis of studies conducted by FAO revealed that 10–37% of rice is lost in Southeast Asia due to poor post-harvest procedures. In China, the losses that occurred on account of improper PHM were estimated in the range of 8–26%. It has been reported that the food loss because of improper storage facility were maximum (42%) of all post-harvest processes. Total loss in the wheat supply chain from harvest to the end consumer in India was estimated to the tune of 4.3%. Techniques adopted in the field contribute to the tune of 75.9% of all losses after harvest. To dry the corns, roadsides are generally preferred, particularly in the Philippines. In Vietnam, huge losses have been observed due to rodents and fungal diseases during the storage of maize [5]. In most of the developing countries such as Africa and South Asia, cereals are generally preferred to be stored in bulk containers or in sacks bags, in simple granaries which are often built locally by available materials such as straw, bamboo, and clay. The latest studies show that the climatic conditions and gradually rising of global temperature in China will hamper the yield of corn. In that case, the rise of 20°C temperature will decrease the produce by –10.4%.

### *6.2.4 AUSTRALIAN CONTINENT*

In Australia, food wastage is about 3.2 million tons each year [6]. Food wastage was observed to be occurring at the extreme end of the value chain. To mitigate food wastage, the discarded food could separately be treated and feed to animals, and it can be mixed with fodder. According to a survey of in Australia alone more than 1,600 households in 2004, consumables worth $10.5 were procured, which were not used and discarded.

### *6.2.5 EUROPEAN CONTINENT*

In the United Kingdom, 6.7 MT of food often gets wasted annually, which is around 1/3$^{rd}$ of the 21.7 million tons of total procurements. In Germany, about 22% of food wastage is observed at the processing and distribution levels, to the tune of 2,400,000 tons per year. However, the per capita wastage in Europe and North America is 95–115 kg/year [5]. Most of the food wastage (4.1 MT or 61%) can be avoided by providing foolproof management [3].

The unavoidable losses are mainly from PH scenario or little in augmenting proper technique and infrastructure.

## 6.3 TECHNOLOGICALLY EQUIPPED INTERVENTIONS TO FACILITATE THE IDEA

Improper preharvest handling, unsuitable, and inappropriate harvesting, post-harvest handling, packaging, and transportation practices coupled with poor logistical support in the supply chain. These incorrect and inadequate practices squeezed the market opportunities and resulted in lesser income for small-scale operators in the supply chain. Fresh produce safety is compromised by pre-harvesting practice, production, and post-harvesting practices including poor hygienic management in the supply chain. Chemical and microbiological hazards can pose a safety threat to human health. It is indeed, imperative that stakeholders in traditional supply chains must focus on improving post-harvest handling and enforce qualitative safety techniques on fruits and vegetables. Recent innovations in post-harvest technology in various countries have been conducted in response to avoid the usage of costly labor and cosmetically ideal produce. This procedure is not sustainable for the long term. But the use of post-harvest pesticides would reduce the occurrence of defects, but it would be a costly approach (Figure 6.3).

**FIGURE 6.3** Illustration of different steps involved in post-harvest processing.

Cultivation of fruit and vegetables contains new opportunities for the farmers. Small scale handlers should be facilitated with credit investments in post-harvest techniques. Many practices possess the potential to withstand the needs of farmers and marketers. There are a number of steps concerned with the post-harvest system, where produce is handled by a stream of people from the farm end to the Consumers.

Though the practices of operations are varied for each crop still there are few series of steps which are common with regard to post-harvest handling that should be followed for the purpose [4, 13].

### *6.3.1 HARVESTING AND PROPER TRANSPORTATION AND DISTRIBUTION OF FOOD GRAIN*

There are a number of elements which causes food grain damage in the harvest. Therefore, some mechanical system is to be evolved to safeguard against such types of losses. There is a need to devise a method for plucking the fruits. It has been observed that night and early morning harvesting is an option for harvesting. During this period, the internal temperature is relatively low, thereby reducing the energy required for cooling and stay worthy of the produce. Exposure to the sun is harmful and invariably should be avoided.

The mechanical damage during harvesting is very common and poses serious problems. Great care is required to be taken for plucking citrus fruits. Citrus fruits, if harvested early in the morning, there is a great chance that it will release essential oils from flavedo oil glands and leads to oil spotting after de-greening. Occasionally, farmers do early harvest citrus fruit crops, in un-ripe or not fully developed stages. The rough handling during packaging, transportation increases bruising and limits the benefits of cooling. Gradual cooling and warming also deteriorates at a slower rate. Fruit Pickers should harvest the crop with utmost care. Reduced tire air pressure of transporting vehicles reduces the amount of motion to the produce. Roads between the farm and the grading facilities houses should be cleared of furrows, knocks, and holes, to avoid further damage. The produce when prepared for marketing, its finishing and cooling play an important role. For horticultural crops, there is no certainty whether the crop will be of full bloom or not. Produce exposed to sunlight can get warmed by 4–6°C. Delay in cooling often shorten the post-harvest life and reduces the quality of the product [5, 6, 18].

### 6.3.2 *MATURITY STANDARDS*

Maturity standards have been prescribed for all food items, including fruits, vegetables, and floral crops. For better results, a proper maturity index should be determined by the handlers before starting plucking fruit crops for marketing. Early harvest does result in a lack of taste and flavors of the fruits in many ways. Fruit growers and puckers should be well trained to identify whether the fruit crop is ready for harvesting (Table 6.1).

**TABLE 6.1** Vegetables and Their Range of Maturity

| | |
|---|---|
| **Textural Properties** | |
| Firmness | Apples, stone fruits, and pears, etc. |
| Tenderness | Peas |
| Color, external | Majority of fruits and vegetables |
| **Compositional Factors** | |
| Starch percentage | Pears |
| Sugar | Apples, grapes, pears |
| sugar/acid ratio | Pomegranates, papaya |
| Juice | Citrus fruits |
| Oil | Avocados |
| Tannin | Dates |

### 6.3.3 *CURING OF ROOT, TUBER, AND BULB CROPS*

The cure and protection of root crop families such as sweet potatoes, cassava, yarns, and many more is of great importance. These crops are kept for any time length by preserving the product at a relatively sustainable temperature and humidity for a longer span allowing the harvesting injuries to heal so that a protective layer is formed. Though this practice is a bit costly to start with but this procedure is beneficial for long time storage, which makes this practice economically viable.

Curing techniques when applied to flower bulbs, garlic, and onions after harvest, utmost care is required to be taken with reference to skin and neck tissues prior to the storage. The dried parts of plants which are invariably covered are shaded during the treatment process. The skin layers further protect the produce from water loss. If conditions do not allow field soaking, then Tent canopies can be used alternatively [4].

### 6.3.4 PACKING HOUSE OPERATIONS

Packaging house operations involve post-cleaning, waxing, sizing, and quality grading. Packaging should be done under the shade alongside sufficient space and provides an easy link for transportation. Produced material is also delivered in packing containers, soon after the harvest to packers. The packers' sort out, do size grading, and then made the appropriate packing to transport the containers [4, 6].

#### 6.3.4.1 GENERAL OPERATIONS

Cleaning can be carried out by various methods, including washing with chlorinated water or dry brushing. Waxing is carried out after washing and soaking of moisture and then grading is done. The premium quality produce is marketed in both the National or International level markets.

#### 6.3.4.2 DUMPING

Dumping is done mildly using watering methods or by dry dumping. Wet dumping reduces the bruising and abrasions, chlorinated (100–150 ppm) water can be potentially used to carry delicate produce.

#### 6.3.4.3 PRE-SORTING

Pre-sorting of the stock is done to minimize damage to the products before cooling.

#### 6.3.4.4 CLEANING

For some of the fruit stocks, for example, kiwifruits or avocadoes, dry brushing is enough to clean the produce whereas, bananas, and carrots, etc., do require washing. Additionally, tomatoes, cucumbers, leafy greens are washed before cooling.

### 6.3.4.5 WAXING

Waxing of unripened fruits, vegetables such as Cucumbers and summer squash, eggplant, peppers, and tomatoes, apples, and peaches is common. Waxing is helpful to safeguard against water loss during handling.

### 6.3.4.6 WASHING

For simple washing, Metal drums can be used by cutting those drums into two halves with drain holes. Metal edges are generally covered with split rubber or either with plastic pipes. The drums are placed on a sloped wooden table, which is Chlorinated and can reduce the spread of contamination at the washing stage. The pH of the wash water needs to be maintained between 6.5 and 7.5 for the best results.

### 6.3.4.7 PACKING CONTAINERS

Useful information on the characteristics of different types of material used for packaging is tabulated in Table 6.2.

**TABLE 6.2** Characteristics of Sacks as Packaging Units

| Sacks | Tearing and Snagging | Impact | Protection Against | | Contamination | Notes |
|---|---|---|---|---|---|---|
| | | | Moisture Absorption | Insect Invasion | | |
| Jute | Good | Good | None | None | Contamination through sack fibers | Bio-deterioration. Insect harborage. Odor retention. |
| Cotton | Fair | Fair | None | None | Fair | High re-use value. |
| Woven plastics | Fair-good | good | None | Some protection | Fair | Adverse effects through ultra-violet light |
| Paper | Poor | Fair-poor | Good-WFP sacks | Protected, if treated. | Good | Consistent quality |

## 6.3.5 DECAY AND INSECT CONTROL

Good management by the proper handling of harvested products acts as a defense wall against insects. The storage environment plays an important role in maintaining high-quality produce. Since free flow water on the surface of produce causes germination and pathogens penetration on nuts, dried fruits, and vegetables, these commodities can be saved for a long period by freezing at less than 5°C. Heat treatment, exclusion of oxygen by 0.5% or minimum use of nitrogen can also be applied to prolong the decay. Leaves of Cassava protect the produce from pests when used as packing material for short-term storage. Pesticidal properties of Neem seeds are becoming more prominent to guard against the attacks by being non-toxic to humans, mammals. Though the Natural pesticide which are known to be safe for humans yet is of paramount importance to regulate the use of such practices by the competent authorities. Chemicals such as chlorinated water, Calcium hypochlorite and along with this sodium hypochlorite are applied to control microbial growth. To control the crown rot fungi, active ingredient sulfur is applied as a paste (0.1%) on bananas to prevent their further damage [4, 5].

### 6.3.5.1 COLD TREATMENT

In the coldest temperature, a certain commodity such as apple and pear can withstand it without incurring damage from fungi and bacterial infections. *Rhizopus stolonifer* and *Aspergillus niger* (black mold) can be targeted by treatment at 0°C for 2 days. Likewise, vegetables can also be stored in moderate cold temperature but cannot last long as compared to certain fruit categories. However, a few vegetables, such as beans, etc., can be kept in frozen conditions.

### 6.3.5.2 HEAT TREATMENT

Hot water or heated airflow is used for direct control on post-harvest insects depending upon the variety of fruits and vegetables, location of country of origin. Brief hot water dips or forced-air heating is found to be highly effective for control of microbial load on various types of fruits. Though efforts are made to market the perishables items on priority (Table 6.3).

**TABLE 6.3** Heat Treatment of Fruits and Vegetables

| Commodity | Pathogens | Temperature (°C) | Time(min) | Possible Injuries |
|---|---|---|---|---|
| Grapefruit | *Phytophthora citrophthora* | 48 | 3 | – |
| Green beans | *Pythium butleri sclerotinia sclerotiorum* | 52 | 0.5 | – |
| Lemon | *Penicillium digitatum Phytophthora sp.* | 52 | 5–10 | – |
| Mango | *Anthracnose collectotrich umgloeosporioides* | 52 | 5 | No stem rot control |
| Melon | *Fungi* | 57–63 | 0.5 | – |
| Peach | *Monoliniafruticola rhizopus stolonifer* | 52 | 2.5 | Motile skin |
| Pepper(bell) | *Erwinia sp.* | 53 | 1.5 | Slight spotting |
| Grapefruit | *Phytophthora citrophthora* | 48 | 3 | – |
| Green beans | *Pythium butleri sclerotinia sclerotiorum* | 52 | 0.5 | – |
| Lemon | *Penicillium digitatum Phytophthora sp.* | 52 | 5–10 | – |
| Mango | *Anthracnose collectotrich umgloeosporioides* | 52 | 5 | No stem rot control |
| Melon | *Fungi* | 57–63 | 0.5 | – |
| Peach | *Monoliniafruticola rhizopus stolonifer* | 52 | 2.5 | Motile skin |
| Pepper(bell) | *Erwinia sp.* | 53 | 1.5 | Slight spotting |

### 6.3.6 TEMPERATURE AND HUMIDITY CONTROL

During the time of harvest and consumption, temperature control is a major key for maintaining product quality. Maintaining the products at their lowest temperature. Fruit and vegetable crops are often more susceptible to cold attack when cooled below 13 to 16°C, i.e., why utmost care is required to avoid chilling injuries to vegetable stocks other the commodity shall perish despite 4 taking all ought efforts for preservation. When the produce is stored, it is imperative to begin with a high-quality product. The produce should be free from damaged or diseased units. In general, proper storage practices include temperature control, relative humidity control, air circulation and adequate ventilation. However, the commodities which are stored together must be able to tolerate the same temperature and relative humidity. So far, the Rabi and Kharif crops are concerned, these crops preserved in dry condition. Utmost care is needed to avoid moisture in the godowns. Regular flow of fresh air is assured, and gases are exhausted by means of proper ventilation and exhaust fans (Table 6.4).

Storage of crops both grain, horticulture, and vegetables are a very technical and far-sighted project. It requires a thorough knowledge with regard to the employment of infrastructural assets and storage areas with proper passages of air, sunlight together fixing of leakages, avoidance of dampness, pest control, guard against rats and other attackers. Storage facility should be built in such a way so that it could serve for the longest period without any alteration. Adequate measures are needed to be taken to manage the storage area, such as by keeping the surrounding area clean, devoid of trash and weeds. Concrete floors could help prevent rodent entry [3].

### 6.3.7 HORTICULTURAL CROPS TRANSPORTATION

Long-distance transportation temperature management is a vital component, i.e., stock must be properly loaded in such a manner to enable proper air circulation to shun heat. Transport vehicles should invariably be ventilated to allow free air movement. The produce should be insulated to maintain cooling all around. Mixed loads have remained to be a serious concern in the situation where temperatures are not compatible. High ethylene producers can also induce certain injuries and can potentially detoriate its quality or may cause certain undesirable changes in terms of color. Lightweight insulated covers can also protect the harvested commodities from excessive heat for longer durations.

**TABLE 6.4** Humidity, Recommended Temperature, Approximate Transit and Storage Life for Fruits and Vegetable Crops

| Product | Temperature (°C) | Relative Humidity (%) | Storage life |
|---|---|---|---|
| Amaranth | 0–2 | 95–100 | 10–14 days |
| Anise | 0–2 | 90–95 | 2–3 weeks |
| Apples | –1–4 | 90–95 | 1–12 months |
| Apricots | –0.5–0 | 90–95 | 1–3 weeks |
| Artichokes, globe | 0 | 95–100 | 2–3 weeks |
| Asian pear | 1 | 90–95 | 5–6 months |
| Asparagus | 0–2 | 95–100 | 2–3 weeks |
| Atemoya | 13 | 85–90 | 4–6 weeks |
| Bananas, green | 13–14 | 90–95 | 14 weeks |

### 6.3.8 MANAGEMENT AT DESTINATION

On the arrival of the produce in markets, it is recommended that utmost care should be maintained such as requisite temperature should be provided. If produce is to be stored, the cold storage units and godowns needed to be well clean and maintained with plenty of air circulation, well-insulated space. As a variety of commodities are handled simultaneously, it is essential to remember and index not to mix the stocks of different temperature requirements altogether. Commodities like bananas, papaya, stone fruits, etc., are left at ambient temperature. Shopkeepers, wholesale marketers should also be educated with regard to the importance and values of cleanliness of the products to fetch a good price. It is a well-known fact that there exist different ways for protecting the produce. No common formulae exist because of variation in geographical features and climatic problems. Anyhow, in some developed countries, attention is given to carry the products from the farms to the Storage facilities and then to the wholesale markets. Those countries have good infrastructure in place, and further, more sellers and buyers are vigilant on this score. Efforts are afoot to make such types of infrastructural facilities around the world by the corporate houses to safeguard against the losses in-between.

#### 6.3.8.1 TEMPORARY STORAGE HOUSE

The foodstuff and other horticultural products, which are grown and consumed in plenty worldwide do not find suitable storage facility. Such items are either stored by the side of farms in temporary shelters and left in open because of lack of place in cold Storage or otherwise. Such items are stocked in an ill manner; transportation facilities are not up to the mark from the farmlands to the storage.

It was also noticed that the cold stores are either constructed near or around the metropolitan cities, and because of improper transportation, plenty of food products go haywire, putting losses. Though the harvested produce is kept at pre-defined locations before transporting to the markets, the handlers can further ensure its quality and therefore can reduce losses by storing products at optimum temperature. For storage of stocks for a lesser period around relative humidity is assessed and maintained between 85 and 95%, and the ethylene level is kept below 1 ppm, by ventilation or using a scrubber to protect the produce from any further damage.

Depending upon the commodity, produce should be packaged properly to avoid the loss [4, 3, 20, 22]. Post-harvest processing normally starts in a simple shed, by shade and running water, or a large-scale, sophisticated, mechanized facility with conveyor belts, automated sifting, packing platforms, and so on. Different crops exhibited varied ranges of temperature and humidity for storage. So far, post-harvest handling is concerned this technique enthused scientists to discover and display the knowledge of fundamental principles and mechanisms of respiration, with regard to post-harvesting. Most often food grain is lost in the pre-harvest and post-harvest stages. Pre-harvest losses occur before the start of the process, which of course is due to hailstorm, insects, weeds, rusts, and dust. Harvest losses occur prior and at the completion of the harvesting period. The chain of losses includes on-farm, transportation, storage, and processing (Table 6.5).

Comparison between properties of non-perishable and perishable crops indicating their storage capacity is indicated below:

- Seasonal harvesting requires for longer duration;
- Products with low level moisture contents (10–15%) or even less;
- Small fruits of less than 1 g;
- Low respiratory activity of the product, limited heat;
- Hard tissues against injuries and good protection;

- Good natural disposition for longer storage and span;
- Losses in storage due to exogenous factors comprised of moisture, insects, and rodents, etc.

➢ **Perishable Food Crops:**

- Products with high level of moisture in between 50 and 60%
- Soft tissues, highly vulnerable.
- Heavy fruits from 5 grams to 5 kilograms or more.
- Permanent or semi-permanent production or short-term storage.
- High/very high respiratory activity of products including heat emission in a particular tropical climate.
- Losses due to endogenous factors.
- Early perishable products, natural disposition in-between weeks and months [26, 27].

**TABLE 6.5** Storage Life of Commodities based on Their Temperature and Humidity Requirement

| Articles | Relative Perishability | Potential Storage Life (Weeks) |
|---|---|---|
| Asparagus, broccoli, cauliflower, green onion. | Very high | <2 |
| Guava, loquat, mandarin, mango, melons, plum, celery, eggplant, avocado, banana, grape (without $SO_2$ treatment). | High | 2–4 |
| Beet, radish, carrot (unripe), grapefruit kiwifruit, apple, orange, lime, pomegranate, pear | Moderate | 4–8 |
| Potato (ripe); pear, apple, lemon, garlic, pumpkin, yam; ornamental plants and bulbs, etc. | Low | 8–16 |
| Vegetables, fruits (dried), nuts tree. | Very low | >16 |

## 6.4 KEY OBSTACLES IN THE IMPLEMENTATION OF PHMP

It is very important to mention here that even after employing suitable techniques responsible for food loss owing to implantation of PH, there had been a wider gap between the growers and the end-users of the foodstuff irrespective of fact whether it is fruit or grain. The main reason for this loss is due to non-existence of a sole authority who could devise proper rules and regulations to quiche the key obstacles in the implementation of PHMP. Geographical features, climatic conditions also play a vital role in this regard. The annual trend of food losses due to post-harvest (PHL) is more than sufficient to suffice the hunger of undernourished people globally. As per estimates, around 870 million people suffer from starvation and faces chronic undernourishment. Africa alone accounts for 27% of malnourished people. This challenge is aggravating at an average of 1.1% the world over due to the gradual growth of population. Food security and squeezing resource scarcity is posing a potential threat towards food loss and wastage (FLW) as it has not a foolproof mechanism which could meet the far-reaching social, economic, and environmental implications. FLW are of particular concern globally. Hence control on food wastage should be a main factor to contribute in abating interlinked sustainability with regard to food insecurity, climate change, and water shortages. As per FAO, about 1/3$^{rd}$ food products worldwide is lost or wasted annually, which comes to a total of almost 1.3 billion tons [5, 10, 15, 28]. Food losses do take place differently at unreliable locations, including food supply chain, geographical features, and social/economic conditions all around. Developing countries are worstly affected due to agricultural production (during harvest, transport, and storage of foodstuffs), whereas the developed countries are losing foodstuff at retail and consumer levels (Figure 6.4).

Under-development all over hampers investment in infrastructure is the main cause of wastage. Management in PHL can contribute towards larger economic, health, and environmental favorable impacts, which will also have the primary results in reducing losses and improve livelihoods of farmers and other members of value chains or provide an opportunity for nutritional security and production diversity.

The majority of efforts have primarily been focused on yields. Whereas, efforts should be intensified to reduce the PHL so that more food that is available for utilization. This will ensure improvements in crop production (via improved yields) will have the positive impacts on food availability.

Raising awareness among industries, retailers, and consumers and finding ways and means for the use of food that is being thrown away should be prioritized. A food loss worth $680 billion in industrialized countries and around $310 billion in developing countries is estimated. The industrialized and developing countries are wasting substantial amounts of food (670–630 million tons). Usually fruits and vegetables, as well as tubers and root crop, also constitutes the highest portion of wastage as compared to other food groups. FLW is around 30% per year

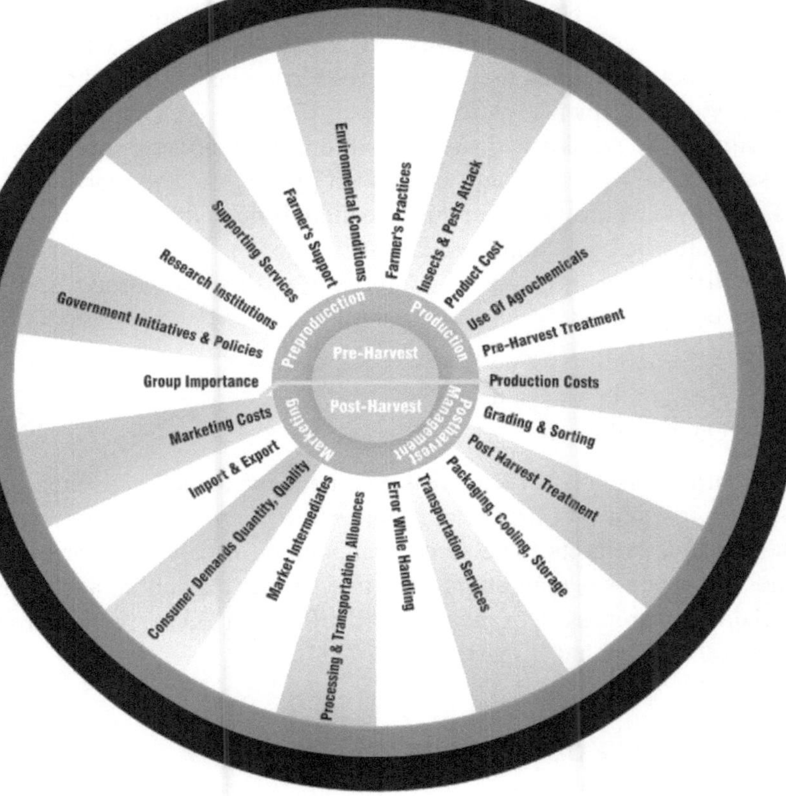


**FIGURE 6.4** Production, share, estimated, and global loss overview of grains, fruits, and vegetables due to proper post-harvest management.

globally. 40–50% root/fruit and vegetables crops are lost, whereas 20% of oilseeds, meat, and dairy products and 35% of fish go waste throughout the world. In developed countries wastage of food items is expected to be around 220 million tons. A significant decrease in post-harvest loss can put checks on food insecurity subject to proper and well-planned management of post-harvest systems, which is certainly a backbone in resolving various social and economic problems. Food safety can be guaranteed by protecting commodities from mold growth and contamination. This would have favorable results in good health and empower weaker communities in lieu. While food loss is happening at different stages of the food value chain wherein around 24% of losses occur during post-harvest handling, transportation, and storage. The main reason for wastage of food products is due to ignorance by the masses. The people who do not have sufficient meal for both the times they cry for food whereas on the other hand the class, who have plenty of food dam care for its preservation. They do not think if the surplus is distributed amongst the poorest strata, they will be serving the society. There is no such program in the world by which masses could be motivated in this direction. If done then, there is no doubt all the living beings would be healthy.

Food security however is comprised of three important and inter-related components:

- Affordability of Food
- Availability of Food
- Quality of Food

In rural areas, food insecurity is prevalent due to lack of adequate infrastructure, particularly for agricultural productivity, inadequate resources, and unstable markets. The three components involved in making any scheme a success is planning in advance, proper execution, and means of spreading awareness. Despite energetic efforts made by the government, India still lags behind (Figure 6.5) [4, 7, 11, 14].

## 6.5 PROBABLE OUTCOME OF PLANNED PHMP

For a sustainable future and hunger-free world, reduction of food wastage is one of the most wanted steps to be taken on priority. It requires a fool-proof system for making farmers' life better. It involves most attentive works, right from commencement of agro-farming till harvesting of crops. When crops are damaged during post-harvest, the income of farmers is also dipped. Since women are widely employed in post-harvest seasons; women can be empowered to excel in farming, provided they are trained thoroughly about the technologies and strategies in reducing PHL. If, food produced is utilized and wastage is minimized, there would be a lesser load on production machine/implements, which would lead to lesser greenhouse gas (GHG) emissions (Figure 6.6).

As per FAO, wastage of food items results in approximately 4.4 gigatons of GHG emissions (4.4 Gt $CO_2$) annually and if reduced through better PHM, two goals viz smart climate and agriculture (CSA) can be achieved:

(i) food security; and (ii) lesser GHG emissions. If implemented, it would add to food production, which in turn will be available for unprivileged people and protect the ecosystem from further damage caused by greenhouse emissions [8, 9, 17, 24].

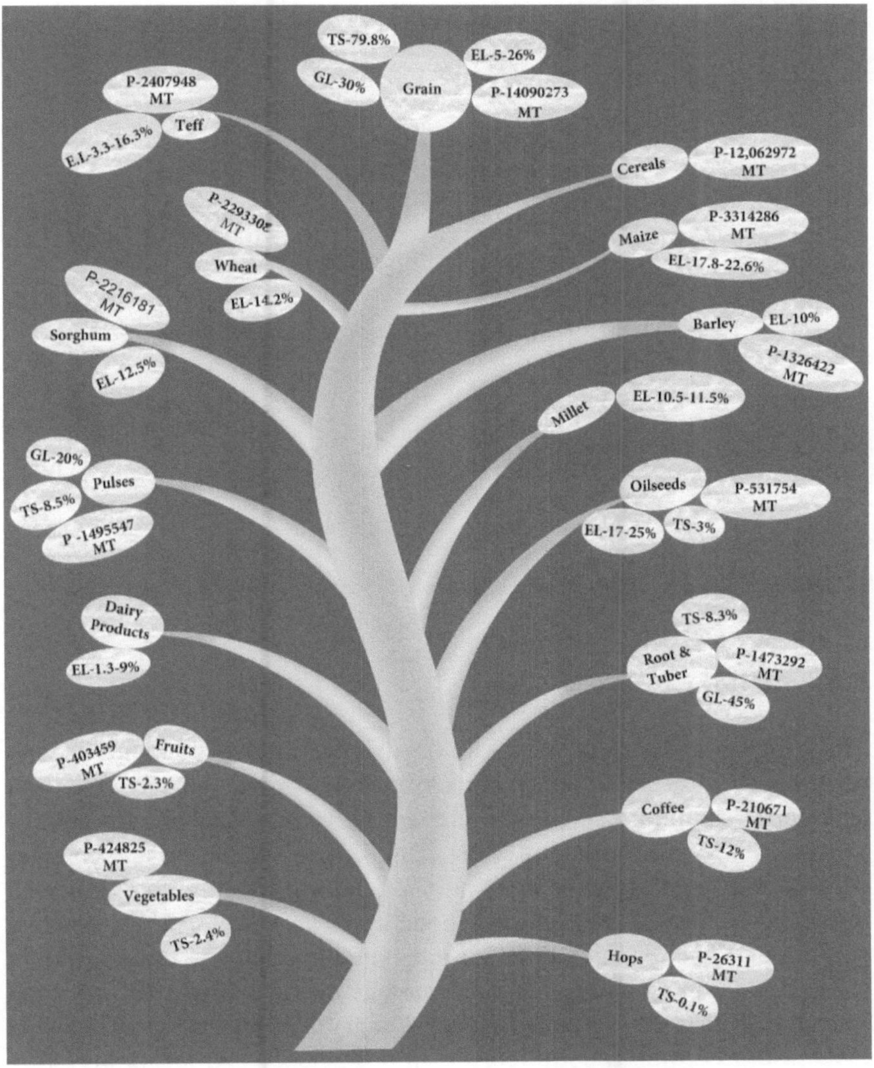


**FIGURE 6.5** Factors affecting post-harvest management practices.

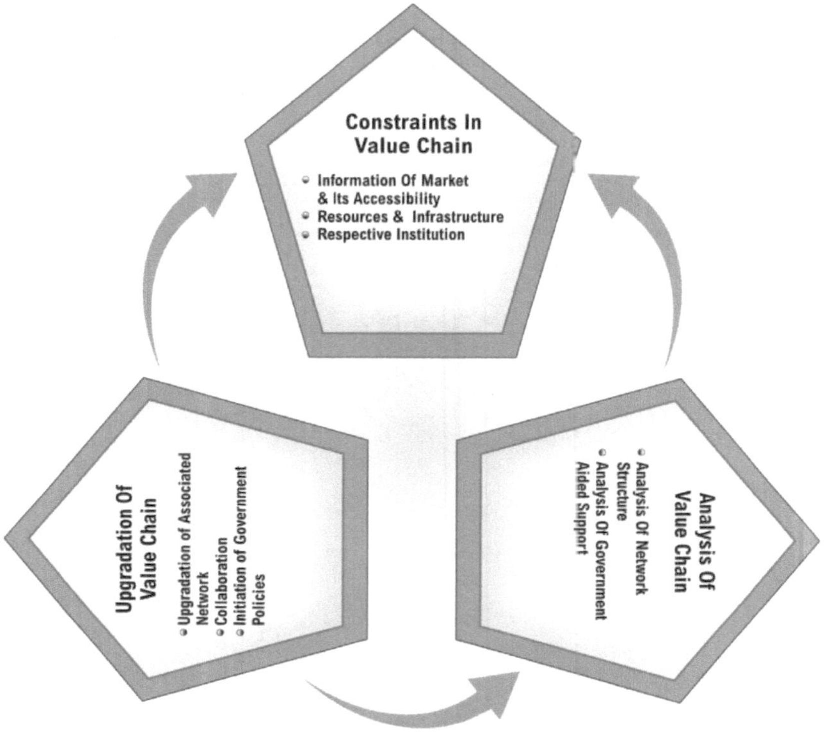


**FIGURE 6.6** Impact of the supply chain over post-harvest management practices.

## 6.6 SUMMARY

This chapter has been drafted to explore and find out ways and means with regard to the causes and safety and losses of food grain, what methods and techniques should be adopted to prevent wastage. Likewise, the significance of PHM in handling of crops, fruits, green vegetables and creating of infrastructure plus transportation of agricultural commodities have been discussed in breadth and length. Though it is an uphill task for creating such an environment and adopt an integrated approach in ensuring more food production globally, minimize wastage and utilization of food grain properly so that no one should sleep hungry and how to grow food in the world according to requirement. In this science and technology Era, nothing is impossible, so therefore to achieve the tasks provided, the task force is determined to get rid of solo approaches. To achieve and to attain success, we are required to bow down ourselves collectively. Though the population of the world is increasing day by day but no concrete steps have been taken to arrange food for one and all. Aspects with reference to strengthening food security and to reduce PHL remained neglected till date. The utmost challenge for food security is how to minimize the food wastage and implement long-term sustainable plan, when the world is over flooding with population disproportionately. It is also imperative to emphasize that if the population growth is not checked in time; there would be a shortage of food production. To meet the food requirements, the need would arise to grow food products by at least 70%. Therefore, Global food production must increase by 60% up to 2050 to provide both times meal. In some country's food distribution is also not up to the mark, which is required to be checked in and honest efforts are made to provide food to the deprived ones. In big hotels, plenty of cooked food is wasted and find the way to drains. In some big kitchens, no headcount is done, and as such, surplus food is cooked without estimation. However, these are little efforts but are of much significance. Roundabout 25% of total grain loss is estimated in most of the countries. To put a check on these losses an eminent role can be played by technologically equipped innovations to prevent improper pre-harvest handling and inappropriate harvesting methods, post-harvest handling, packaging, and transportation coupled with logistical operations. Harvesting and PHM is linked to each other. The value chain in PHM of horticultural crops play a main role and course of which is comprised of pre-harvesting factors, marketing followed by pre-cooling, sorting of produce, grading, packaging, and on-farm storage, transportation, storage, value addition/processing, and by-product waste management. Obstacles on the way cannot be erased unless the farmers are educated and understand the matter of PHLs meticulously on all fronts such as better infrastructure and gradual invasion of all odds starting from the agro-farms, transportation, carriage of products, financial incentives, marketing (wholesaler and retailers) and the end-users.

## ACKNOWLEDGMENTS

The authors are thankful to Dr. Anurag Dabas and Mr. Anand Singh for their technical support. We express our sincere thanks to Mr. Pramod Kulshreshtha and Mr. Sunil Kumar for designing the graphics of the chapter.

## KEYWORDS

- **Consumer**
- **Food and Agriculture Organization**
- **Horticultural crops**
- **Nigerian Naira**
- **Post-harvest losses**
- **Post-harvest management**
- **Threshing**

## REFERENCES

1. Abass, A. B., Ndunguru, G., Mamiro, P., Alenkhe, B., Mlingi, N., & Bekunda, M., (2014). Post-harvest food losses in a maize-based farming system of semi-arid savannah area of Tanzania. *Journal of Stored Products Research, 57*, 49–57.
2. Affognon, H., Mutungi, C., Sanginga, P., & Borgemeister, C., (2015). Unpacking postharvest losses in sub-Saharan Africa: A meta-analysis. *World Development, 66*, 49–68.
3. Arah, I. K., Ahorbo, G. K., Anku, E. K., Kumah, E. K., & Amaglo, H., (2016). Post-harvest handling practices and treatment methods for tomato handlers in developing countries: A mini-review. *Advances in Agriculture, 2016*.
4. Artiuch, P., & Kornstein, S., (2012). *Sustainable Approaches to Reducing food Waste in India*. Retrieved from: http://news.mit.edu/2012/sustainable-approaches-to-reducing-food-waste-in-india (accessed on 9 April 2021).
5. Bardner, R., & Fletcher, K. E., (1974). Insect infestations and their effects on the growth and yield of field crops: A review. *Bulletin of Entomological Research, 64*(1), 141–160.
6. Brecht, J. K., (1988). Evaluating precooling methods for vegetable packinghouse operations. In: *Proc. Fla. State Hort. Soc.* (Vol. 101, pp. 175–182).
7. Chauhan, C., (2013). *India Wastes More Farm Food than China: UN*. Hindustan Times. Retrieved from: http://www.hindustantimes.com/delhi/india691wastes-more-farm-food-than-china-un/story-m4QiWkxAXtTIzlWMkHT4CN.html (accessed on 9 April 2021).
8. Dunning, R., (2016). Collaboration and commitment in a regional supermarket supply chain. *Journal of Agriculture, Food Systems, and Community Development, 6*(4), 21–39.
9. Emerson, R., (2013). *The Food Wastage and Cold Storage Infrastructure Relationship in India*. Retrieved from: file:///C:/Users/test/Downloads/Food%20wastage%20&%20cold%20storage%20industry.pdf (accessed on 28 April 2021).
10. Faqeerzada, M. A., Rahman, A., Joshi, R., Park, E., & Cho, B. K., (2018). Post-harvest technologies for fruits and vegetables in South Asian countries: A review. *Korean Journal of Agricultural Science, 45*(3), 325–353.

11. *Food and Agriculture Organization of the United Nations* (2014). Retrieved from: http://www.fao.org/3/ae075e/ae075e02.htm (accessed on 9 April 2021).
12. *Food and Agriculture Organization of the United Nations,* (2014). Retrieved from: http://www.fao.org/3/a-i3989e.pdf (accessed on 9 April 2021).
13. *Global Hunger Index.* (2019). Retrieved from: https://www.globalhungerindex.org/india.html (accessed on 9 April 2021).
14. *GrainPro*, (2018). Retrieved from: https://news.grainpro.com/post-harvest-storagethat-can-help-save-the-planet (accessed on 20 April 2021).
15. Gustavsson, J., Cederberg, C., Sonesson, U., Van, O. R., & Meybeck, A., (2011). *Global Food Losses and Food Waste.* http://www.fao.org/3/i2697e/i2697e.pdf (accessed on 20 April 2021).
16. *International Crop Research Institute for the Semi-Arid Tropics* (2020). Retrieved from: http://exploreit.icrisat.org/profile/postharvest%20management/124 (accessed on 9 April 2021).
17. Johnson, L. K., Bloom, J. D., Dunning, R. D., Gunter, C. C., Boyette, M. D., & Creamer, N. G., (2019). Farmer harvest decisions and vegetable loss in primary production. *Agricultural Systems, 176*, 102672.
18. Kimatu, J. N., McConchie, R., Xie, X., & Nguluu, S. N., (2012). The significant role of post-harvest management in farm management, aflatoxin mitigation and food security in Sub-Saharan Africa. *Greener Journal of Agricultural Sciences, 2*(6), 279–288.
19. Lipinski, B., Hanson, C., Lomax, J., Kitinoja, L., Waite, R., & Searchinger, T., (2013). *Reducing Food Loss and Waste: Creating a Sustainable Food Future, Installment Two.* World Resources Institute, Washington, DC, USA. [online] URL: http://www.wri.org/publication/reducing-food-loss-and-waste (accessed on 9 April 2021).
20. Madakadze, R., Masarirambi, M., & Nyakudya, E., (2004). Processing of horticultural crops in the tropics. In: *Production Practices and Quality Assessment of Food Crops* (pp. 371–399). Springer, Dordrecht.
21. Pantenius, C. U., (1988). Storage losses in traditional maize granaries in Togo. *International Journal of Tropical Insect Science, 9*(6), 725–735.
22. Perera, C. O., & Perera, A. D., (2019). Technology of processing of horticultural crops. In: *Handbook of Farm, Dairy and Food Machinery Engineering* (pp. 299–351). Academic Press.
23. Sawicka, B., (2019). *Post-harvest Losses of Agricultural Produce.* https://www.researchgate.net/publication/332551213_Postharvest_Losses_of_Agricultural_Produce (accessed on 9 April 2021).
24. UNFAO, (2013). *Global Food Report.* Retrieved from: http://www.imeche.org/docs/default-source/reports/Global_Food_Report.pdf (accessed on 9 April 2021).
25. Yahia, E. M., (2019). *Postharvest Technology of Perishable Horticultural Commodities.* Woodhead Publishing. http://documents1.worldbank.org/curateden/282291468321230375/pdf/multi-page.pdf (accessed on 28 April 2021).
26. Yahia, E. M., Barry-Ryan, C., & Dris, R., (2004). Treatments and techniques to minimize the post-harvest losses of perishable food crops. In: *Production Practices and Quality Assessment of Food Crops* (pp. 95–133). Springer, Dordrecht.
27. Yahia, E. M., Fonseca, J. M., & Kitinoja, L., (2019). Post-harvest losses and waste. In: *Postharvest Technology of Perishable Horticultural Commodities* (pp. 43–69). Woodhead Publishing.

CHAPTER 7

# Food Processing, Branding, Retailing: An Industrial Notion towards Monetary Benefits of Agriculture

ASHISH DHYANI, PALLAVI THAKUR, and SWAMI NARSINGH DEV

## ABSTRACT

The agricultural sector is the view as one of the valuable pillars in the growth of the economy across the globe and in food safety for people. It represents an immense segment for the total manufacturing value addition, offers a huge number of employment, and takes a significant part in the gross domestic product (GDP) of most countries. The demand for the agricultural product can be affected by several means such as increasing demand of food, rising food safety awareness, commercial boundaries, change in environmental conditions and customer choice, etc. Due to these factors' food processing, branding, and retailing play a key role in the growth of a healthy civilization and the economy of the nation. Consumers believe that the properly labeled is quite good in freshness, quality; branded agriculture produce will provide a guarantee relates to its freshness, quality, stability, traceability, and durability. The present study focused on the methods of food processing, branding as well as retailing at the global platform. The proper processing branding and marketing strategies deserve a prime focus for the development of the easily available market platform and new market policies.

## 7.1 INTRODUCTION

In the current scenario, the population of the world is rising rapidly with a 1.2%, annual rate (7.7 crore individuals every year) and is projected to extend up to 900 crores in 2050. It requires 60% more food production and the formation of a balanced diet, which is a real challenge to ensure global food security [17, 66]. At the global level, the agriculture sector plays a tremendous part in the food supply and is regarded as a precious support system for the development of the economy as well as food safety for people. It represents a huge fraction for manufacturing, value addition, offers a number of jobs, and has a crucial role in the upliftment of GDP [36]. The demand for agriculture-based products can be affected by several factors such as increasing food requirement, food security awareness, commercial margins, fluctuation in climate conditions, and customer choice [10, 18, 36]. These unpredictable factors in the agricultural segment can cause mistakes in high cost and numerous decision-making power [10]. These problems stimulate for proper processing of food, and it is branding as well as retailing strategies that contribute to the development of healthy human civilization and economy of the nation. Figure 7.1 represents the workflow of agriculture trade.

### *7.1.1 FOOD PROCESSING*

Food processing includes various approaches under which raw food products are converted into edible goods with high commercial value, safety, quality, and suitability for human as well as animal consumption. Processing has numerous advantages for the supply of finished goods such as food availability during the off-season, processed food is good in taste, attractive, high durability, value addition, create a platform for agribusiness, help in climate change mitigation, provide raw resources for further research and industrial uses, rise farmers income to improve his livelihood, improve foreign exchange earnings, provides employment, helps in rural areas development and by-products of processing can be recycled for the formulation of animal feed [1].

It acts as an important bridge between agriculture producers and consumers at national as well as international platforms and is classified into two groups, namely primary processing and value-added processing. Primary processing of agriculture products involves the transformation of raw agriculture products into intermediate consumables products by several activities such as crushing, hulling, milling, packing, polishing, and shelling, etc. Primary processing is required for only certain farm products such as cereals, oilseeds, and pulses, etc. Value-added processing includes the transformation of raw as well as intermediary agriculture product to value-added products by using numerous processes such as baking, flour milling, fortification, refining, etc. Examples of value-added foods are aerated and malted beverages, beer, and biscuits, etc. Processing of agricultural products can be done by biological, physical, mechanical, and biochemical manipulation with scientific as well as traditional means. This technique involves a set of actions that alter raw agriculture material in consumer goods [1]. The common techniques are discussed in the following subsections.

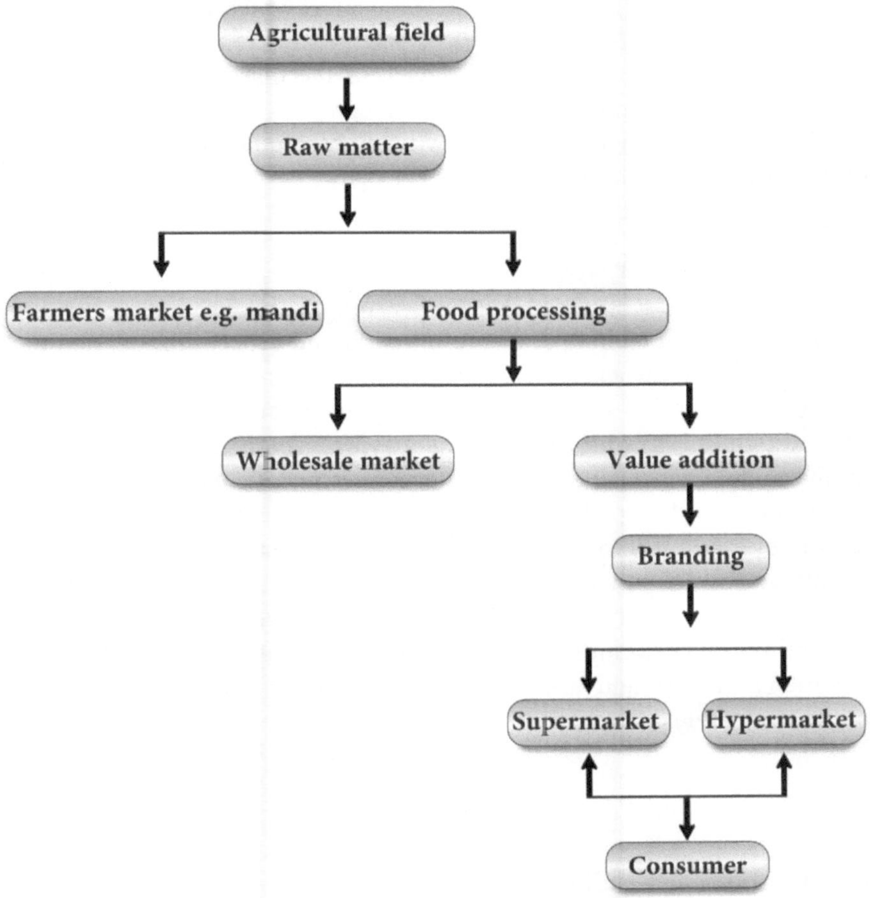


**FIGURE 7.1** Workflow of agriculture trade.

### *7.1.1.1 BAKING*

Baking is a technique of food processing and is defined as a technique that utilizes dry heat in food preparation. It is one of the important processes of a bakery in which dough is converted into the desired finish product with increased volume, color, texture, and flavor. It can alter certain physical, chemical, biochemical, and rheological properties of the products, such as volume expansion, water vaporization, development of porous structure, protein denaturation, gelatinization, and color formation due to the browning reaction [46]. Under high temperature, heat progressively penetrates from the food surface to the center, which offers a firm, dry crust and a soft center such as bread and cookies [54]. It is considered a complex process because heat, as well as the mass transfer, occurs at the same time and the conversion of dough into rigid products. High-quality bakery product is obtained by using high quantity ingredients are optimization of it with the baking process [46].

### *7.1.1.2 FERMENTATION*

Fermentation is a technique of preservation and processing of food that can alter the major as well as minor raw food products into consumer goods by controlled growth of microorganisms and enzymatic process [34, 56]. During this process, enzymes hydrolyze protein molecules into amino acids and low peptides, sugar fermented into alcohol carbon dioxide and lactic acid. It expends the shelf life of fresh foods, aroma, taste, and texture of food [56]. In early civilizations, fermentation is developed as a technique for the preservation of fresh agricultural products, and now it is also used for the preparation of food with desired organoleptic profiles and refining its deliciousness. It also plays a role in the removal of anti-nutritional factors and toxins from raw food materials with increased nutritional values [55].

### *7.1.1.3 DRYING*

Drying is an old technique of food processing in which water fraction removes from the food through evaporation. The core objective of drying is to expand the shelf life of food by reducing its water content. Drying reduces around 75–90% of the water from fruits, vegetable, and finally provide resistance against various deteriorative agents such as micro-organisms. The water can be removed from food by using solar energy or artificially generated hot air. Recently, several new drying methods have been introduced for the transformation of raw farm products into healthy and nutritious food products, such as partial impregnation soaking process, dewatering, and osmotic dehydration. Partial dewatering can be done by osmosis and impregnation soaking process, and operation energy is the process that improves the quality of the dried food products. Nowadays, osmotic dehydration is gaining attention because it provides the stability of the dehydrated product during storage which is due to the low water activity in solute. The low water activity, decreases the rate of the chemical reaction and protects food from deterioration. In various processes, osmotic dehydration is also applied to raise sugar and acid ratio in acidic fruits, which helps in the improvement of appearance, taste, and texture of dried products [47, 54].

#### *7.1.1.4 FREEZING*

Freezing maintains the novel quality of agricultural products over a long storage period. Due to the retention of food quality and nutritive properties, this method is considered a superior technique for fruits as well as vegetable preservation over canning and dehydration methods. In association with several preservation methods such as blanching or dipping, freezing can be applied to increase the shelf life of food products [28]. Commercially frozen food is stored at a temperature range between-10°F and 20°F [54].

#### *7.1.1.5 NANOTECHNOLOGY*

Nanotechnology is seen as the revolution in food industries, and all developed, as well as developing countries, are actively adopting this technology [54]. At present, Nanocarriers are used as a means of transportation for the diffusion of food flavoring substances in specific food products without altering their original properties [53]. This technique is applied during the cultivation, processing, production, and packaging of food. The nanostructure of the food is being developed to enhanced consistency, taste, and texture of food [12]. It helps in upgradation of shelf life of food materials and reduction in wastage of food causes by microorganisms [28]. This technique provides heat resistance, active anti-microbial properties as well as signaling microbiological and biochemical changes [31]. Nonacid is a nanotechnology-based product having anti-bacterial property and seems like a potential applicant in the food industry [31].

#### *7.1.1.6 RADIATION PROCESSING*

Food processing by radiation method is applied to boost safety as well as shelf life of food products by eliminating microbial and insect growth [50]. In this process, ionizing radiations like gamma rays, electrons, and X-rays are applied in a controlled manner against target food product to complete the desired goals. Short wavelength radiations such as Gamma and X-rays are used in this process. Radioisotopes Cobalt-60 and Caesium-137 are the accepted sources of gamma radiation, electron beam up to 10 MeV and X-rays up to 5 MeV for processing as well as preservation of food [22]. This technique can be applied for the storage of bulk and final product for stocking and retail distribution [32].

### *7.1.2 BRANDING*

The word brand originated from "brandr" which means 'burn' and defined as a name, design, logo, sign, or combination of all, which is used to recognize a specific organization product [3]. Branding of agriculture products acts as a bridge between the farm and their customers. Branding of agriculture products increase its value because consumers usually considered that the branded products have better quality, brand loyalty as compares to unbranded products and paid a higher price for the branded items [48]. Producer earns reasonable prices for their product as per the product but by the proper labeling of a brand ensure the quality as well as help in the value addition in product and also support customer against adulteration of products [48]. At present, due to more competition and over saturation is recorded in the market, which is due to the oversupply of products in the domestic market and the inflow of imported products increase competition between producers and sellers. This led to high branding demand in the agriculture sector. Branding added competitive advantage in products and can sell farm produce through hypermarkets and supermarkets at the country and international level [24]. It can be done at a different level as below.

#### *7.1.2.1 BRANDING AT CORPORATE*

At the corporate level, branding is undoubtedly very easy, and various branding methods are available for the agricultural products. At this level government also take various initiatives for the branding of agriculture products, for example, in India, the Spices Board of India established a corporation that promotes spice cultivation and that deals with the marketing of spices under the favorite brand name to ensure both cooperatives and farmers are get the finest price for their farm produces [48, 52].

#### *7.1.2.2 BRANDING BY VALUE ADDITION*

Value addition of a farm product increases the economic value and consumer demand. Farmer may generate price by focusing on the profits associated with the agricultural amenity that can be achieved from quality, functionality, place, and time. Branding of agricultural products can be done by fulfilling necessary guidelines and standers of the authorized organization, which helps in the value addition of the end product. Branding through value addition provides necessary information about the producer or processor and quality of the agricultural produce [48]. It supports the income generation and sustainable development of farmers.

#### *7.1.2.3 RETAIL BRANDING*

Retail branding will consider as the superior method of branding for agricultural products. Retailers support farmers or producers in the branding and enhance the value of the products. Branding of agricultural products at the store level is increasing day-by-day, and retailer selling their food and food-related items in an organized manner [48].

#### *7.1.2.4 GEOGRAPHICAL INDICATORS BRANDING*

Geographical Indicators point out the origin of agricultural products from a specific geographic location. It is used to recognize the manufactured as well as natural products. This branding can be applied to the agricultural produce with good quality which is originated from the place of production, soil type and weather condition [48].

#### *7.1.2.5 VARIETAL BRANDING*

In this branding method, brand name can be allotted to patented crop having good quality as compared to existing varieties. The produces allow its production as well as marketing under license and execute strict quality standards [48].

### ***7.1.3 RETAILING***

The retail sector is the final step of the food supply chain, which deals with the trade of food as well as non-food products for individual or domestic level with few or no processing. They sell their product at different levels, such as wholesale markets, hypermarkets, supermarkets, discounters, convenience stores, independent grocers, free-standing food specialists, and international platforms. It accounts for nearly 40% of sales in the world, but in current times, most of the traditional food retailers expand their businesses to include non-food retailing [18].

#### *7.1.3.1 WHOLESALE MARKETS*

Wholesale markets of agricultural products mostly come under the management of local governments, and at present, they are in poor situations due to poor infrastructure, communication services, and management. In this retailing system, local authorities provide stalls to the wholesaler on a lease basis for 1 year and in turn collects the rent or taxes, etc. These market platforms can be managed by a public-private partnership (PPP), which will invite investment by private partners for refining the infrastructure and other amenities. Policies related to the producer as well as buyers are must be liberal, which builds a healthy environment of their free interaction [38]. Ideal management principles have a crucial part in the development of market functions that can be accomplished by PPP mode and privatization of the wholesale market [43].

#### *7.1.3.2 SUPERMARKETS*

A new method, "farmer-supermarket direct-purchase," is developed for agriculture goods. In this methods, supermarket directly purchases desired agriculture products from the farmers. It maintains the proper circulation of agricultural products and optimization of the supply chain. In this supply chain, supermarkets involve directly in food processing and circulation. "Farmer-supermarket direct-purchase" has several advantages such as sustains the freshness of agricultural products, remove intermediate associations in distribution, reduce procurement costs and decrease the cost of processed products. This system exerts a positive effect on the supermarket business structure and supply of agricultural products [25].

## 7.2 AN ANALYSIS OF GLOBAL MARKET OF AGRICULTURE PRODUCE

Changes in lifestyle, safety, and convenience increase the demand for processed food globally that is due to several benefits over unprocessed food, for example, easy cooking, handling, immediate consumption, preservation, and storage, etc. Based on processing, food can be categorized as minimal, high base on product type is classified as plant-based such as Fruits and vegetables, etc. Based on the product application, processed food can be baked goods and baby foods, etc. [14].

### *7.2.1 OVERVIEW ON CURRENT STATUS OF PROCESSED FOOD MARKET*

After the Green revolution, the global market of the agriculture sector has transformed, and a number of industrial players started their investment in agriculture products. The agriculture sector focused on various agricultural products such as agriculture inputs, seed, and traits, application of biotechnology, and several other amenities have been gaining consideration in the form of organic farming and precision farming all across the globe to generate more economy [35]. In 2014, global agricultural trade is near about $1 trillion, which has been rising around 3.6% per year in the last two decades which is due to the technology transformation, increment in productivity, liberalization, alteration in trade patterns and development of new trade policy. However, the biggest traders and exporters of agricultural produce are remain untouched over the previous 20 years, but Brazil, China, India, Indonesia, and Russia grew much faster as compared to the global average [21]. Developing countries continuously attain a higher portion in the world agriculture market, and the trading involves between developing countries is known as 'South trade,' which is increasing at a higher rate as compared to trading between the developing and developed countries (North-South trade). An international organization, i.e., World Trade Organization (WTO) is authorizing for the policy, plane, and rule preparation for smooth, predictable, and free as possible trade at the international level [58]. The larger part of agricultural food market is made up of processed food products. According to Coherent Market Insights (2020), due to the increasing trend of snack consumption, North America is the world's largest market for processed food. Europe has a great demand for high-quality and nutritious food. The growth of the processed food market is very fast in the Asia Pacific region, which may be due to increment in population and consumer purchasing ability. In India, customers spend on processed food is approximately US$ 3.6 trillion by 2020. In Latin America, the demand for processed food also increases due to urbanization and modernization.

The rank of "classical" food products such as coffee and tea has been changed by the processed food [3, 29]. The total world food exports were increased from 44% in 1980 to approximately 63% in 2006, which was part of processed food. In which, a portion of developing countries increases three times while the share of developed countries was quite stable. The increasing demand for processed food exports also changes the world's agricultural trade. Processed food export is fast growing as compared to other exports of agricultural products, especially primary food [30]. The annual growth rate of global food exports for processed foods during 1981–2006 is summarized in Table 7.1.

**TABLE 7.1** The Annual Growth Rate of Global Exports for the Processed Foods During 1981–2006

| Year | Agriculture (%) | Agriculture Raw Matter (%) | Unprocessed Food (%) | Processed Food (%) | Manufacturing (%) |
|---|---|---|---|---|---|
| | | **Developing Countries** | | | |
| 1981–1990 | 8.2 | 6.8 | 4.8 | 12.9 | 18.9 |
| 1991–2000 | 6.3 | 6.1 | 5.3 | 6.6 | 15.4 |
| 2001–2006 | 10.8 | 12.1 | 10.4 | 10.6 | 15.3 |
| | | **Developed Countries** | | | |
| 1981–1990 | 4.5 | 5.0 | – | 6.3 | 8.4 |
| 1991–2000 | 2.0 | 2.0 | 1.1 | 2.1 | 5.5 |
| 2001–2006 | 7.8 | 7.0 | 6.6 | 8.4 | 9.1 |

## 7.3 HIGHLIGHTING THE WORLDWIDE AVAILABLE MARKETING PLATFORMS

Uneven supply and inefficient marketing channels of agriculture products will lead to disturbance and problems between the producers (farmer cooperatives, agriculture industrialization leading enterprises, planting, and raising households) and consumers (supermarkets, community markets and grocery stores, etc.). A well-managed market platform can provide tighter corporations and valuable information to both the producers and consumers. The primary purpose of market platform in agriculture product supply is to assist in unbiased deal between clients, which means either the producers as well as end-user, could competently find each other and the utility of one side will increase with the growth of number on another side [11]. Several marketplaces and online systems are available worldwide to help farmers, food buyers, agribusiness, and leaders at various levels.

### *7.3.1 AGRICULTURE MARKET PLATFORM IN BRAZIL*

The agriculture sector plays an utmost role in the increment of the Brazilian economy and is expected to register a compound yearly growth rate (CAGR) is about 5.9% during 2020–2025. This sector contributes near 14% GDP, 28% for the whole agricultural chain and employment of almost 18 million people. The growth and development of agricultural products as business extremely relies on exports which is around 31% of total agricultural production. Currently, Brazil is known as the world's third agri-food exporter country followed by European Union (EU) and the United States [8] and become the world's biggest exporter of beans, cassava, coffee, cotton, corn, crop-based ethanol, maize, onion, potato, rice, soybeans, sorghum, sugarcane, tobacco, tomato, watermelon, and wheat [7, 22]. As per Euromonitor, the sale of the packaged food market in Brazil had been projected to reach up to US$94.9 billion in 2019, which was the largest in Latin America and the 5th in the world. Since 2015, it also represents 24.3% or US$18.5 billion growth rate. In 2024, the sale of the packaged food market in Brazil is probable to reach up to US$131.1 billion with a 30.8% or US$30.8 billion growth rate. High growth rates in the forecast included.

The agricultural activities in Brazil are governed by the two distinct ministries namely the Ministry of Agriculture, Livestock, and Food Supply (MAPA) which deals with the development of agribusiness and market integration and the Ministry of Agrarian Development shaped in 1999, which is meant for the land reorganization, growth, and promotion of family farming system. As per the Agricultural Census 2006 report, 84% family farm was established all over the country, which contributes nearly 38% gross value of agricultural production [18, 20]. Ministries of Social Development, Education, and Health, now developing new integrated strategies for sustainable growth in the agriculture sector with social inclusion. Due to which significant improvements in the reduction of poverty and inequality have been recorded, and as per the Gini index, discrimination has fallen up to 0.52, which is the lowest in 50 years [42]. Moreover, in 2012 Human Development Index of Brazil is 0.730, which is 40% higher than in 1980 [26]. Agricultural policies of Brazil depend on the rural credit that provided approximately 80% market interest rates, followed by market price support and rural insurance.

#### *7.3.1.1 RURAL CREDIT*

Rural Credit system was formed in 1965 to accelerate wealth development to support farmers [15, 49]. Rural credit help in the investment of crop production and market-related produces. Government of Brazil provide subsidy for rural credit which represents around third part of total credit needs in the agricultural sector. From 1994, private sectors such as brokers, farmers, input providers, and private banks, along with the Government of Brazil, provides credit needs in the reformation of rural agriculture. Agricultural credit focused on financing crop and livestock production, capital investments in infrastructure development and equipment. Capital investments include machinery for planting, harvesting, and processing of livestock products as well as expansion of pastureland. Between 1985 and 2006, the investment credit expended and issued to the livestock sector and boost its 4.7% credit annually, which is two times large investment credit distributions for crops during the same period [4]. Investment credit indicates the higher growth in the livestock sector, and Brazil takes priority in modernizing livestock production [45].

#### *7.3.1.2 MARKET SUPPORT*

This is a regional and market-oriented approach. The main goal of this policy was to reduce price uncertainty and help in the planning and trade decision of farmers, which includes providing information about estimated futures market prices to support agriculture. The Government of Brazilian buys surplus production, balance price, finance storage, and recommends public and private sales contract when the market prices are below than lowest prices. These procedures help in the maintenance of proper market supply, suppress price instability and work to improve farmer's income. Every year minimum support price is revised in 33 different crops such as castor beans, jute, and mallow, etc. Along with this, the production of these crops is quite valuable in terms of environment protection and sustainable development of the North and Northeast regions due to the direct benefit of traditional communities and farmers [15]. Since 2003 government promoted revision in product promotion and revenue assurance agricultural policy instruments. The moto of this revision is to improve public resources and develop coordination between the fellows of productive chains to control financial limits in the implementation of the programs for price and income assurance [6]. The market support policies are:

1. Minimum Prices Guarantee;
2. Federal Government Purchase (AGF);
3. Federal Government Loan (EGF);
4. Special Credit Line for Marketing (LEC);
5. Sale Option Contract;

6. Repurchase or Transfer of Sale Option Contract;
7. Premium for Commercial Buyers Program (PEP) and Product Value of the Commercial Buyers Program (VEP);
8. Rural Product Note (CPR);
9. Rural Promissory Note and Rural Trade Note (NPR and DR);
10. Equalizing Premium Paid to Producer (PEPRO);
11. Private Sale Option Contracts and Private Option Risk Premium (PROP);
12. Agricultural Deposit Certificate (CDA) and Agricultural Warrant (WA);
13. Agrinote or Commercial Agribusiness Note (NCA).

#### *7.3.1.3 RURAL INSURANCE*

Rural insurance is the private sector policy in Brazil. However, the Government of Brazil provided up to 70% subsidy in premium insurance which includes agriculture, livestock, forest, and fishery. In 2012, Resources assigned under the Rural Insurance schemes support the southern region of Brazil and grain crops, especially soy, which covers 5.24 million hectares of area and 43.538 producers benefited from this. In the 2013–2014, subsidy resources available for Rural Insurance scheme in crops are increases up to 75% and convert 10 million hectares' land eligible for the subsidy. In 2010, the Government started a Catastrophe fund for extra coverage of this scheme in the favor of re-insurance organizations. Ministry of Agriculture started an Agricultural Livestock Guarantee Program (Proagro) which provided the same coverage of insurance to the farmer under a payment of premium fee and farmers' compliance with the agricultural zoning on the basis of climatic risk. Brazil's re-insurance market was controlled by a state-owned company until 2007 when the government opened it to international re-insurers with the aim of stimulating competition and reducing premium values [15]. Brazil continuously implemented its fully open trade policy for agricultural products with the aim of increasing its remarkable presence in world markets. Brazil kept free or privileged trade agreements with most of Latin America and Caribbean and MERCOSUR countries members, i.e., Argentina, Brazil, Paraguay, Uruguay, and Venezuela. Moreover, Brazil also established agreements with two new countries, namely India (2009) and Israel (2011) [67].

### 7.3.2 AGRICULTURE MARKET PLATFORM IN CHINA

China developed a stable agriculture sector which is due to the major portion of its population is depends on farming, and this can be better for the constancy of society. The China government is most concerned for their agriculture sector and rural economies. They continually implement their agricultural policies to improve farmer income, local support, and promoting long-term food security goals. Recently, they make great effort to reduce its large stocks, especially for cotton and corn, and reduce the price gap between national as well as international markets by altering their agricultural-related policies, particularly price support policies for corn, cotton, and soybeans [27]. To develop and improve agriculture policies in China, the government sincerely concentrated on numerous strategies such as the reformation of the rural economic and institution, filtering incentives for farmers and encourages local governments for investment in the agriculture field, increasing the rate of revival for renewable resources, regulating the use of non-renewable resources, providing input as well as output price to increase the multiple cropping index, application of technical methods and tools in the agricultural sector, regulating rural financial arrangements, improving associations of agricultural production, strengthening anti-poverty programs, opening the agricultural section for foreign investment.

In China, online trading of agriculture products is in the growing phase, and various organizations are energetically adopting e-commerce. Many provinces of China are established their own "agricultural information website" and build several online market platforms, e.g., "Fuzhou peak," "Nanjing BaiYunting," etc. At present, more than 10,000 agriculture websites are available in China. These websites offer valuable data related to aquaculture, fruits, livestock, poultry, trees, vegetables, agricultural demand and supply, economy, and investment facts. It plays a crucial role in the promotion, circulation, improvement in agricultural products and farmer income. Some important websites having high influence in the agriculture sector are China's agricultural science and technology information network (www.caas.net), China agricultural information network (www.agri.gov.cn), seed group co., LTD. (www.chinaseeds.com.cn), the North China seed industry information center (www.northseed.com.cn), etc. [19]. For example, in 2018 'Alibaba' launched "ET Agricultural Brain" program that provides data by which farmers can implement new strategies in their effective planting methods and maximize crop yields [2]. Agricultural e-commerce provides the following platform for agriculture-related marketing in China:

- Enterprise's own e-commerce platform;
- Based on third-party e-commerce public platform;
- Collaborative e-commerce platform;
- International agricultural e-commerce platform.

### 7.3.3 AGRICULTURE MARKET PLATFORM IN INDIA

About 58% of India's population rely on the agriculture sector for their livelihood. Between 2000 and 2014, India's agricultural production has raised from US$101 billion to about US$367 billion mainly due to high-value-added segments such as dairy, inland aquaculture, horticulture, and poultry. India is a food as well as non-food agriculture bases country and can become a principal participant in the world agricultural trade. The food processing industry accounts for 32% and deliberated as the biggest industrial sectors in India and ranked 5$^{th}$ in terms of manufacture, consumption, export, and growth. It contributes around 8.80 and 8.39% of Gross value addition in the manufacturing and agriculture sector, respectively. At present, the value of the food market in India is US$ 1.3 billion with a rising 20% CAGR and projected to increase up to three times in 2020. In India, online food ordering trade is in the budding phase but showed exponential development day by day. The online food delivery sector grew at 150% per year with a projected US$ 300 million Gross Merchandise Value, as recorded in the year 2016 [40]. There are mainly three methods of agricultural product distribution in India.

- Direct method;
- Indirect method; and
- Online method.

#### 7.3.3.1 DIRECT METHOD

The direct method involves direct selling of farm products into the market and requires a high level of customer interaction. Commonly, the farmer receives a price similar to grocery store charge and is more entrepreneurial like than wholesale marketing. In a fashion, the farmer grows a "product" more than a crop and consumers directly interact with farmers and purchase their goods. For this type of marketing, farmers required proper transportation and storage facilities, the ability to accept multiple forms of payment, and develop innovative strategies to sell their merchandise on a target day. Farmer also wants to keep in mind the required vendor fees and other requirements for the market [44]. The direct marketing platform in India are as follows:

1. **Roadside Markets:** This type of marketing system permits the farmer to remain close to their farm and minimizing transport to market. This option required minimal time and less infrastructure. A more substantial roadside stand requires an investment in infrastructure, directional signage, marketing, and staffing. Location is also playing a major role in these outlets [44].
2. **Farmers Markets/Mandies:** This marketing system considers as a satisfactory way to build up a good relationship and loyalty with their target population, get direct feedback and promote farm business. Farmer's market sales can make notoriously long days, being off the property or investing in staff, running trucks, and dealing with weather-related uncertainty on market days [44].

#### *7.3.3.1.1 Government Schemes to Promote Direct Marketing*

In India, the Ministry of Finance construct plans for smooth direct selling of fruits and vegetable, without the involvement of middlemen and food inflation due to the re-emergence if monsoon falters. Ministry is working on the valuable variations in the Agricultural Produce Marketing Committee (APMC) Act, which manage the marketing of agriculture products and permit producers to sale their farm products directly to consumers and lead to saving high intermediation costs [51].

1. **Apni Mandies in Punjab and Haryana:** *Apni Mandis* (Our Market) program was 1st established in the mid-1990s in Punjab and Haryana state and a 1st initiative which directly linked farmers and consumers. Farmer sell their products directly to its customer. The APMC of *Apni Mandi* area provides all essential amenities

such as space, watershed, counters and weighing balances to the farmer [41]. In 1988, 10,278 farmers participated in apni mandies which increase up to 353,610 in 2010. The approximate sale was recorder ₹ 29,624,761 in 1988, which increased to ₹ 629,347,474 in 2010.

2. **Rythu Bazaars in Andhra Pradesh:** Government of Andhra Pradesh India, introduced the *Rythu Bazaars* on the occasion of republic day in 1999. These Bazaars are situated on government lands which is marked by District Collectors on the basis of convenience of both farmers and consumers. The standards criteria for the inauguration of new Rythu Bazaars are it requires a minimum of one-acre land in the specific location and identifies 250 farmers with 10 groups. The price should be fixed by the committee of farmers and the Estate Officer of the Bazaar. Generally, the price of *Rythu Bazaars* is 25% higher than the wholesale rate and less than 25% from local retail price. The maintenance of this Bazaar is being observed by the financial sources of APMC [41].
3. **Uzhavar Santhai in Tamil Nadu:** This Market is governed by the administrative control of the State's 16 Agricultural Marketing Committees, a part of the Department of Agricultural Marketing. This committee has a Marketing Committee officials and farmers representative which are responsible for the fixation of price vegetables per day. Committee members have to collect price details from the central and retail markets before 3.00 a.m. and by 6.30 a.m. The maximum selling prices is fixed at 15–20% more than the night sale price at the central market and 20% less than retail markets in the Farmers' Market is [41].
4. **Hadaspar Vegetable Market in Pune:** This market is a role model market for the direct selling method in Pune, Maharashtra. It is located 9 km far from Pune city and governed by Pune Municipal Corporation, which is responsible for collecting rent from the farmers for consuming the space in the market. In this market, the farmer sells their product to target consumers without any agents or middlemen. This market is equipped with modern weighing machines [41]. The common problem of the direct market method is the involvement of middlemen in bazaars However, identity cards have been introduced and periodically checked, but the problem remains in many bazaars [41].

### *7.3.3.2 INDIRECT METHODS*

Indirect methods involve different channels for retailing agriculture produces. Some representative marketing channels for different product groups are given below:

- **Marketing Channels for Cereals:** Marketing channels for different cereals are less or more similar in India except for rice. For pulse crops, dal mills appear prominently in the channel. For example:
  - **Networks for Rice:**
    - Producer-miller->consumer (village sale);
    - Producer-miller->retailer-consumer (local sale);
    - Producer-wholesaler->miller-retailer-consumer;
    - Producer-miller-cum-wholesaler-retailer-consumer;
    - Producer-village merchant-miller-retailer-consumer;
    - Producer-govt. procurement-miller-retailer-consumer [58].

### *7.3.3.3 ONLINE METHOD*

The e-commerce introduced by information technology (IT) led to improve the market access and information. It exhibited considerable potential for integration with agricultural markets. Directorate of Marketing and Inspection, Government of India, has launched an internet-based system (AGMDRKNET) to spread market information across the country. The market research and information network (MRIN), previously linked 735 wholesale markets, is aimed at catering to all investors. In this system, data from 154 different commodities were collected by the Directorate of Economics and Statistics and made accessible on the internet, however, this information only helps to the policymakers and researchers [39]. To overcome from this problem, The Government of India launched some of the following online programs for awareness of farmers regarding their agriculture crops and their market.

#### ***7.3.3.3.1 E-Choupals***

The world "E-choupal" is derived from the Hindi word 'choupal' which means "village meeting place" and is defined as a meeting market place where traders and consumers come together for trading. The India Tobacco Company (ITC) launched E-Choupal initiative in 2000 and at the end of 2003 up to 2700 Choupals, covering 18000 villages and 1.2 million populations in five states, namely Andhra Pradesh, Madhya Pradesh, Maharashtra, Karnataka, and Uttar Pradesh was established. Internet act as a backbone for communication in this platform (Figure 7.2) [5]. E-choupal is an online trading platform for farmers where they selling their farm products directly and can recovered valuable rate for their farm produce [39]. The commodities covered are coffee, pulses, rice, shrimp, soybean, and wheat. ITC is going to extent it into 15 states with 20,000 Choupals covering 10 million populations [39]. The core charm of this platform is that it can be used to connect small as well as large producers and small or large users, thereby eradicating the requirement of brokers and geographic distances cannot restrict participation of farmers [5]. It facilitates the movement of information, knowledge, and supports market transactions online:

- It transfers information about farm, prices, risk management and weather;
- It assists sales of farm input (for quality screening);
- It provides substitute output-marketing channels (convenience and lower transaction costs) to farmers right at his doorstep;
- It is an interlocking network partnership platform (ITC + Met Dept + Universities + Input COs + Sanyojaks, the erstwhile Commission Agents) bringing the "best-in-class" in information, knowledge, and inputs.

E-Choupal platform reduced procurement, transit, and material handling costs. Acquisition of business prices are decreases form the industry 8% (farmer earns 3% and the processor earns 5%) to 2% (with farmer saving all his 3% and the processor-ITC-saving 3%) as shown in (Table 7.2). This platform has come a long way since its inauguration and is documented as India's largest Internet-based initiative, which covers 1,300 choupals, 7,500 villages and helping approximately 1 million farmers. In Madhya Pradesh, 1,045 E-choupals are present in more than 6,000 villages which cover 600,000 farmers. E-choupal also recognized its valuable presence in other states such as Andhra Pradesh, Karnataka, and Uttar Pradesh. ITC exports US$140 million agricultural products in the world out of the US$15 million worth was collected from the commodities of E-choupals in 2001. 120,000 metric tons of various farm products have already been obtained through this channel and resulted in more than US$1 million worth of savings. This money is distributed between ITC and farmers from different states. Price offered by ITC are not more than the mandi, the farmer chooses this platform because trades are done closer to home with practices of weighing and quality assessment in an efficient and transparent way. Farmers save their travel expenses and experience less wastage. An estimated at ₹ 400 to 500 (US$8 to US$10) per ton of soybeans was recorded from farmer ends. The beauty of this market platform is the final decision related to selling farm crops to the mandi or ITC is resting with the farmers.

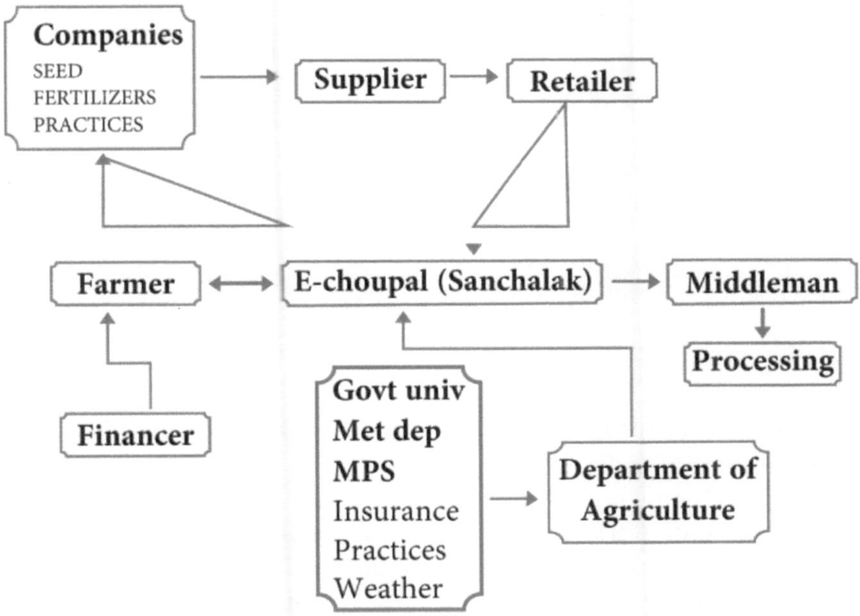


**FIGURE 7.2** E-Choupal, new business model.

Farmers can directly sell their produces in ITC with orders on the Internet. In this process, farmers save around ₹ 250–500 per ton which purely depends on the distance between their geographic locations and ITC recognize collection center. ITC also assured their supply and save more than ₹ 200 per ton by reducing transport cost from mandi to collection center and other costs in the supply chain. Overall, more than ₹ 1 billion or US$21 million transactions have occurred so far, and ITC plans to cover 15 states in upcoming few years [5].

**TABLE 7.2** Comparison between Conventional Transaction vs. E-Choupal Costs

| Cost Element | Conventional Market | E-Choupal |
|---|---|---|
| Trolley freight to mandi | 100 | Nil |
| Filling and weighing Labor | 70 | Nil |
| Labor Khadi Karai | 50 | Nil |
| Handling loss | 50 | Nil |
| Sub Total | 270 | Nil |
| Processor Incurs | | |
| Commission to agent | 100 | 50 |
| Cost of Gunny bags | 75 | Nil |
| Labor (Stitching and Loading) | 35 | Nil |
| Labor at factory (Unloading) | 35 | 35 |
| Freight to factory | 250 | 100 |
| Transit losses | 10 | Nil |
| Sub Total | 505 | 185 |
| Grand Total | 775 | 185 |
| As % of Beans Value | 8% | 2% |

### *7.3.3.3.2 Electronic-National Agricultural Markets (E-NAM) [Government of India, 2019]*

The Ministry of Agriculture (Government of India) started this portal with an electronic market pilot project across the country. Launched in 14 April 2016 [16], this portal is operated by small farmers' agribusiness consortium (SFAC) under the Department of Agriculture, Cooperation, and Farmers' Welfare technology provider, NFCL's iKisan division. The Congress government also introduced a similar project in Karnataka with tremendous success, and the NDA government has applied it on a national level. It connects 585 markets (APMCs) in 16 states and 2 union territories with more than 45 lakh farmers. This online market platform provides help to sellers and exporters to obtain quality products in bulk at the same place and assure transparent transactions (Table 7.3) [16].

**TABLE 7.3** Number of Mandis Undertaking Online Agribusiness in India

| Sr. No. | State | Number of APMC | Mandis Doing Online Trade |
|---|---|---|---|
| 1. | Andhra Pradesh | 22 | 16 |
| 2. | Chandigarh | 1 | 1 |
| 3. | Chhattisgarh | 14 | 9 |
| 4. | Gujarat | 79 | 22 |
| 5. | Haryana | 54 | 6 |
| 6. | Himachal Pradesh | 19 | 11 |
| 7. | Jharkhand | 19 | 7 |
| 8. | Madhya Pradesh | 58 | 30 |
| 9. | Maharashtra | 60 | 15 |
| 10. | Odisha | 10 | 3 |
| 11. | Puducherry | 2 | 1 |
| 12. | Punjab | 19 | 10 |
| 13. | Rajasthan | 25 | 17 |
| 14. | Tamil Nadu | 23 | 8 |
| 15. | Telangana | 47 | 12 |
| 16. | Uttar Pradesh | 100 | 58 |
| 17. | Uttarakhand | 16 | 14 |
| 18. | West Bengal | 17 | 15 |

### *7.3.4 AGRICULTURE MARKET PLATFORM IN THE UNITED STATES*

Agriculture industry is the net exporter of food in the United States [33]. As per the census data of agriculture 2007, there were 2.2 million farms are located in the US which covers around 922 million acres (1,441,000 sq mi) area with an average of 418 acres (169 hectares) per farm [59]. The trading of agriculture products is governed under the supervision of the United States Department of Agriculture (USDA), and it involves in policy planning and program implementation. Under the Wholesale Market and Facility Design group, it provides technical assistance to farmers for the construction of new structures or upgrade old one platform and improve the quality as well as a variety of foods across the world. USDA continuously working for the establishment of international morals and rules to improve predictability and accountability for agricultural business [64]. It mainly focused on the functioning of wholesale markets, public markets, farmer's markets, and food hubs services which are necessary for the distribution of food at the national platform. It started several programs for the development of world-class agriculture trade such as farmers market promotion program (FMPP), market access program (MAP), foreign market development (FMD) program, emerging markets program (EMP), technical assistance for specialty crops (TASC), agricultural trade promotion program, and quality samples program (QSP).

#### *7.3.4.1 FARMERS MARKET PROMOTION PROGRAM (FMPP)*

FMPP provides financial assistance to develop, coordinate, and increase direct producer-to-consumer markets and also support to increase accessibility of local agriculture products by evolving, increasing, managing, and providing better outreach, training, and technical assistance to local farmer's markets, roadside stands, community-supported agriculture programs, agritourism activities and online sales or direct producer-to-consumer market opportunities [61].

#### *7.3.4.2 MARKET ACCESS PROGRAM (MAP)*

In MAP, a joint association of Foreign Agricultural Service (FAS), US agricultural trade organizations, state regional trade groups, and small businesses helps to build commercial export markets for US agricultural products and merchandises by providing financial support for external marketing and promotional activities. FAS provides financial help to authorized US organizations in business-related activities such as consumer advertising, public relations, point-of-sale demonstrations, participation in trade fairs and exhibits, market research, and technical assistance. Every year, FAS declares the MAP application period and conditions for federal register. Eligible candidates apply in this program through the unified export strategy (UES) process that permits qualified organizations to demand funding from multiple USDA market development programs [63].

#### 7.3.4.3 FOREIGN MARKET DEVELOPMENT (FMD) PROGRAM

This program is also known as the Cooperator program which helps to generate, develop, and sustain long-term export markets for US agricultural products. FMD mainly focused on the promotion of agriculture products rather than consumer-oriented branded product. Fund for this program is usually allotted for the long-term prospects to decrease import or increase export growth such as reduce infrastructural market impairments, develop processing proficiencies, modify codes and ethics, and identify the new market platforms for the agricultural product [62].

#### 7.3.4.4 AGRICULTURAL TRADE PROMOTION PROGRAM (ATP)

This program helps exporters to form new market platforms and reduce negative effects of other countries' tariffs as well as non-tariff barriers. ATP provides financial support for suitable organizations agribusiness-related activities such as consumer publicity, public associations, demonstrations of point-of-sale, participation in trade fairs, market research, and technical assistance. ATP mainly works through partnerships of non-profit national and regional organizations [60].

#### 7.3.4.5 AGRICULTURE MARKETING RESOURCE CENTER (AGMRC)

The USDA Agricultural Marketing Service collects data from suppliers, brokers, and buyers to analyze market conditions for smooth agriculture product supply. This center provides valuable information to analyze the food industry in terms of value addition and industries. In this agriculture-based system, there are two types of direct market:

1. Community supported agriculture (CSA) markets;
2. Farmers' markets.

##### *7.3.4.5.1 Community Supported Agriculture* **(CAS)** *Markets*

CSA markets mainly focused on the community that helps farm operations by providing support and sharing risks as well as benefits of growers and consumers for food production. There are two models of the CSA market system, namely traditional CSA and current business model. In the traditional CSA model, community fellows share hazards and profits in food production with producers. Members purchase their desired product from the farm before the growing season, and in return for this, they obtain steady deliveries of products throughout the season. The producer receives advance prices for their produces, improves financial safety, receives better prices, and benefits from the direct marketing plan. In Current business models, producers fit a variety of developing direct marketing opportunities such as institutional health and wellness programs; Multi-farm systems to increase scale and scope; Season extension technologies; and Incorporating value-added products, offering flexible shares, and flexible electronic purchasing and other e-commerce marketing tools [40].

#### *7.3.4.5.2 Farmers Markets*

Farmers markets are located in public places and permit farmers to sell their farm products directly at a particular place. In the US, this marketing system is established in 1994 by the USDA's Agricultural Marketing Service, and there are up to 8,720 markets were recognized in 2013. These markets are operated based on the season and managed by a manager who is responsible to build coordination between vendors and market promotion. Consumers have the opportunity to provide direct feedback to its grower, which helps in the improvement of product quality. This form of direct marketing is also considered as agritourism. Consumers have a chance to buy locally-generated products and are promoted as being "fresh." Food and Agriculture Organization of the United States, develop Virtual Farmers' Market (VFM) e-platform for agriculture products [65].

- **Maano-Virtual Farmer's Market (VFM):** The World Food Program's (WFD) Innovation Accelerator develops an e-commerce platform, i.e., Maano-Virtual Farmers Market (VFM). This app is designed to help smallholder farmers access markets via mobile and provide a transparent, open, and reliable place to negotiate fair prices and deals. It facilitates the sale of farm products by online payment method, and United Nations WFD acts as a barrier between farmers and buyers. In July 2016, WFP provided funding to a pilot study of the prototype of VFM with farmers from rural

areas located in three districts of Zambia; hence this pilot study is also known as Zambia pilot. In the initial stage, the report indicated that the app progress is good, and 70 buyers from the national and international levels express interest in purchasing crops through the Maano. The Maano-Virtual Farmers Market app was launched in May 2017 with targeting 2,500 Zambian farmers during 2017 [65].

## 7.4 GAP AND CHALLENGES ASSOCIATED WITH CURRENT SUBJECT MATTER

The global agricultural market is expected to expand continue to meet the demand for food safety with new endeavors, however, facing several challenges. The challenges for producers in agricultural produce marketing are the unorganized food processing and branding sector, shortage of financial resources, inadequate market platform, an inappropriate storage facility for their crops and shortages of the storage facility.

### *7.4.1 UNORGANIZED FOOD PROCESSING AND BRANDING SECTOR*

Food processing is a series of organized technique which required at a different level in industries. In developing countries, the food processing sector faces several challenges that affect the performance and growth of products by value addition. The main challenges are lack of new technology and equipment, training, access of institutional credit, shortage awareness on quality of products and lack branding and marketing skills [13]. To resolve these issues, the government must be focused on new policies to reframe the infrastructure by conducting awareness programs, training in each district, state, and national level. Governments should involve in the building of partnerships between the banks or financial institutions and producers for the development of marketing of food at the global level.

### *7.4.2 SHORTAGE OF FINANCIAL RESOURCES*

At the global level, around 500 million smallholder farming households are located, which signify 2.5 billion people depend on agricultural activities for their livelihoods. The financing as well as insurance in agriculture sectors are the important for strategies for the eradication of extreme poverty level in these households. This is the major problem of the agriculture sector, and the need for investment in agriculture is increasing day by day which is due to the growth of global population and altering dietary patterns towards higher-value agricultural products. The growth and trading of agriculture products require financial assistance for smooth functioning. A significant challenge is to address systemic risks through insurance and other risk management procedures and low operating costs in dealing with smallholder farmers. In developing countries, financial institutions not or less share their portfolios to agriculture as compared to the GDP of the agriculture sector which can be due following factors [57]:

- Insufficient or ineffectual policies;
- High transaction cost to reach remote locations;
- Disputes in product production, marketing, and price risks;
- Inadequate mechanisms for risk management;
- Low levels of demand due to disintegration and incipient growth of value chains;
- Lack of skills of financial institutions in managing agricultural loan portfolios.

### *7.4.3 INADEQUATE MARKET PLATFORM*

In the supply of agriculture produce market platform has an important position because it provides a place to both traders and consumers. The disorganized infrastructure of their target wholesale markets which causes a long chain of intermediaries, damage of quality, and rise of gap between the producer and consumer price. Therefore, the growth of the market gets decreases and to solve this problem, standardization of an adequate market platform is required.

### *7.4.4 LACK OF STORAGE*

Agricultural goods are fragile with seasonal production. To supply of agriculture product throughout the year is need to be stored in warehouses as per the crops. Producers or farmers may not have their storage facilities. Therefore, they sell their produce at the earliest at very low prices. At each level government should provide a specific storage facility at a minimum price and can aware farmers for the new storage facility and technique [37].

## 7.5 REMEDIAL MEASURES FOR IMPROVEMENT OF PROMOTIONAL ASPECTS

The agriculture sector is acting as an important part of human civilization and the economy. Certain actions should be taken for their better functioning and growth. Consolidation of holding in the direction of modernization can be done by passing proper legislation. Uneconomic small farms and small fragmented holdings also consolidated by forming co-operative farming societies. Necessary action should be taken to resolve problems caused by natural factors such as include the creation of adequate irrigation facilities, the use of a good quantity of pesticides, insecticides, and flood control measures. Awareness programs can be conducted at each level to provide training related to modern technique, current market demands, and use of a high-yielding variety of seeds and proper crop planning. Economic is the most important factor which must be accepted to make agriculture more remunerative. Adequate plans and laws are must be applied for the extension of farm association, land management, and formation of agro-based industries. Furthermore, the government must implement minimum price support and crop insurance policy.

## 7.6 SUMMARY

The agriculture sector is an integral part of the world economy and approximately 11% (150 crore hectare) of the world is used for agriculture production. The demand for agricultural products can be affected by several means like increasing demand of food, rising food safety awareness, commercial boundaries, change in environmental conditions and customer choice, etc. Due to these factors' food processing, branding, and retailing play a key role in the improvement of healthy human civilization and the economy of the nation. If products are appropriately labeled, then it helps the consumer in reducing the search, which leads to obtaining a price premium. Production of branded agriculture produce will provide a guarantee relates to its freshness, quality, stability, traceability, and durability. Along with this, it helps in the growth and upliftment of producers. The production and marketing infrastructure in this sector deserves the attention of all the stakeholders and new market policies. Proper and unbiased marketing played an effective role in the global platform to generate high economy and sustainable development of farmers.

## ACKNOWLEDGMENTS

The authors are thankful to Mr. Saurabh Saini and Mrs. Ekta Arya for the IT support. We are also thankful to Mr. Sunil Kumar for handling the graphics of the chapter.

## KEYWORDS

- **Agricultural Trade Promotion**
- **Agriculture Sector**
- **Branding**
- **Electronic-National Agricultural Markets**
- **Food Processing**
- **Market Platforms**
- **Retailing**

## REFERENCES

1. Adiaha, M. S., (2017). Complete guide to agricultural product processing and storage. *World Scientific News, 81*(1), 1–52.
2. *Agricultural Industry in China.* https://www.1421.consulting/2020/03/agricultural-industry-in-china (accessed on 28 April 2021).
3. Baker, M. J., (1999). *The Marketing Book* (3rd edn.). Butterworth Heinemann, Oxford.
4. Banco Central do Brazil (BACEN), (2009). *Anuário Estatístico Do Crédito Rural.* www.bcb.gov.br (accessed on 28 April 2021).
5. Bowonder, B., Gupta, V., & Singh, A., (2003). *Developing a Rural Market E-Hub the Case Study of E-Choupal Experience of ITC.* Planning Commission of India. http://www.planningcommission.gov.in/reports/sereport/ser/stdy_ict/4_e-choupal%20.pdf (accessed on 28 April 2021).

6. Brazil Agricultural Policies, (2008). *Ministry of Agriculture, Livestock and Food Supply Secretariat of Agricultural Policy.* http://extwprlegs1.fao.org/docs/pdf/bra167590.pdf (accessed on 28 April 2021).
7. Calmon, P., (1939). *Rei Filosofo, Vida D* (Vol. 120). Companhia Editora Nacional.
8. Chaddad, F. R., & Jank, M. S., (2006). The evolution of agricultural policies and agribusiness development in Brazil. *Choices, 21*(316-2016-6401), 85–90.
9. Chen, L., Madl, R. L., Vadlani, P. V., Li, L., & Wang, W., (2013). Value-added products from soybean: Removal of anti-nutritional factors via bioprocessing. *Soybean: A Review* (pp. 161–179). Intech Open.
10. Chen, W., Li, J., & Jin, X., (2016). The replenishment policy of agri-products with stochastic demand in integrated agricultural supply chains. *Expert Systems with Applications, 48*, 55–66.
11. Chiu, Y. T., & Chen, C. K., (2014). Linear structural cross-group relationship between social capital and national competitiveness. *Journal of Statistics and Management Systems, 17*(3), 213–234.
12. Cientifica Report, (2006). *Nanotechnologies in the Food Industry*. http://www.cientifica.com/www/details.php?id47 (accessed on 28 April 2021).
13. Civilsdaily, (2020). *Food Processing Industry: Issues and Developments*. https://www.civilsdaily.com/story/food-processing-industry-issues-and-developments/ (accessed on 28 April 2021).
14. Coherent Market Insights, (2020). *Processed Food Market Analysis*. https://www.coherentmarketinsights.com/ongoing-insight/processed-food-market-735 (accessed on 28 April 2021).
15. De Moraes, A. L. M., (2014). Brazil's agricultural policy developments. *Revista De Politica Agricoia*, 55–64.
16. *e-NAM Portal This Month is a Landmark Achievement in e-NAM History*, (2019). Retrieved from: https://pib.gov.in/Pressreleaseshare.aspx?PRID=1561146 (accessed on 28 April 2021).
17. Espolov, T., Espolov, A., Kalykova, B., Umbetaliyev, N., Uspanova, M., & Suleimenov, Z., (2020). Asia agricultural market: Methodology for complete use of economic resources through supply chain optimization. *International Journal of Supply Chain Management, 9*(3), 408.
18. FAO, (2017). *The State of Food and Agriculture, Leveraging Food Systems for Inclusive Rural Transformation*. http://www.fao.org/3/a-i7658e (accessed on 28 April 2021).
19. FAOSTAT, (2014). China Internet Network Development Statistic Report. 2014; https://www.cnnic.com.cn/IDR/ReportDownloads/201411/P020141102574314897888.pdf (accessed on 28 April 2021).
20. Food and Agriculture Policy Decision Analysis, (2014). *Socio-Economic Context and Role of Agriculture*. http://www.fao.org/3/a-i4911e.pdf (accessed on 28 April 2021).
21. Gale, F., (2015). China's growing participation in agricultural markets: Conflicting signals. *Choices, 30*(2), 1–6.
22. Gautam, S., (2020). *Recent Developments and Challenges in Radiation Processing of Food*. http://aujournals.ipublisher.in/File_upload/90180_77626060.pdf (accessed on 28 April 2021).

23. Government of India-Ministry of Commerce and Industry, (2020). *Agriculture Export Policy*. https://commerce.gov.in/writereaddata/uploadedfile/MOC_636802088572767848_Agriexport_policy.pdf (accessed on 28 April 2021).
24. Haimid, T., Rixki, D., & Dardak, R. A., (2012). Branding as a strategy for marketing agriculture and agro-based industry products. *Economic and Technology Management Review, 7*, 37–48.
25. Han, X. Y., (2011). Fresh agricultural products logistics under "farmer-supermarket direct-purchase": Problems and suggestions analysis. In: *Applied Mechanics and Materials* (Vol. 97, pp. 1046–1049).
26. HDI, (2020). *The HDI Measures National Development Through Health, Education, and Income Levels*. http://hdr.undp.org/sites/default/files/reports/14/hdr2013_en_complete.pdf; (accessed on 28 April 2021).
27. Hejazi, M., & Marchant, M. A., (2017). China's evolving agricultural support policies. *Choices, 32*(2), 1–7.
28. Ioannou, I., (2013). Prevention of enzymatic browning in fruit and vegetables. *European Scientific Journal, 9*(30), 310–341.
29. Jaffee, S., & Gordon, S., (1993). *Exporting High-Value Food Commodities: Success Stories from Developing Countries*. World Bank Discussion Papers 198, World Bank, Washington, DC.
30. Jongwanich, J., & Magtibay-Ramos, N., (2009). Determinants of structural change in food exports from developing countries. *Asian-Pacific Economic Literature, 23*(2), 94–115.
31. Joseph, T., & Morrison, M., (2006). *Nanotechnology in Agriculture and Food* (Vol. 2, pp. 2, 3). Nano forum Report.
32. Kume, T., Furuta, M., Todoriki, S., Uenoyama, N., & Kobayashi, Y., (2009). Status of food irradiation in the world. *Radiation Physics and Chemistry, 78*(3), 222–226.
33. Stephen, M., (2018). *Latest U.S. Agricultural Trade Data*. USDA Economic Research Service. https://www.ers.usda.gov/data-products/foreign-agricultural-trade-of-the-united-states-fatus/us-agricultural-trade-data-update/ (accessed on 28 April 2021).
34. Marco, M. L., Heeney, D., Binda, S., Cifelli, C. J., Cotter, P. D., Foligné, B., Ganzle, M., et al., (2017). Health benefits of fermented foods: Microbiota and beyond. *Current Opinion in Biotechnology, 44*, 94–102.
35. Markets and Markets Research, (2020). *Agriculture Industry Market Research Reports and Consulting from Markets and Markets*. https://www.marketsandmarkets.com/agriculture-market-research-173.html (accessed on 28 April 2021).
36. Miranda, J., Ponce, P., Molina, A., & Wright, P., (2019). Sensing, smart, and sustainable technologies for agri-food 4.0. *Computers in Industry, 108*, 21–36.
37. Money Matters, (2020). *Top 10 Problems Faced in Marketing Agricultural Goods*. https://accountlearning.com/top-10-problems-faced-in-marketing-agricultural-goods/ (accessed on 28 April 2021).
38. Murthy, M., Reddy, G., & Rao, K., (2013). Agricultural marketing and organized retailing for changing agri-food system in India. *International Journal of Business Management and Research, 3*(3), 169–178.
39. Naik, S. T., (2011). *Emergence, Growth, and Impact of Agricultural Commodity Marketing Center: A Study of Gadhinglaj*. (Published Doctoral Thesis). Shivaji University, Maharashtra, India. http://hdl.handle.net/10603/111579 (accessed on 28 April 2021).

40. National Agricultural Library, (2020). *Community Supported Agriculture*. https://www.nal.usda.gov/afsic/community-supported-agriculture (accessed on 28 April 2021).
41. National Institute of Agricultural Extension Management (MANAGE), (2020). *Training Program on Agricultural Marketing: the New Paradigms*. https://www.manage.gov.in/studymaterial/Am.pdf (accessed on 28 April 2021).
42. OECD Publishing, (2013). *Agricultural Policy Monitoring and Evaluation 2011: OECD Countries and Emerging Economies*. https://www.oecd.org/agriculture/topics/agricultural-policy-monitoring-and-evaluation/ (accessed on 28 April 2021).
43. Planning Commission, (2012). *Government of India, Working Group, Agricultural Division, Agricultural Marketing Infrastructure and Policy Required for Internal and External Trade*. https://niti.gov.in/planningcommission.gov.in/docs/aboutus/committee/wrkgrp11/wg11_agrpm.pdf (accessed on 28 April 2021).
44. Pujara, M. K., (2016). *F2C (Farmer to Consumer)-Farm Direct Marketing to Get More Share from Consumer Rupee*. https://www.linkedin.com/pulse/f2c-farmer-consumer-farm-direct-marketing-mandeep-pujara (accessed on 28 April 2021).
45. Rada, N., & Valdes, C., (2012). *Policy, Technology, and Efficiency of Brazilian Agriculture* (p. 137). USDA-ERS Economic Research Report.
46. Sai, M. R., (2014). Baking. *Conventional and Advanced Food Processing Technologies*, 159–196.
47. Sakif, A. S., Saikat, N. M., & Eamin, M., (2018). Drying and dehydration technologies: A compact review on advance food science. *Journal of Mechanical and Industrial Engineering Research, 7*(1), 1–10.
48. Salokhe, S., (2017). Branding of agricultural commodities/products for adding value. *International Conference on Recent Innovations in Engineering, Applied Sciences and Management*, 48–58.
49. Schnepf, R., Dohlman, E., & Bolling, C., (2001). *Agriculture in Brazil and Argentina: Developments and Prospects for Major Field Crops*. WRS-01-3, Economic Research Service, US Department of Agriculture.
50. Sharma, A., (2010). *Recent Advances in Radiation Processing of Food and Agricultural Commodities*.
51. Sikarwar, D., & Pandey, V., (2014). *The Economic Times: Farmers May be Allowed to Sell Directly to Consumers*. https://economictimes.indiatimes.com/news/economy/agriculture/farmers-may-beallowed-to-sell-directly-to-consumers/articleshow/32623139.cms (accessed on 28 April 2021).
52. Singh, B., (2016). Enhancing income of farmers through branding of agricultural produce. *International Journal of Applied Research, 2*(4), 168–179.
53. Singh, T., Shukla, S., Kumar, P., Wahla, V., Bajpai, V. K., & Rather, I. A., (2017). Application of nanotechnology in food science: Perception and overview. *Frontiers in Microbiology, 8*, 1501.
54. Soni, S., & Dey, G., (2014). Perspectives on global fermented foods. *British Food Journal, 116*(11), 1767–1787.
55. Soni, S., Gupta, M., Chaudhary, M. S., & Garg, A., (2013). Updates on agro-based food processing industry in India. *International Journal of Scientific and Engineering Research, 4*(9), 1303–1308.
56. Terefe, N. S., (2016). *Emerging Trends and Opportunities in Food Fermentation*. Elsevier.

57. The World Bank Group, (2020). *Agriculture Finance and Agriculture Insurance.* https://www.worldbank.org/en/topic/financialsector/brief/agriculture-finance (accessed on 28 April 2021).
58. TNAU Agritech Portal, (2015). *Agricultural Marketing and Agri-Business: Agricultural Marketing.* http://agritech.tnau.ac.in/agricultural_marketing/agrimark_India.html (accessed on 28 April 2021).
59. *U.S. Census of Agriculture*, (2007). Agcensus.usda.gov. https://www.nass.usda.gov/Publications/AgCensus/2007/Full_Report/Volume_1,_Chapter_1_US/usv1.pdf (accessed on 28 April 2021).
60. U.S. Department of Agriculture, (2020). *Agricultural Trade Promotion Program.* https://www.fas.usda.gov/programs/agricultural-trade-promotion-program-atp (accessed on 28 April 2021).
61. U.S. Department of Agriculture, (2020). *Farmers Market Promotion Program.* https://www.ams.usda.gov/services/grants/fmpp (accessed on 28 April 2021).
62. U.S. Department of Agriculture, (2020). *Foreign Market Development Program (FMD).* https://www.fas.usda.gov/programs/foreign-market-development-program-fmd (accessed on 28 April 2021).
63. U.S. Department of Agriculture, (2020). *Market Access Program (MAP).* https://www.fas.usda.gov/programs/market-access-program-map (accessed on 28 April 2021).
64. U.S. Department of Agriculture, (2020). *Trade.* https://www.usda.gov/topics/trade (accessed on 28 April 2021).
65. World Food Program, (2020). *Maano-Virtual Farmers Market.* https://innovation.wfp.org/project/virtual-farmers-market (accessed on 28 April 2021).
66. WTO, (2013). *World Trade Policy Review: Brazil.* http://www.wto.org/english/tratop_e/tpr_e/s283_e.pdf (accessed on 28 April 2021).
67. Wu, M., (2017). The development of the agri-food market in China: Opportunities and challenges. In: *China, New Zealand, and the Complexities of Globalization* (pp. 169–201). Palgrave Macmillan, New York.

# CHAPTER 8

# Food Adulteration: Havoc Leading to an Extorted Economy

ARUN KUMAR KUSHWAHA, PALLAVI THAKUR, SHIVAM SINGH, and TANUJA GAHLOT

## ABSTRACT

Food adulteration is a very common business practice to earn unfair profits or extorted economic gain. The increasing population has been turned up into a large business market for different industries and entrepreneurs. The international food business/trade has been increasing at a very fast rate in adopting the food habits and tastes of different regions of the world. In this race, the chances of ignorance or intentional addition of different materials/substances into the food have been increased many folds. Entry of microbiological, chemical, and physical hazards materials in our foods, causing foodborne illnesses and sometimes deaths. Food adulteration, which is presented in its crudest form generally achieved by several methods, which affect the natural quality, quantity, and composition of food ingredients. Analysis of current situations on food adulteration is increasing day by day. It has been noticed that several health problems in humans, as well as animals, arise due to adulteration in food products. The role of advertising companies in exaggerating false health benefits of food items endorsed by them have been increasing day by day. The strict laws and orders are required by every county in order to control and stop all the malicious activities, which are responsible for food adulteration knowingly/unknowingly. This study is a critical review on food adulteration. Every aspect of adulteration has been touched by the reviewer to provide a detailed understanding about food adulteration and its spreading business worldwide.

## 8.1 INTRODUCTION

According to the Indian Food Safety and Standards Act, 2006 [29, 30], "adulterant" is any material or substance which is used or employed in food to make the food unsafe, unhealthy, substandard, containing malnutrition causing matter or misbranded. According to the Federal Food, Drug, and Cosmetic Act, 1938, of USA, food is "adulterated" if it contains or bear; any "poisonous or deleterious substance" which may render it injurious to health or if it contains or bear a chemical pesticide residue that is not safe to consume [3]. Another term "adulteration" used in the ninth edition of Encyclopedia Britannica defined as the act of degrading the pure or genuine commodity for obtaining commercial gain by adding inferior quality of articles, or by extracting one or more of its constituents to increase the quantity or weight of the article and improving its physical appearance in order to present it better for marketing purposes [18].

Many times, food contamination may happen naturally, for example, biogenic amines may contaminate alcoholic beverages in the process of fermentation while production of alcohol without the intentional motive of the producer [27].

The growth of fungus producing aflatoxin on red pepper or chilies during the storage without any intention of such contamination is another example of natural food contamination [6, 19, 46].

In this era of the 21st century, every human being desire to have access to safe and healthy food. Generally, the foods that taste good are considered good foods since a common man is not much aware and educated to distinguish between tasty and healthy foods and often considers tasty foods as healthy or nutritious food. The food quality is not only affected by taste, color, texture, shape, and size, etc., but also by other reasons like chemicals, microbes, and other physical factors which are usually invisible. Food adulteration normally present in its most crude form generally done by several methods in which the subtraction or partly or exclusively substituted of any substance to food that the natural quality, quantity, and composition of food ingredient are affected. It is a venerable problem, especially where there is a challenge between the physical availability of food, either for financial gain or due to carelessness and lack of proper hygienic conditions of processing, storing, transportation, and marketing [56]. Adulteration of food reduces not only its nutritional value but also may disturb its safety issue, which may adversely affect the health of consumers [3]. Due to food adulteration, very hazardous impacts are reported suffering from heart disease, kidney failure, skin diseases, asthma, stomach ache, body ache, anemia, paralysis, diarrhea, ulcers, increase in the incidence of tumors, pathological lesions in vital organs, abnormalities of skin and eyes and other chronic diseases [7]. Food adulteration is quite common in emergent republics or backward countries. The proportion of adulterated food items on the market varied between 70 and 90% [67]. Humans consume many types of food, more the types of food more the adulteration in them, so it is very difficult to estimate the incidence of foodborne diseases globally. In the industrialized countries, more than 30% of the population suffers from foodborne disease every year. To avoid food contamination, WHO has given the following keys to Safer Food such as Keep Clean, Separate raw and cooked food, Keep food at safe temperatures. According to a report of the WHO website, about 600 million people in the world fell sick in April 2020 after the consumption of contaminated food. Every year about 420,000 people deaths are caused because of contaminated food. Approximately every year because of food adulteration, 33 million healthy lives come at risk. Consumption of such contaminated or unhealthy foods have been reported more in children. Children under the age of 5 years carry about 40% burden of the foodborne disease, which results in 125,000 deaths every year. The most common food contaminated disease is diarrhea, because of this disease, about 550 millions of people fall ill and 230,000 die every year. The numbers were based on 4.6 billion cases of diarrhea and 1.6 million deaths due to diarrhea that occurred worldwide in 2010, similar to the numbers occurring in later years [75]. The best way to avoid health problems due to contaminated food is prevention. Here we give some suggestions such as self-alertness is mandatory to procure any foodstuff, the examination of food before purchase make sure to ensure the absence of insects, visual fungus, foreign matters, etc. The label of declaration on packed food is very important for knowing the nutritional value and ingredients it contains, it also helps the consumer in checking the freshness of the food item and the period up to which it is best before use. The consumer awareness is also important; he/she should avoid taking food from the unhygienic places. Food stores which are usually present in unhygienic condition should be avoided completely. Many times, people consume food being prepared under unhygienic conditions and consume cut fruits which are being sold in unhygienic conditions. It is better to buy certified food from a reputed company and shop, during purchasing please check hallmarks on items for quality recommendations.

On the one hand, the world talks about good food and good health, while on the other hand, food adulteration has been speeded up to earn unfair profits or extorted economy. To gain this type of profits, sellers mix additional substances in main products to increase weight, quantity, durability, artificial freshness and cleanliness such as sand mix with cereals, water or chemicals mix in milk, Rasin or gum scent and color in Asafetida (Hing), Sawdust, and color in Chilles, Cassia bark which resembles cinnamon in taste and color in Cinnamon (Dalchini), Iron fillings to add weight in Rawa, Charcoal or brick powder in Shahajira, powder, and artificial color in Tea, etc. In this way, the amount of Food adulteration increases from one seller to another and finally to the buyer. Many times, adulterants are being intentionally added to more expensive substances to increase visible qualities, quantities, and to reduce manufacturing costs, or for some other false or malicious purposes. In this whole event, some people are getting a lot of benefits, but large groups suffer from many serious health problems. WHO in 2020 [35, 84] stated that unsafe food usually contains harmful microbes such as viruses, bacteria, parasites, or chemical substances, causes more than 200 serious diseases from diarrhea to cancers. In low- and middle-income countries about US$110 billion is lost every year in productivity and medical expenses, resulting from the consumption of unsafe food.

Although there is a vast possibility for improvement and development in the food adulteration field, some quick detection methods of adulterants in food, developed by various government and private organization have been presented. Although food testing laboratories are being established everywhere to test the quality of food, still an independent third-party testing is required by them to provide reliable and accurate information to researchers, regulators, industries, and consumers about the quality/worth of food products. Several organizations should be activated-supporting civil society organizations, sensitizing policymakers, issues, and promoting food safety as a culture. It is prudent that every physician should take a special interest in this subject and educate our friends and families about this menace. The proper and timely detection of adulterants in food is a vital requirement of food safety controls in any country. In order to ensure food quality, a quick observation at the household level provides enormous help in reducing the threat of food adulteration in society. Education on Food nutrition, food safety, and food fraud must be incorporated in the syllabus of pupils at both primary and secondary school levels. It will help both the consumers and workers in the food industry at a very initial age. So that appropriate, animal health, welfare, and protection of the mother earth could be taken care of from the beginning [3].

### 8.1.1 TYPES OF FOOD ADULTERATION

Due to adulteration, food characterization is necessary to ensure quality and safety along with food production, which is cost-effective, safe, nutritious, desirable, and help the consumers to make healthy selections about their food habits [56]. Food adulteration can be categorized into four categories (Figure 8.1).

1. **Intentional Adulteration:** It can be defined as the addition of lower ingredients having similar properties to the foods in which they are being added. It is intentionally or knowingly added by a person or group of people externally or internally to a food business. This type of adulteration is difficult to detect normally. There are many examples of intentional adulteration cases from around the globe, the most common are: addition of water to liquid milk, superfluous matter to ground spices, or the substitution or removal of milk solids from the natural product, strawberries were implicated in deliberate sabotage involving sewing needles, sand, marble chips, stones mud, other filth, chalk powder, mineral oil, and harmful color, etc., [7].
2. **Unintentional Adulteration:** It is the insertion of undesirable substances due to obliviousness or lack of proper facilities during the processing of food. Most of the time, this is of an acquired type such as contamination of foods by droppings of rodents, larvae in foods and pesticide residues, bacteria or fungi, dust, and stones, presence of certain chemicals or organic compounds or radicals naturally present in foods, for example, toxic varieties of pulses, vegetables, mushrooms, fish, and seafoods, also the harmful residues from packing material, etc. Contamination of crop unintentionally is another category of unintentional adulteration. Adulterants are usually contaminated the food articles in the process of their production like operations usually carried out in animal

husbandry, crop husbandry and use of veterinary medicine, during the manufacturing process, preparation, treatment, packing, and packaging of food [7]. Any time any kind of addition of adulterant in food before it consumes becomes a harmful action after being consumed.

3. **Metallic Contamination:** It is the direct or indirect addition of various type of metals or metallic compounds into the food. In the metal compounds, lead, mercury, arsenic, and cadmium are considered as the most toxic among the metals [7, 17, 70].
4. **Microbial Contamination:** It is the rotting or spoilage of food due to fermentation/infusion of food by different microbes through numerous sources. Microorganisms like bacteria, fungi, protozoa, and viruses cause food disorders. The microbiological health risks in fowl consumption and its raw products include contamination by the above food pathogens [7, 32].

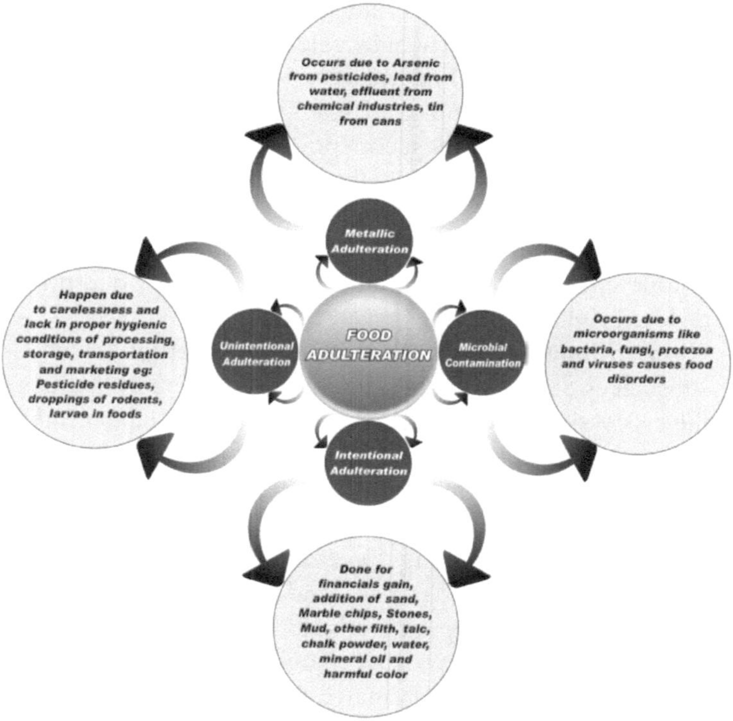


**FIGURE 8.1** Diagram showing types of food adulteration.

### 8.1.3 METHODS FOR DETECTING ADULTERANTS

Detection of adulteration in food is a vital requirement for ensuring the food safety in the modern era. The ancient history of toxicology, medical records on poisons of Indian Vedas, and Chinese knowledge of medicinal systems are the pioneer to check contaminated food to the next generations continuing his work. An adulterant can be demonstrated by the presence of an extraneous/foreign substance or a marker in the commodity, representing that a component is deviated from its normal form [7, 85]. Many techniques, based on anatomical and morphological characterization, chemical testing, and organoleptic markers have been developed to authenticate to check adulterants.

These techniques comprise of physical, chemical, analytical, and most popular DNA-based molecular techniques. Although physical and chemical techniques are common, easy, and more convenient for routine detection of food adulterant, although in certain instances, they may not provide exact qualitative and quantitative results. Physical methods for adulterant detection include microscopic and macroscopic techniques [31, 55, 63, 73]. Many chemical and biochemical methods for the detection of food adulterants have designed by researchers, which are categorized as spectroscopy-based, chromatography based, electrophoresis based and immunology based [34]. Among different analytical techniques of adulterant detection, the most commonly used is-high-performance liquid chromatography-(HPLC) [10, 58, 60, 74]. Gas chromatography (GC) along with Fourier transform infrared spectroscopy (FTIR) and mass spectroscopy (MS) is also widely used for adulterant detection in food [11, 62]. Near-infrared spectroscopy is the most popular among the spectroscopic techniques, in the rapid adulterants detection in raw material but is not very useful in identifying the contaminant [39]. The nuclear magnetic resonance (NMR) technique is very important in the detection and identification of adulterants. For the adulterant detection in ground roasted coffee containing roasted barley, the solid-phase micro-extraction-gas chromatography-mass spectroscopy (SPME-GC-MS) has also been employed successfully [64]. The electrophoresis technique is not only being utilized in detection of contaminant but also in the detection of food frauds, for example, the electrophoretic analysis had the potential to identify and quantify additional whey present in milk and dairy products [12]. Adulteration of milk can also be detected by urea-PAGE electrophoretic technique. This indirect enzyme-linked immunosorbent assay (ELISA) technique was developed for the quantification and detection of adulteration of bovine milk in goat's milk [86].

DNA-based methods for the detection of food adulterants have the potential to complement these techniques [53], and hence the food analysis laboratories are benefitting from the rapid development of DNA techniques in the field of adulteration. However, not all methods have been proved robust enough to be applied. If adulterant and original food items show the physical resemblance, then the molecular markers are being used for discrimination of adulterants from original food items. Primarily three strategies are followed for adulterants detection utilizing DNA based methods: PCR based, hybridization-based, and sequencing-based [13, 14, 16, 24, 25, 51, 77, 88]. Food Safety and Standards Authority of India (FSSAI) conducts rapid tests for establishing tentative authentication of food products by sensory evaluation (Table 8.1).

For the control of recent ongoing process of food quality deterioration or food adulteration, the laboratory services need to work in association. The national food control system must be integrated with them in order to obtain the required scientific evidence regarding various quality and safety issues indirectly affecting the public health and to solve the trade-related issues [28]. United Nations Industrial Development Organization (UNIDO), which helps the countries in economy and industrial development, has proposed three objectives for the economic and industrial development: (i) reducing poverty through productive activities, (ii) trade capacity building, and (iii) environment and energy conservation [43].

Many tools and guidelines have been proposed by UNIDO to identify the essential requirements needed for the development of national quality infrastructure, which clear the way to achieve sustainable development goals along with fulfilling the technical requirements of developing countries in getting the multilateral trading system [43]. In achieving the goal, many organizations have been collaborated such as "Fast Forward," a joint effort by UNIDO and International Organization for Standardization, " LabNetwork," Portal, a collaborative attempt by the World Association of Industrial and Technical Research Organization and UNIDO, with International Laboratory Accreditation Cooperation, International weights and Measures Offices, ISO, and Vimta Labs Ltd, India, for developing a global laboratory network in the arena of testing and calibration.

**TABLE 8.1** Sensory Evaluation Method for Food Adulteration

| Food Product | Adulterant | Method of Sensory Evaluation |
|---|---|---|
| Milk | Synthetic milk | • Synthetic milk gives bitter after taste.<br>• If adulterated, it gives a soapy feeling on rubbing between the fingers. |
| Black pepper/cloves | With mineral oil coating | • It gives kerosene-like smell. |
| Chili powder | Salt powder, brick powder, or talc. powder | • Take one teaspoon chili powder in a glass of water and examine the residue.<br>• When the residue is rubbed and if any grittiness is felt, it indicates the presence of brick powder/sand.<br>• When the white residue is rubbed, soapy, and smooth feel indicates the presence of soapstone. |
| Cloves | Volatile oil extracted cloves (exhausted cloves) | • It can be identified by its shrunken appearance and small size. The characteristic pungency of genuine cloves is less distinguishing in exhausted cloves. |
| Sugar | Urea | • Rub little sugar on palm and smell. If adulterated with urea, it will smell of ammonia.<br>• Dissolve a small amount of sugar in water.<br>• If adulterated, urea in sugar gives a smell of ammonia |
| Wheat, Rice, Maize, Jowar, Bajra, Channa, Barley, etc. | Kernel Bunt | • Separate out the non-characteristic grains and examine.<br>• Kernel bunt has a dull appearance, blackish in color, and rotten fish smell. |
| Atta | Resultant atta/Maida | • When the dough is prepared from resultant atta, less water is needed.<br>• The normal taste of chapati prepared out of atta is somewhat sweetish whereas those prepared out of adulterated will taste insipid (tasteless). |
| Sago | Sand or talcum | • Put a little quantity of sago in mouth.<br>• If adulterated, it will have a gritty feel. |

**TABLE 8.1** *(Continued)*

| Food Product | Adulterant | Method of Sensory Evaluation |
|---|---|---|
| Powdered spices | Common salt | • Taste for addition of common salt.<br>• If present, it will taste salty. |
| Sweetmeats | Artificial sweetener | • Taste small quantity of sample.<br>• Artificial sweetener leaves a lingering sweetness on the tongue for a considerable time and leaves a bitter after taste. |

## 8.2 A PURVIEW OF FOOD ADULTERATION ACROSS THE GLOBE

Too many records reported on food adulteration from all over the world. Since prehistoric periods people have altered the state of food to prolong its longevity or improve its taste. About 300,000 years ago, people had harnessed fire to cook and conserve food like meat and, later, determined that salt could be added to preserve meat without cooking. In ancient Rome and Greece, the wine was often mixed as a preservative with herbs, spices, honey. During the Middle Ages, spices were quite valuable, high priced and limited supply. Merchants sometimes combined spices with ground nutshells, juniper berries, pits, and stones. Since the ancient times, manufacturers or producers of food products have tried to alter their product in an attempt to get higher price for cheaper goods. Thus, earlier peoples used to get some materials increase in the quantity of food, but now it is done for the taste, color, huge quantity, more income, and more sales. Adulteration causes food poisoning, and it can be well versed in food toxicology. Shen Nung (ca. 2696 BC), known as the father of Chinese medicine, was famous for tasting numerous herbs in order to determine their qualities related to poisonous properties. In 994 BC, about 40,000 people passed away because of the gangrene which developed after the consumption of ergot contaminated wheat [47, 75]. Some notable incidents reported after ancient and middle ages times on food contamination are reported by many authors listed in Table 8.2.

Food adulteration probably peaked in 1850–1950 [18]. The industrial revolution created a drastic change during the 18th and 19th centuries, as the United States shifted from an agriculturally based economy to an industrial economy. Urbanization disconnected people from food production and the debasement of food for profit became rampant. Some major examples such as milk was often watered down and colored with chalk or plaster substances which were also added to bulk up flour. Wine and beer were contaminated with lead, coffee, tea, and spices were routinely mixed up with dirt, sand, or other leaves. Although a number of laws strictly forbade the mixing of harmful substances to food, however they were tough to enforce since there were no dependable tests or visual recordings available to prove the existence of pollutants. In 1820, numerous cases of food tampering "A Treatise on Adulterations of Food, and Culinary Poisons" was seen. In 1902, the effects of ingesting some of the most common food preservatives in use at the time, such as borax, copper sulfate, sulfuric

**TABLE 8.2** Some Notable Incidents Due to Food Contamination

| Time | Notable Incidents | References |
|---|---|---|
| 19th Century | Swill milk scandal in New York; Adulteration of bread with alum in London, causing rickets; Esing Bakery incident. Sweets poisoned with arsenic in Bradford, England. | [52, 76] |
| 1900 to 1949 | Severe and widespread neurological disorder due to bread flour bleached with agene in South Africa; OPIDN Organophosphate poisoning due to drinking of Ginger jack in America. | [26, 41] |
| 1950 to 2000 | Pont-Saint-Esprit mass poisoning in Southern France; Minamata disease in Japan; Morinaga Milk arsenic poisoning incident; Chicken feed and thence chickens were contaminated with dioxins from polychlorinated treated cowhides in U.S.A.; Yushō disease in Japan; Iraq poison grain disaster, wheat contaminated with weed seeds that contain pyrrolizidine alkaloids in Afghanistan; Seveso disaster in Italy; Spanish Toxic oil syndrome; Rajneeshee bioterror attack in USA; Milk contamination with dioxin in Belgium; adulterated mustard oil poisoning in India; Meat and milk were found with elevated dioxin concentrations in Germany and the Netherlands; Pyrrolizidine alkaloids poisoning in Afghanistan. | [1, 9, 21, 22, 29, 33, 38, 46–46, 54] |
| 2001 to 2010 | Spanish olive pomace oil was contaminated with Polycyclic aromatic hydrocarbons in Spain; Nitrofurons were detected in samples of chicken in Northern Ireland; The banned antibiotic Chloramphenicol was found in honey from China in UK and Canada; The banned veterinary antibiotic nitrofurans were found in chicken from Portugal in U.K.; Detection of chloramphenicol honey in Canada; Soy milk manufactured with added kelp contained toxic levels of iodine in New Zealand; EHEC O104:H4 contamination of hamburgers as a possible cause in South Korea; Worcester sauce was found to contain the banned food coloring, Sudan in U.K.; Pork, in China, containing clenbuterol when pigs were illegally fed the banned chemical to enhance fat burning and muscle growth in china; Pet food recalls in North America, Europe, and South Africa; Baby milk scandal in china; Irish pork crisis of 2008; Pork containing the banned chemical clenbuterol when pigs were illegally fed it to enhance fat burning and muscle growth in china; Hoola Pops contaminated with lead in Mexico; Edible Snakes were contaminated with clenbuterol when fed frogs treated with clenbuterol in China. | [36, 49, 59, 69, 79] |

**TABLE 8.2** *(Continued)*

| Time | Notable Incidents | References |
|---|---|---|
| 2011 to present | Poor-quality illegal alcohol in West Bengal in India; German *E. coli* O104:H4 outbreak was caused by EHEC O104:H4 contaminated fenugreek seeds; American black licorice products recalled due to high lead levels in America; A batch of 1800 almond cakes with buttercream and butterscotch from the Swedish supplier in china; Bihar school meal poisoning incident; Taiwan food scandal; Mozambique funeral beer poisoning; Caraga candy poisonings in the Philippines; United States *E. coli* outbreak; CRF Frozen Foods recalled over 400 frozen food products due to listeria; Fipronil eggs contamination in Europe; Australian strawberry contamination. | [15, 23, 37, 71, 81, 87] |

acid and formaldehyde which were lethal for humans [72]. In May 2009, after the incidence of melamine in infant and pet food scandals, a public meeting was held by the Food and Drug Administration (FDA) to discuss the methods to "better predict and prevent economically motivated adulteration (EMA)" of all FDA-regulated products [20]. The second-lowest burden of foodborne diseases globally is found for the Region of America. An average of 77 million people still falls ill annually from the contaminated food, with an estimation of 9000 deaths every year in the region. Out of those who fall ill, about 31 million are under 5 years of age, results in more than 2000 deaths of these children annually. European Region has the lowest estimated burden in terms of foodborne diseases globally; it was found that above 23 million people in this region fall ill every year by consuming unsafe food, results in about 5000 deaths [83] annually. When it comes to China, which is one of the main animal food producers in the world, the last three decades in China, as industrialization and urbanization hastens, China's food industry has resulted in increasing animal-based food production along with the growing capacity of food processing. The consumption of processed and packaged food has been increased by many folds in recent times [89]. The economically motivated food fraud documented in Australia such as the use of pesticides, food additives, and preservatives. In Australia, research scholars, institutions, and Food Standards Australia New Zealand (FSANZ) play a key role in managing food safety [20]. According to WHO [83], the Eastern Mediterranean Region has the third highest estimated burden of foodborne diseases per population, after the African and South-East Asia regions. Every year more than 100 million people living in the Eastern Mediterranean Region are estimated to become ill with foodborne diseases every year and of them, 32 million are children under 5 years of age. In addition, the South-East Asia Region (after the African Region) has the second-highest burden per population of the foodborne diseases. Although, in terms of absolute numbers, more than 150 million cases, of which 175,000 deaths occur from foodborne diseases every year living in the WHO South-East Asia Region than in any other WHO Region. According to WHO [83] in the South-East Asia Region, some 60 million children usually below 5 years of age fall ill and almost 50,000 die from foodborne diseases every year. In China, the use of melamine was (a flame-retardant generally used in the furniture industry in the form of plastic "melamine-formaldehyde resin") found in increasing the nitrogen content in food. However, the use of melamine in food is prohibited by the WHO or any national authorities [3]. In China, an estimated 300,000 children were tested positive suffering from problems in kidney, of which 54,000 babies were sent to hospitals and of them 6 children died from kidney stones, in July 2008. In response to the several ongoing scandals in China, on 28 February 2009, China announced the Food Safety Law of the People's Republic of China to replace its Food Hygiene Act [20]. Food adulteration has become rampant in India it starts from the agriculture field itself, where fertilizers and pesticides are over-used. More and more people are moving away to readymade fast foods and eating regularly at restaurants and nearly half of them are adulterated. However, the food should be safe and healthy. Unfortunately, there is little awareness among the Indian public. In the National Survey conducted by the food safety and standards authority, India to ascertain the quality of milk in the country, 68% of samples were found to be non-conforming to food safety and standards regulations (FSSR). The government has released state-wise details of non-conforming samples in the descending order of percentage with respect to the total samples collected in different states [67]. The highest burden of foodborne disease per population was estimated in WHO African Region. Each year about 91 million people fall ill and more than 137,000 people die from foodborne diseases. The traditional food control systems in most African countries are very poor in assessing the food quality. They do not provide the concerned agencies a clear mandate and authority to prevent food safety-related problems. A major problem with food production, processing, and marketing in most of the countries is a large number of small producers and handlers lack the adequate knowledge and expertise in the field of modern practice and food hygiene, which make the food more prone to contamination [61]. Thus, so many examples are available regarding food adulteration over the world. Barilla Centre for Food and Nutrition [8] of the European Foundation Centre, was given the food sustainability index of the country of the world. The overall score is calculated from a weighted average of the three category scores, such as food loss and waste, sustainable agriculture, and nutritional challenges. A country that attains a higher score means that it is on the right path concerning a sustainable food and nutrition system. The greater score showing better sustainable food and nutrition system, the lower the score denotes adulteration/contamination and bad nutrition system (Figure 8.2).

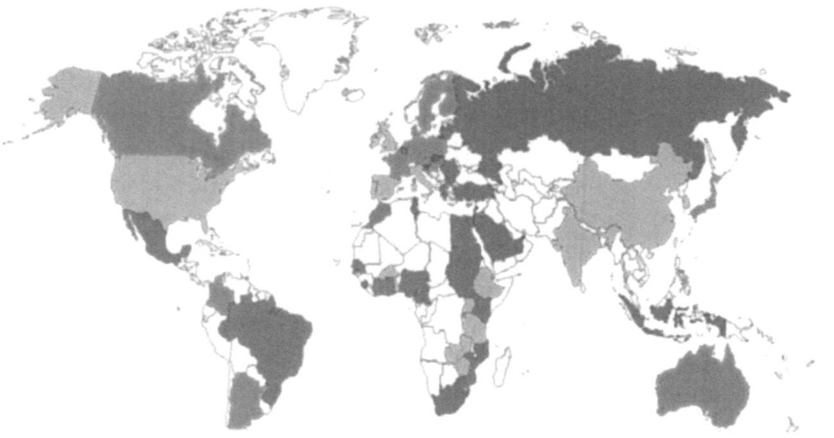

| VERY HIGH Score 71.2 to => 76.1 | | # | HIGH Score 66.5 to => 71.1 | | # | MEDIUM Score 62.7 to => 66.4 | | # | LOW Score 52.3 to => 62.6 | |
|---|---|---|---|---|---|---|---|---|---|---|
| France | **76.1** | 1 | Ireland | **71.0** | 1 | Belgium | **66.2** | 1 | Lebanon | **62.4** |
| Netherlands | **75.6** | 1 | Spain | **70.9** | 1 | Cote d'Ivoire | **65.9** | 1 | Tunisia | **62.3** |
| Canada | **75.3** | 1 | Estonia | **70.8** | 1 | Cyprus | **65.8** | 1 | Latvia | **61.9** |
| Finland | **74.1** | 1 | Portugal | **70.6** | 1 | Senegal | **65.8** | 1 | Malta | **61.5** |
| Czech Republic | **74.0** | 1 | South Korea | **70.5** | 1 | Mexico | **65.6** | 1 | Slovakia | **61.4** |
| Japan | **73.8** | 1 | China | **70.2** | 1 | Brazil | **65.5** | 1 | Slovenia | **60.9** |
| Denmark | **73.5** | 1 | United Kingdom | **70.0** | 1 | Lithuania | **65.3** | 1 | Sudan | **60.9** |
| Sweden | **73.4** | 1 | Uganda | **68.7** | 1 | Israel | **64.6** | 1 | Turkey | **60.1** |
| Austria | **73.3** | 1 | United States | **68.6** | 1 | Greece | **64.5** | 1 | Cameroon | **59.7** |
| Hungary | **72.5** | 1 | Ethiopia | **68.5** | 1 | Kenya | **64.4** | 1 | Indonesia | **59.1** |
| Australia | **71.8** | 1 | Italy | **68.1** | 1 | Romania | **64.4** | 1 | Sierra Leone | **58.8** |
| Rwanda | **71.6** | 1 | Luxembourg | **67.9** | 1 | Nigeria | **63.7** | 1 | Ghana | **57.6** |
| Argentina | **71.5** | 1 | Tanzania | **67.4** | 1 | Morocco | **63.5** | 1 | South Africa | **56.4** |
| Croatia | **71.4** | 1 | Zimbabwe | **67.3** | 1 | Egypt | **63.0** | 1 | Saudi Arabia | **56.2** |
| Poland | **71.3** | 1 | Zambia | **67.2** | 1 | Mozambique | **63.0** | 1 | Russia | **56.1** |
| Colombia | **71.2** | 1 | Burkina Faso | **66.4** | 1 | Jordan | **62.8** | 1 | Bulgaria | **54.5** |
| Germany | **71.2** | 1 | India | **66.4** | 0 | | | 1 | UAE | **52.3** |

**FIGURE 8.2** Food sustainability index showing sustainable food and nutrition system of the world.

China and India are among the major producers and consumers of food; however, due to their huge populations, the large gap between their production and consumption is a major concern for food producers. Milk adulteration is often being done through water, synthetic milk, skim milk powder, starch, cane sugar, melamine, ammonium sulfate, fat, etc. These additives are used to increase its volume and give a natural look while maintaining its specific gravity, protein, and fat content, etc. However, to increase the shelf life of adulterated milk formalin, antibiotics, caustic soda, salicylic acid, benzoic acid, hydrogen peroxide, carbonates, bicarbonates, etc., are frequently used [3].

## 8.3 SITUATIONAL ANALYSIS OF ADULTERATED FOOD PRODUCTS AND THEIR IMPACT OVER LIVELIHOOD

Livelihood is a set of activities essential to everyday life. Such activities include securing water, food, medicine, shelter, clothing, and fodder. One of the basic needs for every living being is food. It is also a very important aspect for life. But now day's foods are affected by different adulterants. It is the duty of every "Welfare Country" to ensure their citizens to have the access to affordable, nutritious, and safe food. Risk for carcinogenic and non-carcinogenic effects of contaminated foods are related to the exposure to contaminants through three specific mechanisms on human and animals due to water pollution, food adulteration, and bio-magnification. In India, about 3000 persons fell ill and 54 deaths because of dropsy triggered by the consumption of adulterated mustard oil. It is caused by a toxin called sanguinarine, present in the *Argemone mexicana* [82]. Due to Adulterated 'Dal' mixed with 'Kesari Dal' hundreds of people were disabled in Madhya Pradesh, India after consuming this adulterated dal. Some common adulterants are mineral oils, castor oils, and argemone oils, in edible oils; vanaspathi, mashed potato in ghee; invert sugar or jaggery in honey, lead chromate in pulses, etc., which can lead to epidemic dropsy, glaucoma, cardiac arrest, lathyrism, paralysis, anemia, abortion, brain damage, cancer, etc. [4]. An increased incidence of kidney stones and renal failure among infants was identified in China due to milk adulteration [56]. In China, farmers in Fujian provinces were detained for selling carcasses as fake meat in roadside shops. In these cases, suspects are accused of using gelatin, red pigment, and nitrates to alter the dead pigs, ducks, and rats [51]. In a random survey by the Public Health Laboratory of Dhaka Municipal Corporation in 2004, more than 76% of food items on the market were found adulterated [67] are also responsible for cancers of oral cavity, pharynx, larynx, esophagus, liver, and breast in Bangladesh [40]. Reports from recent newspapers say, the problem is now alarming and people are suffering from food phobia. They are eating fewer fruits and vegetables, resulting in undernutrition, on the other hand, they are putting them vulnerable to dreadful diseases by eating adulterated food [67]. Consumption of artificially ripened or adulterant fruits can be the reason of developing skin rashes, mouth ulcers, diarrhea, and many times gastric problems. It is very harmful for pregnant women because consumption of these artificially ripened fruits can cause miscarriages and developmental abnormalities in the fetus [3] Some of the common adulterants in food and their impact on livelihood are mentioned in Table 8.3. Adulterants of various animal feedstuff are harmful to animals and indirectly to humans also. The edible products from animals, eggs, meat, milk, etc., are obtained from animals and these things are an essential part of the health of human beings. The pure quality of these edible items depends on the pure, unadulterated, and uncontaminated feed which the animals consume. Some common feedstuff adulterants are cobs, cob dust, sand in maize; marble, grit in rice kani; water in molasses; urea, raw soybean, hulls in soybean mineral mixture; ground rice husk, sawdust in de-oiled rice bran and wheat bran; groundnut husk, urea, non-edible oil cakes in groundnut cake; common salt, marble powder, sand, limestone in soybean meal mineral mixture; *Argimone maxicana* seeds, fibrous feed ingredients, urea in mustard cake; common salt, urea, sand in fish meal [66]. The adulteration of crude drug is a burning problem. It may be deliberate or accidental substitution of crude drug partially or completely with other harmful substances which are either free or inferior in chemical or therapeutic properties. The unwanted aspect of substitution lies under sub-standard conditions [2] which sometimes become life-threatening caused a variety of adverse effects from mild (allergic reactions, mood disturbances or muscle weakness, fatigue, gastrointestinal upset, pain, nausea, and respiratory complaints) to moderate (confusion, convulsions, seizures or lethargy or, dermatitis, leucopenia, vomiting, and sensory disturbances) to severe (poisoning, metabolic acidosis, carcinomas, cerebral edema, intracerebral hemorrhage, coma, nephrotoxicity, perinatal stroke, multi-organ failure, renal or liver failure or death) life-threatening effects [65].

**TABLE 8.3** Some Injurious Adulterants/Contaminants and Their Types in Foods and Their Health Effects

| Sr. No. | Food Article | Adulterant | Harmful Effects |
|---|---|---|---|
| 1. | Bengal gram (toor dal) | Kesari dal | Lathyrism cancer |
| 2. | Tea | Used tea leaves processed and colored | Liver disorder |
| 3. | Coffee powder | Tamarind seed, date seed powder | Diarrhea |
| | – | Chicory powder | Stomach disorder, giddiness, and joint pain |
| 4. | Khoa | Starch and less fat content | Less-nutritive value |
| 5. | Wheat and other food grains (Bajra) | Ergot (a fungus containing poisonous substance) | Poisonous |
| 6. | Sugar | Chalk powder | Stomach disorder |
| 7. | Black powder | Papaya seeds and light berry | Stomach, liver problems |
| 8. | Mustard powder | Argemone seeds | Epidemic dropsy and glaucoma |
| 9. | Edible oils | Argemone oil | Loss of eyesight, heart diseases, tumor |
| | – | Mineral oil | Damage to liver, carcinogenic effects |
| | – | Karanja oil | Heart problems, liver damage |
| | – | Castor oil | Stomach problem |
| 10. | Asafetida | Foreign resins galbanum, colophony resin | Dysentery |
| 11. | Turmeric powder | Yellow aniline dyes | Carcinogenic, anemia, epilepsy, neurotoxicity |
| | – | Non-permitted colorants like metanil yellow | Highly carcinogenic |
| | – | Tapioca starch | Stomach disorder |
| 12. | Chili powder | Brick powder, sawdust | Stomach problems |
| | – | Artificial colors | Cancer |
| 13. | Sweets, juices, jam | Non-permitted coal tar dye (Metanil yellow) | Metanil yellow is toxic and carcinogenic |
| 14. | Jaggery | Washing soda, chalk powder | Vomiting, diarrhea |
| 15. | Pulses (Green peas and dhal) | Coaltar dye | Stomach pain, ulcer |

**TABLE 8.3** *(Continued)*

| Sr. No. | Food Article | Adulterant | Harmful Effects |
|---|---|---|---|
| 16. | Suapari | Color and saccharin | Cancer |
| 17. | Honey | Molasses sugar (sugar plus water) | Stomach disorder |
| 18. | Carbonator water beverages | Aluminum leaves | Stomach disorder |
| 19. | Liquid Milk | Water or urea | Loss of vision and heart diseases |
| | – | Detergent | Hypertension and heart ailments |
| | – | Starch | Stomach diseases |
| | – | Caustic soda | Hypertension and heart ailments. |
| 20. | Mustard seed | Argemone seed | Loss of vision and heart diseases |
| 21. | Ice creams | Washing powder | Appendicitis and small intestine problems |
| 22. | Groundnut Oil | Argemone oil | Loss of vision and heart diseases |
| 23. | Fruits/Vegetables | Calcium carbide/metallic lead/oxytocin | Vomiting, diarrhea, kidney failure, brain damage |
| 24. | Alcoholic liquors | Methanol | Blurred vision, blindness, death |
| 25. | Cold drinks | Brominated vegetable oils | Anemia, enlargement of heart |
| 26. | Drinking water, seafoods, tea, etc. | Fluoride | Mottling of teeth, skeletal, and neurological disorders |
| 27. | Spinach, amaranth | Oxalic acid | Renal calculi, cramps, failure of blood to clot |
| 28. | Cottonseed flour and cake | Gossypol | Cancer |
| | | **Foodborne Illnesses (Bacterial)** | |
| 29. | Rice | *Bacillus cereus* | Nausea, vomiting, abdominal pain, diarrhea |
| 30. | Raw milk, goat cheese | *Brucella abortus*, *B. melitensis*, and *B. suis* | Fever, chills, sweating, weakness, headache, muscle and joint pain, diarrhea, bloody stools |

**TABLE 8.3** *(Continued)*

| Sr. No. | Food Article | Adulterant | Harmful Effects |
|---|---|---|---|
| 31. | Honey, home-canned vegetables and fruits, corn syrup | *Clostridium botulinum* | Lethargy, weakness, poor feeding, constipation, hypotonia, poor head control, poor gag, and sucking reflex. |
| 32. | Unrefrigerated or improperly refrigerated meats, potato, and egg salads, cream pastries | *Staphylococcus aureus* | Nausea and vomiting, abdominal cramps, diarrhea, fever |
| 33. | Cereal products, custards, puddings, sauces | *Bacillus cereus* | Nausea, vomiting, abdominal pain, diarrhea |
| 34. | Undercooked or raw seafood, such as fish, shellfish. | *Vibrio parahaemolyticus* | Watery diarrhea, abdominal cramps, nausea, vomiting |
| 35. | Undercooked or raw shellfish, especially oysters, other contaminated seafood | *Vibrio vulnificus* | Vomiting, diarrhea, abdominal pain |
| 38. | Aspergillus flavus contaminated foods such as groundnuts, cottonseed, etc. | Aflatoxins | Liver damage and cancer |
| 39. | Meats, poultry, gravy, dried or pre-cooked foods, time-, and/or temperature-abused food. | *Clostridium perfringens* | Watery diarrhea, nausea, abdominal cramps; fever |
| 40. | Undercooked beef especially hamburger, unpasteurized milk and juice, raw fruits, and vegetables | *Escherichia coli* | Diarrhea that is often bloody, abdominal pain and vomiting. |
| 41. | Water or food contaminated with human feces | *E. coli* | Watery diarrhea, abdominal cramps, some vomiting. |

**TABLE 8.3** *(Continued)*

| Sr. No. | Food Article | Adulterant | Harmful Effects |
|---|---|---|---|
| 42. | Fresh soft cheeses, unpasteurized milk, inadequately pasteurized milk, ready-to-eat deli meats, hot dogs | *Listeria monocytogenes* | Fever, muscle aches, and nausea or diarrhea.<br>Pregnant women may have mild flu-like illness, and infection can lead to premature delivery or stillbirth |
| 43. | Unrefrigerated or improperly refrigerated meats, potato, and egg salads, cream pastries | *Staphylococcus aureus* | Sudden onset of severe nausea and vomiting, abdominal cramps, diarrhea and fever |
| 44. | Undercooked pork, unpasteurized milk, tofu, contaminated water. Infection has occurred in infants whose caregivers handled chitterlings. | *Yersinia enterocolytica* and *Y. pseudotuberculosis* | Appendicitis-like symptoms (diarrhea and vomiting, fever, and abdominal pain) occur primarily in older children and young adults |
| | | **Foodborne Illnesses (Viral)** | |
| 45. | Shellfish harvested from contaminated waters, raw produce, contaminated drinking water, uncooked foods and cooked foods that are not reheated after contact with infected food handler | Hepatitis A | Diarrhea, dark urine, jaundice, and flu-like symptoms, i.e., fever, headache, nausea, and abdominal pain |
| 46. | Shellfish, fecally contaminated foods, ready-to-eat foods touched by infected food workers (salads, sandwiches, ice, cookies, fruit). | Noroviruses (and other caliciviruses) | Nausea, vomiting, abdominal cramping, diarrhea, fever, myalgia, and some headache. Diarrhea is more prevalent in adults and vomiting is more prevalent in children |
| 47. | Fecal contaminated foods. Ready-to-eat foods touched by infected food workers (salads, fruits) | Rotavirus | Vomiting, watery diarrhea, low-grade fever, temporary lactose intolerance |

**TABLE 8.3** *(Continued)*

| Sr. No. | Food Article | Adulterant | Harmful Effects |
|---|---|---|---|
| | | **Foodborne Illnesses (Parasitic)** | |
| 48. | Raw or undercooked intermediate hosts (e.g., snails or slugs), infected paratenic (transport) hosts (e.g., crabs, freshwater shrimp) | *Angiostrongylus cantonensis* | Headaches, nausea, vomiting, neck stiffness, paresthesias, hyperesthesias, seizures, and other neurologic abnormalities |
| 49. | Various types of fresh produce (imported berries, lettuce) | *Cyclospora cayetanensis* | Diarrhea (usually watery), loss of appetite, substantial loss of weight, stomach cramps, nausea, vomiting, fatigue |
| 50. | Any uncooked food or food contaminated by an ill food handler after cooking, drinking water | *Entamoeba histolytica/Giardia lamblia* | Diarrhea (often bloody), frequent bowel movements, lower abdominal pain |
| 51. | Accidental ingestion of contaminated substances (e.g., soil contaminated with cat feces on fruits and vegetables), raw or partly cooked meat (especially pork, lamb, or venison) | *Toxoplasma gondii* | Cervical lymphadenopathy and/or a flu-like illness |
| 52. | Raw or undercooked contaminated meat, usually pork or wild game meat (e.g., bear or moose) | *Trichinella spiralis* | Acute: nausea, diarrhea, vomiting, fatigue, fever, abdominal discomfort followed by muscle soreness, weakness, and occasional cardiac and neurologic complications. |

## 8.4 NORMS AND REGULATIONS OF FOOD SECURITY AT INTERNATIONAL LEVEL

Too many norms, regulations, and agencies are working against adulterated food. Food legislations had started to regulate food safety to overcome food fraud issues among all the nations in the world. The laws relating to adulteration of food lacked sufficient means of enforcement and remained largely in affective before the 70th century through the general regulatory laws. Due to the increasing number of food adulteration incident, several agencies/organizations have been set up by the Government to remove adulterants from foodstuffs. Different food acts/law from time to time in different countries kept coming such as in England Adulteration of Food or Drink Act (1860), Sale of Food and Drugs Act (1875), Food and Drugs (Adulteration) Act (1928), Food and Drugs Act (1938), the third was the Food and Drugs (Milk, Dairies, and Artificial cream) Act (1950) and the last and the existing full consolidating Act is the Food and Drugs Act (1955) [68]. During British Rule in India, the first attempt was made to deal with poisonous or noxious food under the Indian Penal code, 1860 [42]. The four categories were distinguished by the Indian Penal Code. viz.: (i) spreading of infections (Sec.269 to 271); (ii) adulteration of food, drinks, and drugs (Sec. 272 to 276); (iii) fouling of water (Sec. 277); and (iv) making atmosphere noxious to health (Sec. 273). The Prevention of Food Adulteration Act (1954) (Act 37 on of 1954) received the assent of the President of India, and came into force on 1.6.1955 [36]. In the United States, the Federal Food and Drug Act (1906) defined food adulteration and misbranding of products and continuing with the Food, Drugs, and Cosmetics Act (1938) it was amended in 1958 and 1962 to define and regulate food and food coloring. Imported goods that violated the provisions of the Act may be denied admittance to the United States [79], etc.

Some of the national and international agencies have been set up by the Governments to check and remove adulterants from foodstuffs such as ISO (The International Organization for Standardization), FSMS (Food Safety Management System), FSSR, FCI (The Food Corporation of India), FSSAI, FSMS, FAO (The Food and Agriculture Organization), WHO (The World Health Organization), WTO (The World Trade Organization), QMS (quality management system), FSA (Food Standards Agency), FSIS (Food Safety and Inspection Service), USDA (The United States Department of Agriculture), FSSA (Fitness and Sports Sciences Association), FDA, FSSAI and herewith some of the NGOs such as The Daily Star and RDRS working on Hazards of food contamination on national life are contribute in this mission.

The campaign *Jago Grahak Jago* was started in 2005 by the ministry of consumer affairs. The main motive of this program was too aware the consumers about the consumer rights. It also educates the consumer about the rules of purchase and sales [5]. This campaign has now become a common household name because of its publicity. The best part of the campaign was that through this program, the rural, distant, and isolated areas have been given the main priority in order to getting legal information of their consumer rights.

The first question raises in the mind of a consumer before doing complaint of the product that do I have right to file the complaint against such a big brand? Therefore, the answer is "Yes" a consumer or one or more consumers in relation to any kind of goods or services are eligible to file a complaint. The central and state government can also file a complaint. Any voluntary consumer association, which is registered under the Companies Act (1956), can file a complaint. In case of death of any consumer after consuming the food product, a complaint can be filled. However, a power of attorney holder has no right to file a complaint under the act. A complaint against the product can be filed by writing its plea on plain paper containing the name, and proper address of the petitioner and the opposite party. All the information and facts associated to complaint should be mentioned in a document. Allegations supporting documents need to be attached with the complaint. Along with this information, the complainant also needs to write about relief, which he is seeking. A very nominal court fee is charged in such type of complaints, and there is no lawyer required for this purpose. Government ensures consumer to take very strict action in favor of consumes if the opposite party is proved guilty. Depends upon the value or cost of goods and services, a complaint could be filed under three forums: (1) National Commission, (2) State Commission, and (3) District Forum. Thus, all the advertising and communication agencies must realize their professional and social ethics in promoting a healthy lifestyle by focusing on achieving high standards of public health goal.

## 8.5 CHALLENGES AND OPPORTUNITIES

There are many challenges the global market is facing in maintaining food quality in the food industry. The increasing demand of food due to increasing population is one of the major issues which has become a global issue in the context of food production. The ongoing trends which are utilizing harmful substances or adulterant in order to maintain the enormous quantity of food product, are rendering many lives at risk. Food hygiene is the main cause of food quality deterioration. There are different types of adulteration that occur in different types of ways. The identification of food adulterant by the food laboratories has also become a big challenge. The changing form of adulterant day by day is causing a serious issue for testing laboratories. The greed of making more profit at a lower investment has become very common in food industry and due to this even the big industries are not following whatever standards have been proposed by health organizations in making of the food products. It has been noticed that the increasing marketing of the food products through different media open up scope for different food industries to sell their products globally no matter how much they are affecting the health and how much parameters they are following in production of their food items. Advertisement industry plays a very important role in spreading the information regarding the products launched by different companies. Advertising in the food industry means commercial communication of any food item through a food business operator (FBO) to common people [5]. Through advertisement, products get an identification among consumers. The marketing of products through advertisement has become prevalent in national and international market. Awareness is also very important in any field. Due to lack of awareness, many food adulterants are being added on a daily purpose by unaware people.

Some major challenges in the field of food adulteration are:

1. **Greed of Making More Profit by Cooperate Buyers:** The cooperate buyers in greed of making more profit use to play with the health of consumers. They use to buy low quality or cheap raw goods and use them in their production process, due to which the nutritional value of the product is ignored.
2. **Vulnerability Fear:** This can also be a reason in the food industry responsible for deteriorating nutritional value or adding adulterant to the food item. It is well understood now that food has given birth to a number of big band industries. Now the market competition has been increasing day by day. On our TV screen, we see a number of new food products launched under the name of a number of brands. This kind of race in food industries has

caused several loopholes in the quality of food products. Hence to overcome the fear of vulnerability, the food industrialist uses the added substances to increase the taste of their particular foot items by adding harmful substances or adulterant to them.

3. **Challenge for Food Testing Labs:** A big challenge is to detect the food adulterant in food items. A wide range of adulterants are being added to the food day by day by hundreds of methods, and the most challenging task is to identify a different type of adulterant with old developed methodology and technique. Every methodology/technique developed by researcher today expose a number of adulterants in food but every day the increasing cases of food adulterant again create a new challenge for researchers.
4. **Advertisements and Marketing Challenge:** Advertisements and marketing for any food items and beverages should be honestly. No misleading or illusory rumors should be publicized. Sometimes advertiser claims or highlights about particular ingredients in the food item, however, it is not present or present in modified form. All the claims of the advertisement/marketing communications for food and beverages should be supported by scientific proof and should also meet the requirement of basic food standards proposed by the Food Safety and Standards Authority. Advertiser should not discourage the original or good dietary practices of the food product over its commercialization. For example, the dairy products are better when they are purchased from the original milk seller than getting packed dairy items, fresh fruits are better than packed one, etc. Hence advertisements and marketing fields have also turned up into a challenge.
5. **Awareness and Education:** Many times, food adulteration occurs due to unawareness of the food producer. People of rural and remote areas are not so much educated and aware about the food nutrition. The use of fertilizers or illegal substances/chemicals in the agriculture by uneducated farmers is a main problem in food quality deterioration.

Opportunities in food adulteration can be seen in prevention. There are several areas in research which need systematic, long-term, and serious actions. It has been seen that this area of study is still untouched in many respects. Many kinds of adulterant are still available easily because of a lack of research on them. Another opportunity may be seen in education program. There should be a separate subject for the study of food nutrition and adulteration at primary and secondary level in schools. Food laboratories should be opened in every city, village, and district so that a common man could go and check the nutritional value of its food product. The awareness program at the local level should be promoted to aware the local people about food fraud and food adulteration. Food adulteration awareness should also be encouraged at national and international levels.

## 8.6 SUMMARY

Food adulteration has been speeded up to earn unfair profits or extorted economy. To earn unfair profits, human negligence and enter microbiological, chemical, and physical hazards materials in our foods causing foodborne illnesses and deaths. Food adulteration normally presents in its crudest form generally done by several methods in which the subtraction or partly or exclusively substituted of any substance to food that the natural quality, quantity, and composition of food ingredients are affected. There are four types of adulteration in food that have been noticed, i.e., Intentional adulteration, Unintentional adulteration, Metallic, and Microbial contamination. Analysis of current situations on food adulteration increasing day by day causes several health problems in humans as well as animals such as heart disease, kidney failure, skin diseases, asthma, stomach ache, body ache, anemia, paralysis, diarrhea, ulcers, increase in the incidence of tumors, pathological lesions in vital organs, abnormalities of skin and eyes and other chronic diseases. The number of cancer patients, heart, and kidney disease are sharply increasing because of food adulteration and taking a heavy toll on public lives. According to WHO, about 600 million episodes of illness, 420,000 deaths every year due to contaminated food. The growing market of food industry worldwide deteriorating the quality of food by compromising its nutritious value. Advertising and marketing industries have been expanding their business in order to earn more profit by claiming wrong statements regarding the health benefits of the product. This is a serious matter of concern. The advertisers proclaim what is not actually present or present in a very less quantity in a food product by exaggerating their health benefits. Such advertising or marketing agencies are responsible to encourage the consumers to purchase food adulterant containing products under the name of big brands. For overcoming such issues, many initiatives have been taken by government and non-government organizations in all over the world. But still very few changes have been noticed in order to achieve health benefits out of the food products selling in markets. A campaign started by the ministry of consumer affairs, India-2005, "*Jago Grahak Jago*" was one of the major steps by the government of India to aware and educate the people about the purchase and sale of products. This campaign not only aware the consumer about their consumer rights but also create a social responsibility among product selling companies. Livelihood is a set of different activities. The basic requirement of a common person is water, food, shelter, and medicine. Food is one of the most basic needs for a living being. Nowadays, food items are affected by different adulterants. It is the responsibility of a Welfare Country to ensure its public to have access to affordable, nutritious, and safe food. Hence, to achieve the goal of providing safe and healthy food to their citizens, every country in the world needs to implement all the food safety rules and regulations properly by taking strict legal actions against food fraud agencies.

## ACKNOWLEDGMENTS

The authors are thankful to Ms. Divyajyoti Arya for publication assistance and Mr. Pradeep Badola for administrative support. We are also thankful to Mr. Pramod Kulshreshtha for graphical support.

## KEYWORDS

- **Enzyme-Linked Immunosorbent Assay**
- **Food Adulteration**
- **Food Business Operator**
- **Food Safety and Standards Regulations**
- **Gas Chromatography**
- **Harmful effect**
- **Regulatory Bodies**

## REFERENCES

1. Afful, S., Awudza, J. A., Twumasi, S. K., & Osae, S., (2013). Determination of indicator polychlorinated biphenyls (PCBs) by gas chromatography-electron capture detector. *Chemosphere, 93*(8), 1556–1560.
2. Ahmed, S., & Hasan, M. M., (2015). Crude drug adulteration: A concise review. *World J. Pharm. Pharm. Sci., 4*(10), 274–283.
3. Attrey, D. P., (2017). Detection of food adulterants/contaminants. In: *Food Safety in the 21st Century* (pp. 129–143). Academic Press.
4. Awasthi, S., Jain, K., Das, A., Alam, R., Surti, G., & Kishan, N., (2014). Analysis of food quality and food adulterants from different departmental & local grocery stores by qualitative analysis for food safety. *IOSR-JESTFT, 8*(2), 22–26.
5. Bajaj, S., (2017). Regulation of advertisement for food products in India-advertisement for food products. In: *Food Safety in the 21st Century* (pp. 469–477). Academic Press.
6. Banerjee, D., Chowdhary, S., Chakraborty, S., & Bhattacharyya, R., (2017). Recent advances in detection of food adulteration. In: *Food Safety in the 21st Century* (pp. 145–160). Academic Press.
7. Bansal, S., Singh, A., Mangal, M., Mangal, A. K., & Kumar, S., (2017). Food adulteration: Sources, health risks, and detection methods. *Critical Reviews in Food Science and Nutrition, 57*(6), 1174–1189. doi: 10.1080/10408398.2014.967834.
8. BCFN, (2018). *Barilla Center for Food and Nutrition of the European Foundation Center: Food Sustainability Index of the Country of the World.* https://www.barillacfn.com/en/// (accessed on 9 April 2021).
9. Bernard, A., Broeckaert, F., De Poorter, G., De Cock, A., Hermans, C., Saegerman, C., & Houins, G., (2002). The Belgian PCB/dioxin incident: Analysis of the food chain contamination and health risk evaluation. *Environmental Research, 88*(1), 1–18.
10. Blanch, G. P., Mar, C. M., Ruiz, D. C. M. L., & Herraiz, M., (1998). Comparison of different methods for the evaluation of the authenticity of olive oil and hazelnut oil. *J. Agric. Food Chem. 46*, 3153–3157.
11. Cabanero, A. I., Recio, J. L., & Ruperez, M., (2006). Liquid chromatography coupled to isotope ratio mass spectrometry: A new perspective on honey adulteration detection. *Journal of Agricultural and Food Chemistry, 54*(26), 9719–9727.
12. Cancalon, P. F., (2006). Electrophoresis and isoelectric focusing in food analysis. *Encyclopedia of Analytical Chemistry: Applications, Theory, and Instrumentation.* doi: 10.1002/9780470027318.a1007.
13. Carles, M., Cheung, M. K., Moganti, S., Dong, T. T., Tsim, K. W., Ip, N. Y., & Sucher, N. J., (2005). A DNA microarray for the authentication of toxic traditional Chinese medicinal plants. *Planta Medica, 71*, 580–584.
14. Chandrika, M., Maimunah, M., Zainon, M. N., & Son, R., (2010). Identification of the species origin of commercially available processed food products by mitochondrial DNA analysis. *Int. Food Res. J., 17*, 867–876.
15. Chaves, B. D., & Brashears, M. M. Food Safety magazine. https://www.food-safety.com/articles/5134-mitigation-of-ilisteria-monocytogenesi-in-ready-to-eat-meats-using-lactic-acid-bacteria (accessed on 9 May 2021).
16. Cheng, C. Y., Shi, Y. C., & Lin, S. R., (2012). Use of real-time PCR to detect surimi adulteration in vegetarian foods. *J. Marine Sci. Tech., 20*(5), 570–574.

17. Coelho, L. M., Pessoa, D. R., Oliveira, K. M., De Sousa, P. A. R., Da Silva, L. A., & Coelho, N. M. M., (2016). Potential exposure and risk associated with metal contamination in foods. *Significance, Prevention and Control of Food Related Diseases*. doi: 10.5772/62683.
18. Collins, E. J. T., (1993). Food adulteration and food safety in Britain in the 19th and early 20th centuries. *Food Policy, 18*(2), 95–109. doi: 10.1016/0306-9192(93)90018-7.
19. Costa, J., Rodríguez, R., Garcia-Cela, E., Medina, A., Magan, N., Lima, N., & Santos, C., (2019). Overview of fungi and mycotoxin contamination in Capsicum pepper and in its derivatives. *Toxins, 11*(1), 27.
20. Curll, J., (2015). The significance of food fraud in Australia. *Australian Business Law Review, 43*(4), 270–302.
21. Dakeishi, M., Murata, K., & Grandjean, P., (2006). Long-term consequences of arsenic poisoning during infancy due to contaminated milk powder. *Environmental Health, 5*(1), 31.
22. De Marchi, B., Funtowicz, S., & Ravetz, J., (1996). Seveso: A paradoxical classic disaster. *The Long Road to Recovery: Community Responses to Industrial Disaster*, 86–120.
23. Dewey-Mattia, D., Manikonda, K., Hall, A. J., Wise, M. E., & Crowe, S. J., (2018). Surveillance for foodborne disease outbreaks-United States, 2009–2015. *MMWR Surveillance Summaries, 67*(10), 1.
24. Dhanya, K., & Sasikumar, S., (2010). Molecular marker-based adulteration detection in traded food and agricultural commodities of plant origin with special reference to spices. *Curr. Trends Biotech. Pharm., 4*, 454–489.
25. Dhanya, K., Syamkumar, S., Jaleel, K., & Sasikumar, B., (2008). Random amplified polymorphic DNA technique for the detection of plant-based adulterants in chili powder (*Capsicum annuum*). *J. Spices Aromatic Crops, 17*, 75–81.
26. Dyer, P., (2009). The 1900 arsenic poisoning epidemic. *Brewery History, 130*, 65–85.
27. EFSA Panel on Biological Hazards (BIOHAZ), (2011). Scientific opinion on risk-based control of biogenic amine formation in fermented foods. *EFSA Journal, 9*(10), 2393.
28. FAO, (2016). *Food Safety and Quality*. Available from: http://www.fao.org/food/food-safety-quality/a-z-index/laboratory/en/ (accessed on 9 April 2021).
29. Firestone, D., (1973). Etiology of chick edema disease. *Environmental Health Perspectives, 5*, 59–66.
30. FSS, (2006). *Food Safety and Standards*. Rules and Regulations, 2011.F. No. 6/FSSAI/ Dir (A)/Office Order/2011-12.
31. FSSAI, (2012). Manual of Methods of Analysis of Foods (Spices and Condiments). http://www.old.fssai.gov.in/Portals/0/Pdf/15Manuals/SPICES%20AND%20CONDIMENTS.pdf (accessed on 9 May 2021).
32. Gilson, S., & Glennerster, A., (2012). High fidelity immersive virtual reality. *Virtual Reality-Human Computer Interaction*. doi: 10.5772/50655.
33. Giraud, G., & Latour, H., (1952). Mass poisoning by bread at Pont-saint-esprit; systematic analysis of its component syndromes, either independent or associated. *Bulletin De L'academie Nationale De Medecine [Bulletin of the Academic Nation of Medicine], 136*(24–26), 465–476.

34. Gonzalez, M., Gallego, M., & Valcarcel, M., (2003). Determination of natural and synthetic colorants in pre-screened dairy samples using liquid chromatography-diode array detection. *Anal. Chem., 75*, 685–693.
35. Government of India, Legislative Department. (2018). The Prevention of Food Adulteration Act, 1954. https://legislative.gov.in/sites/default/files/a1954-37_1.pdf (accessed on 9 May 2021).
36. Gossner, C. M. E., Schlundt, J., Ben, E. P., Hird, S., Lo-Fo-Wong, D., Beltran, J. J. O., et al., (2009). The melamine incident: Implications for international food and feed safety. *Environmental Health Perspectives, 117*(12), 1803–1808.
37. Gudo, E. S., Cook, K., Kasper, A. M., Vergara, A., Salomão, C., Oliveira, F., & Viegas, S. O., (2018). Description of a mass poisoning in a rural district in Mozambique: The first documented bongkrekic acid poisoning in Africa. *Clinical Infectious Diseases, 66*(9), 1400–1406.
38. Harada, M., (1995). Minamata disease: Methylmercury poisoning in Japan caused by environmental pollution. *Critical Reviews in Toxicology, 25*(1), 1–24.
39. Haughey, S. A., Graham, S. F., Cancouet, E., & Elliott, C. T., (2012). The application of near-infrared reflectance spectroscopy (NIRS) to detect melamine adulteration of soya bean meal. *Food Chem., 136*(3, 4), 1557–1561.
40. Hussain, S. M. A., (2013). Comprehensive update on cancer scenario of Bangladesh. *South Asian Journal of Cancer, 2*(4), 279.
41. Inaba, T., Kobayashi, E., Suwazono, Y., Uetani, M., Oishi, M., Nakagawa, H., & Nogawa, K., (2005). Estimation of cumulative cadmium intake causing Itai-Itai disease. *Toxicology Letters, 159*(2), 192–201.
42. *IPC*, (1974). Indian Penal Code, 186Q, (Sec. 272 to 278) Sec. 269: Whoever unlawfully or negligently does any act India to safeguard public health by making provisions under the Penal Law to deal with contingencies that affects the health of the community.
43. ITF, (2010). *Challenges in Agri-Food Exports: Building the Quality Infrastructure*. International Trade Forum (issue 3/2010). Available from: http://www.tradeforum.org/Challenges-in-Agro-Food-Exports-Building-the-Quality-Infrastructure/ (accessed on 9 April 2021).
44. Kakar, F., Akbarian, Z., Leslie, T., Mustafa, M. L., Watson, J., Van, E. H. P., Omar, M. F., & Mofleh, J., (2010). An outbreak of hepatic veno-occlusive disease in Western Afghanistan associated with exposure to wheat flour contaminated with pyrrolizidine alkaloids. *Journal of Toxicology*, 1–7.
45. Karaali, G., & Khadjavi, L. S., (2019). Unnatural disasters: Two calculus projects for instructors teaching mathematics for social justice. *PRIMUS, 29*(3, 4), 312–327.
46. Karvetski, C. W., Olson, K. C., Gantz, D. T., & Cross, G. A., (2013). Structuring and analyzing competing hypotheses with Bayesian networks for intelligence analysis. *EURO Journal on Decision Processes, 1*(3, 4), 205–231.
47. Khan, M. A., Asghar, M. A., Iqbal, J., Ahmed, A., & Shamsuddin, Z. A., (2014). Aflatoxins contamination and prevention in red chilies (*Capsicum annuum* L.) in Pakistan. *Food Additives and Contaminants: Part B, 7*(1), 1–6.
48. Kirk, M. D., Angulo, F. J., Havelaar, A. H., & Black, R. E., (2017). Diarrheal disease in children due to contaminated food. *Bulletin of the World Health Organization, 95*(3), 233.

49. León-Camacho, M., Viera-Alcaide, I., & Ruiz-Méndez, M. V., (2003). Elimination of polycyclic aromatic hydrocarbons by bleaching of olive pomace oil. *European Journal of Lipid Science and Technology, 105*(1), 9–16.
50. Liu, C. Y., & Hua, J., (1939). Dead pigs scandal questions Chinas public health policy. *Lancet 2013,* 381.
51. Lockley, A. K., & Bardsley, R. G., (2000). DNA-based methods for food authentication. *Trends Food Sci. Tech., 11*, 67–77.
52. Lowe, K., & McLaughlin, E., (2015). 'Caution! The bread is poisoned': The Hong Kong mass poisoning of January 01857. *The Journal of Imperial and Commonwealth History, 43*(2), 189–209.
53. Lum, M. R., & Hirsch, A. M., (2006). Molecular methods for the authentication of botanicals and detection of potential contaminants and adulterants. *Acta Horticulture, 720*, 59–72.
54. Malisch, R., (2000). Increase of the PCDD/F-contamination of milk, butter, and meat samples by use of contaminated citrus pulp. *Chemosphere, 40*(9–11), 1041–1053.
55. Mangal, M., Bansal, S., & Sharma, M., (2014). Macro and micromorphological characterization of different *Aspergillus* isolates. *Legume Res., 37*(4), 372–378.
56. Manning, L., & Soon, J. M., (2014). Developing systems to control food adulteration. *Food Policy, 49*, 23–32.
57. McClements, D. J., (2003). *Analysis of Food Products, Food Science 581.* Department of Food Physico-chemistry, University of Massachusetts Amherst, Chenoweth Laboratory; Holdsworth Way, Amherst, MA, extracted/reproduced with permission from Prof. McClements, D.J. Available from: http://people.umass.edu/~mcclemen/581Sampling.html (accessed on 9 April 2021).
58. McDowell, I., Taylor, S., & Gay, C., (1995). The phenolic pigment composition of black tea liquors; Part I: Predicting quality. *J. Agric. Food Chem., 69*, 467–474.
59. Montet, D., & Dey, G., (2018). History of Food Traceability. CRC Press. file:///C:/Users/test/Downloads/crcpress-chapter1.pdf (accessed on 9 May 2021).
60. Mouly, P., Gaydou, E. M., & Auffray, A., (1998). Simultaneous separation of flavone glycosides and polymethoxylated flavones in citrus juices using liquid chromatography. *J. Chromatogr. A., 800*, 171–179.
61. Ndiaye, C., (2005). National food safety systems in Africa-a situation analysis. In: *FAO/WHO Regional Conference on Food Safety for Africa* (pp. 47–57).
62. Nogueira, J. M. F., & Nascimento, A. M. D., (1999). Analytical characterization of Madeira wine. *J. Agric. Food Chem., 47*, 566–575.
63. Ohe, K. V. D., (1991). Scanning electron microscopic studies of pollen from apple varieties. 6th pollination symposium. *Acta Hortic.,* 405–409.
64. Oliveira, R. C. S., Oliveira, L. S., Franca, A. S., & Augusti, R., (2009). Evaluation of the potential of SPME-GC-MS and chemometrics to detect adulteration of ground roasted coffee with roasted barley. *J. Food Comp. Anal., 22*(3), 257–261.
65. Posadzki, P., Watson, L., & Ernst, E., (2013). Contamination and adulteration of herbal medicinal products (HMPs): An overview of systematic reviews. *European Journal of Clinical Pharmacology, 69*(3), 295–307.
66. Qudoos, A., Bayram, I., Iqbal, A., & Shah, S., & Çetingul, I., (2019). Determination of adulteration in animals feed by using ft-nir technology ft-nir detection of Adulteration in Feed Using Technology.

67. Rasul, C. H., (2013). Alarming situation of food adulteration. *Bangladesh Medical Journal Khulna, 46*(1, 2), 1–2.
68. Rowlinson, P. J., (1982). Food adulteration its control in 19th century Britain. *Interdisciplinary Science Reviews, 7*(1), 63–72.
69. Rumbeiha, W., & Morrison, J., (2011). A review of class I and class II pet food recalls involving chemical contaminants from 1996 to 2008. *Journal of Medical Toxicology, 7*(1), 60–66.
70. Samsahl, K., & Wester, P. O., (1977). Metallic contamination of food during preparation and storage: Development of methods and some preliminary results. *Science of The Total Environment, 8*(2), 165–177. doi: 10.1016/0048-9697(77)90075-4.
71. Schaefer, K. A., & Scheitrum, D., (2020). Sewing terror: Price dynamics of the strawberry needle crisis. *Australian Journal of Agricultural and Resource Economics, 64*(2), 229–243.
72. Schumm, L., (2014). *Food Fraud: A Brief History of the Adulteration of Food.* http://www.realfoodsgrow.com/food-fraud-a-brief-history-of-the-adulteration-of-food.html (accessed on 9 May 2021)
73. Sheorey, R. R., & Tiwari, A., (2011). Rapid amplified polymorphic DNA (RAPD) for identification of herbal materials and medicines. *J. Scientific Ind. Res., 70*, 319–326.
74. Silva, B. M., Andrade, P. B., Mendes, G. C., Valentao, P., Seabra, R. M., & Ferreira, M. A., (2000). Analysis of phenolic compounds in the evaluation of commercial quince jam authenticity. *J. Agric. Food Chem., 48*, 2853–2857.
75. Singh, S. P., Kaur, S., & Singh, D., (2017). Food toxicology-past, present, and the future (the Indian perspective). In: *Food Safety in the 21st Century* (pp. 91–110). Academic Press.
76. Snow, J., (1857). On the adulteration of bread as a cause of rickets. *The Lancet, 70* (1766), 4, 5.
77. Soares, S., Amaral, J. S., Mafra, I., & Oliveira, M. B., (2010). Quantitative detection of poultry meat adulteration with pork by a duplex PCR assay. *Meat Sci., 85*(3), 531–536.
78. William, H. H., & Judith, S. L., (1975). *The New Columbia Encyclopedia*, 973.
79. Thomson, B., Poms, R., & Rose, M., (2012). Incidents and impacts of unwanted chemicals in food and feeds. *Quality Assurance and Safety of Crops and Foods, 4*(2), 77–92.
80. Tlustos, C., (2009). The dioxin crisis in Ireland 2008-challenges in risk management and risk communication. *Organohalogen Compd., 71*, 1152–1154.
81. Tran, T. H., (2018). Critical factors and enablers of food quality and safety compliance risk management in the Vietnamese seafood supply chain. *International Journal of Managing Value and Supply Chains (IJMVSC), 9.*
82. Verma, S. K., Dev, G., Tyagi, A. K., Goomber, S., & Jain, G. V., (2001). Argemone Mexicana poisoning: Autopsy findings of two cases. *Forensic Science International, 115*(1, 2), 135–141.
83. WHO, (2015). *World Health Organization.* https://www.euro.who.int/en/health-topics/disease-prevention/food-safety/news/news/2015/12/more-than-23-million-people-in-the-who-european-region-fall-ill-from-unsafe-food-every-year (accessed on 9 April 2021).
84. WHO, (2020). *Food Safety.* World Health Organization. https://www.who.int/news-room/fact-sheets/detail/food-safety (accessed on 9 April 2021).

85. Wilhelmsen, E. C., (2004). Food adulteration. *Food Sci. and Tech., 138*, 2031–2056.
86. Xue, H., Hu, W., Son, H., Han, Y., & Yang, Z., (2010). Indirect ELISA for detection and quantification of bovine milk in goat milk. *J. Food Sci. Tech., 31*(24), 370–373.
87. Yang, J., Hauser, R., & Goldman, R. H., (2013). Taiwan food scandal: The illegal use of phthalates as a clouding agent and their contribution to maternal exposure. *Food and Chemical Toxicology, 58*, 362–368.
88. Zammatteo, N., Lockman, L., Brasseur, F., De Plaen, E., Lurquin, C., Lobert, P. E., Hamels, S., et al., (2002). DNA microarray to monitor the expression of MAGE-A genes. *Clin. Chem., 48*, 25–34.
89. Zhang, W., & Xue, J., (2016). Economically motivated food fraud and adulteration in China: An analysis based on 1553 media reports. *Food Control, 67*, 192–198. doi: 10.1016/j.foodcont.2016.03.004.

# CHAPTER 9

# Biofortification: A Way Forward

ACHARYA BALKRISHNA, BHOOMIKA SHARMA, RASHMI MITTAL, and SHERRY BHALLA

## ABSTRACT

A comprehensive review of the world's nutritional status highlighted that hidden hunger, popularly known as the triple burden of malnutrition, has greatly affected the life of billions globally. Fortified micronutrients together with supplementation programs have gained popularity among the national and international health agencies to address the issue of micronutrient deficiency amongst the lower and middle-income countries. Dietary supplementation, diversification, paradigm shift towards fortified and bio-fortified crop plants are the major adapted approaches for mitigating hidden hunger. Malnourished communities have started choosing biofortified crops to combat the issue. If consumed regularly, these crops can restore micronutrients in human beings. This strategy may lead to the overall elimination of hidden hunger, but for the triumph of the goal, various strategies at the global level need to be initiated. Three major methodologies named transgenic, crop breeding, and fertilization were observed to be the potent techniques to implement the same. This chapter aimed to provide a brief overview of malnutrition status worldwide and the strategies already adopted or can be adapted to combat the current situation.

## 9.1 INTRODUCTION

At present, the food system is experiencing a great burden globally as the world population is awaited to rise from 7.6 to 8.6 billion by 2030 and is expected to jump to 9.8 billion by 2050. Sustainable food availability has raised global concerns [5]. Even in the 21st century, there still exists inequalities in food availability and affordability. WHO report of 2017 revealed that 462 million adults are still underweight, constituting 45% death rate due to under nutrition issues in children under the age group of 5 years [7]. Micronutrient deficiencies are often referred as 'hidden hunger.' Inadequate intake of micronutrients has led to the:

- Poor overall growth;
- Cognitive impairments;
- Morbidity and mortality.

Eliminating hidden hunger in its all-existing forms have imposed a serious challenge over the world economy. Eradication of hidden hunger have been considered as a primary goal for United Nation's Sustainable Development for 2030. The highest impact of hidden hunger has been observed in developing countries, more prominently in Asia and Sub-Saharan regions [3]. Micronutrient deficiency furthermore negatively impacted National prosperity by limiting:

- Educational attainment;
- Reduces workability.

Ultimately leading to lower-income and ending up in poverty.

International Food Policy Research Institute revealed that on an average 2 billion people across the globe are suffering from micronutrient malnutrition which clearly indicated that a major section of the society is deprived of adequate supply of minerals and vitamins. The impact of micronutrient deficiency can be somewhat reversed by providing them with a micronutrient enriched diet, but few of them are irreversible. Among a long list of micronutrients, deficiency of vitamin A, iron, and iodine are the most prominent ones and can lead to the onset of multiple diseases in the human body [16]. Children below the age group of 5 years and pregnant women are most susceptible to the effect. Adequate supply of micronutrients in a bioavailable form is also highly essential to facilitate its proper absorption. An appropriate diet which includes vegetables, fruits, and animal products is essential to meet micronutrient and energy demands for proper growth and development of the human body. Hidden hunger does not only lead to disease burden but also causes deterioration of economic productivity and social welfare worldwide.

In developing countries, mineral deficiency leads to a reduction in work productivity and ultimately affects Gross National Product. Statistical analysis of micronutrient scenario in United States revealed that 11–38% of children population is currently suffering from anemia whereas in Canada and America, 10% of their population is currently under the threat of Zn deficiency. Zn and Fe deficiency has affected more than 3 billion people across the globe. According to WHO Report, 25% of the world's population is suffering from anemia [9, 17]. 17.3% of the world population is expected to deal with Zn deficiency. 5% of South Asia's population is dependent upon rice-pulse diet which has caused serious health impacts in them. If we consider the status of micronutrient deficiency in developing countries, Zn occupies $5^{th}$ rank in the list of 11 deficient micronutrients.

In developing countries, a large proportion is dependent upon cereal and animal-based food trend due to their affordability issues. Wheat, rice, and maize are considered to be the prominent staple crops. Wheat, Fe, and Zn were observed to be present in an aleuronic layer which were generally lost during milling and processing. Due to the improper processing, highly essential micronutrients get denatured [6].

### *9.1.1 APPROACHES TO IMPROVE THE NUTRITIONAL PROFILE OF CURRENT UPTAKE*

In the scientific community, several methods and techniques are available to mitigate the micronutrient malnutrition by accelerating the micronutrient concentration and their bioavailability in edible regions of staple crop. The following different methods are available for the same:

- Dietary diversification;
- Food supplementation;
- Food fortification;
- Biofortification.

#### *9.1.1.1 DIETARY DIVERSIFICATION*

A food strategy named as dietary diversification includes consumption of different varieties of food products with special emphasis on plant-based food products, for example, fruits, vegetables, and cereal grains to meet the micronutrient requirement essential for proper growth of the human body [29]. The present technique involves different basic steps at household level such as food production preparation by:

i. Soaking;
ii. Fermentation;
iii. Germination.

To enhance the micronutrient concentration and its bioavailability in regular human diet. Vegetables and fruits were observed to be rich in promoter substances, for example, β-carotene and ascorbate, which can remarkably increase its nutrient uptake and therefore must be consumed with regular dietary intake. Anti-nutrient compounds such as polyphenols and phytic acids should be ignored because they can inhibit mineral absorption, thereby decreasing its bioavailability. Some specialized minerals and nutrients such as ascorbic acid which can potentially promotes the Fe absorption must be consumed. Furthermore, the process of fermentation and germination can enhance the bioavailability of iron as these techniques can enhance the functioning of phytase enzyme, which could hydrolyzes the phytic acid in legumes and cereal grains. Intake of folate-rich foods or sprouted foods can increase the folate content in the human body, which is highly essential for their adequate growth and development [34].

#### *9.1.1.2 FOOD SUPPLEMENTS*

- **In the form of powder, pills, and solutions:** Micro-nutrients are consumed as food supplements to meet the nutrient deficiency. Food supplementation can be adopted as a temporary program to support the patient to recover from the particular deficiency; it need not be adopted as a permanent measure. In few countries, programs initiated for the supplementation of vitamin A and zinc have gained success in the recent past. Especially in pregnant women and children; iron, zinc, and folic acid supplementation has gained significant popularity [24]. Being cost-effective in nature, it is approachable for all sections of society, especially for lower income-based countries. In spite of attaining positive results from this approach, it cannot be considered as a permanent solution of the above-mentioned problems. Zinc, iron, and folic acid supplementation exhibited different physiological response and absorption rate in comparison to its original form, which indicated that food supplementation cannot eradicate the root cause of hidden hunger [14]. Furthermore, food supplementation process needs access to medical centers, awareness programs, and proper monitoring of demand and supply chain with proper storage facility. Process of food supplementation can be easily initiated in developed countries but will have to face several hurdles and challenges in developing countries [4, 11].

#### 9.1.1.3 FOOD FORTIFICATION

Addition of vital micronutrients to enhance the nutritional content of food uptake is referred as food fortification. World Food Program (WFD) has initiated several food assistance assistance programs to provide the nutritional benefits with minimum risks factor using pre-cooked and processed cereals, pulses, and likewise other products fortified with micronutrients. Iron, ferric pyrophosphate, ferrous sulfate, electrolytic iron powder is commonly preferred for food fortification [9]. To increase the folate content in diet, biofortification can be carried out with folic acid. Salt iodization is one such example of it where fortification was carried out with iodine to reduce the goiter condition.

#### 9.1.1.4 BIOFORTIFICATION

Biofortification is commonly referred to as a process of enriching the nutritional status of foods obtained from plants through (Figure 9.1):

i. Agronomic interventions;
ii. Genetic engineering;
iii. Plant breeding.

Initially, the technology was initiated as a food-dependent strategy to encounter the iron, zinc, and vitamin A deficiency which has remained to be a major issue of concern in lower-income countries [36]. Several strategies have been implemented to overcome the micro-nutrient deficiency:

i. Proper supply of micro-nutrient supplements in the form of Pharmacological preparations;
ii. Food fortification;
iii. Dietary diversification.

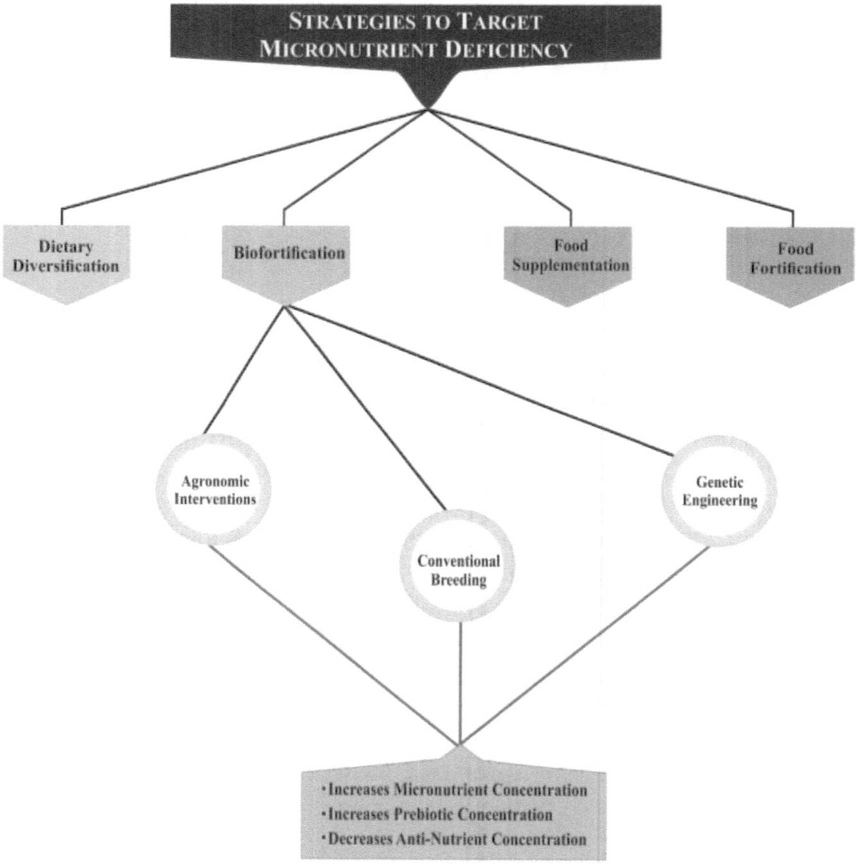


**FIGURE 9.1** Strategic plan to deal with micro-nutrient deficiency.

Biofortification can increase the micro-nutrient level and enhance their bioavailability. The technique is primarily developed to target rural people who heavily depends on stable crops produced in the local farm. Because of the financial crisis and lack of market access to commercially available food supplements, a large proportion of their population is suffering from hidden hunger. Biofortification strategy further possess the vast potential to ensure sustainability as the planting material used during the process, is once obtained can be further preserved, saved, recycled, or can be distributed to other farmers [8]. After attaining the initial development and dissemination stage, the cost needed for their maintenance can be significantly reduced. Moreover, possible micro-nutrients which can be supplied through biofortification are significantly higher in comparison to other approaches currently getting used worldwide to deal with hidden hunger [32]. In the current scenario, biofortification can be considered as a potential approach and a significant complementary approach to overcome the micronutrient deficiency (Figure 9.2).

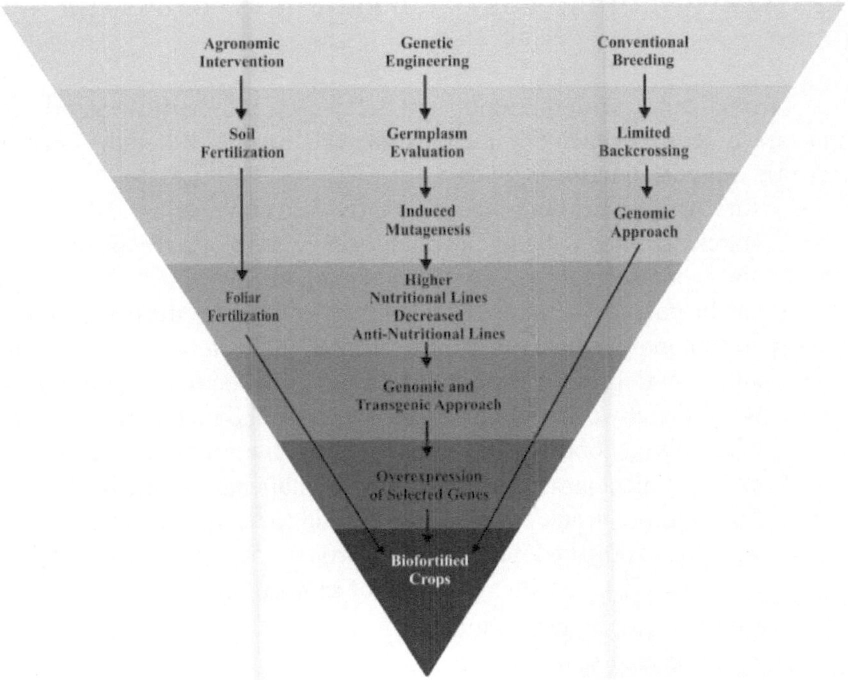


**FIGURE 9.2** An outline of different approaches to deal with bio-fortification.

An International Organization named Harvest Plus, which comprises of a group of distinguished researchers is predominantly involved in biofortification strategy. The named organization is managed by the Consultative Group on International Agricultural Research (CGIAR). This consortium is dealing with all the stages of strategy development starting from agricultural and biotechnological research to food science and even induces economics and policy sciences. Another consortium named as Agro Salud is also running with the similar motive and is currently coordinating with the research related to biofortification with the interventions currently on the run in Latin America. In India, Brazil, and China, different biofortification research programs have also been established to collectively deal with the current situation [26]. In Nicaragua, Panama, and Zambia, biofortification have been included in the list of National Nutrient strategies adopted by the government for the welfare of their community.

## 9.2 A LAYOUT OF BIO-FORTIFIED PRODUCTS ACROSS THE GLOBE

Apart from dealing with micronutrient deficiency, biofortification process may serve several other applications as well in public health nutrition concern. Several functional foods

— such as oilseeds with increased fatty acid content; starch samples with an elevated level of probiotic fructans; and increased calcium content in vegetables

— can be potentially marketed to prevent several chronic diseases such as osteoporosis, cancer, cardiovascular, etc. Many micro-nutrients also function as co-factors for proper functionality of the human body, thereby controlling different cellular and metabolic processes of the human body [15]. Unequal distribution of micro-nutrients in different plant parts has also raised the challenges of hidden hunger. For example, iron content was observed to be higher in rice leaves, whereas in polished rice grains, the iron amount was observed to be very low [18]. Nutritional targets of biofortification include:

- Elevated mineral content;
- Enhanced vitamin content;
- Enhanced amino acid level;
- Raised fatty acid content;
- Escalated antioxidant level.

Biofortification implementation in crop plants can fulfill the calorie requirement of the human body to meet the body's energy needs. Moreover, biofortification of crop plants that are easily affordable by poor people can profoundly improve their nutritional status. Several biofortified varieties of cereals, legumes, pulses, vegetables, oilseeds, fruits, fodder, etc., have been released by different countries to overcome the issue of micronutrient deficiency (Figure 9.3).

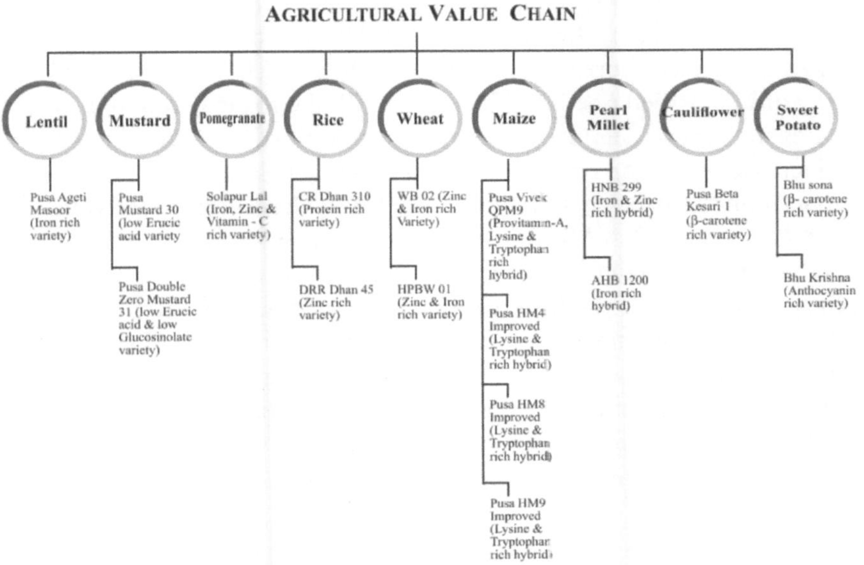


**FIGURE 9.3** Biofortified varieties of lentil, mustard, pomegranate, rice, wheat, maize, pearl millet, cauliflower, and sweet potato.

## *9.2.1 BIOFORTIFIED CEREALS*

Different varieties of biofortified cereals have been developed by distant corners of the world with the objective to deal with acute and chronic diseases caused due to micronutrient deficiency [10]. Biofortified rice, wheat, maize, barley, and sorghum have been initiated and, in this section, we will further discuss their role in improving nutrient deficiency.

### *9.2.1.1 BIOFORTIFIED RICE (ORYZA SATIVA)*

Rice, being the staple crop, is widely targeted to combat the challenge of hidden hunger across the world. Vitamin deficiency has constantly remained to be a major challenge for the underprivileged population due to the financial crisis. Discovery of Golden rice has remained to the major breakthrough in this direction enriched with higher concentration of Vitamin A and can potentially decrease the burden of disease by facilitating the expression of genes coding for *PSY* and carotene desaturase. Phytoene, a precursor of β-carotene has observed to be elevated by 23-fold via targeting the expression of gene encoding for Arabidopsis GTP-cyclohydrolase I (GTPCHI) and amino deoxy chorismate synthase (ADCS) [22]. It was believed that 100 g of biofortified rice is sufficient to fulfill the folate requirements.

Several reports suggested that genetic modifications in rice have led to an increase in its iron content by altering the genes encoding for nicotinamide aminotransferase, iron transporter OSIRTI, nicotinamide synthase 1 (OsNAS1) and 2 (OsNAS2) soybean ferritin along with bean ferritin. Iron-rich biofortified rice can further be synthesized by incorporating multiple genes, which play a significant role in iron nutrition. By eliminating the anti-nutrient compounds, for example, phytic acid, iron bioavailability can be significantly increased in rice. By inducing the overexpression of OsIRT1, zinc content can further be increased in genetically modified rice and by modifying the expression of HvNAS1, Hv NAAT-A, Hv-NAAT-B, IDS3 zinc content can significantly be enhanced in rice (Table 9.1).

**TABLE 9.1** Genes Altered in *Oryza sativa* to Increase Nutrient Content Through Biofortification

| **Altered Gene** | **Commodity** |
|---|---|
| *GmFAD3* | α-linolenic acid |
| Maize *C1*, Phenylalanine ammonia-lyase, R-S regulatory gene [*Myb*-type] and chalcone synthase (*CHS*) gene | Antioxidants |

### *9.2.1.2 BIOFORTIFIED WHEAT (TRITICUM AESTIVUM)*

Worldwide food is considered as to the major staple crop. Different researches were carried out to deal with deficiency of Vitamin A, iron, and other essential nutrients through biofortified wheat. Bacterial *PSY* and *carotene desaturase genes* were expressed in wheat to enhance the pro-vitamin A content. Iron content were further elevated by expressing ferritin gene obtained from soybean. Similarly, iron bioavailability was subsequently enhanced by increasing the phytase activity by targeting phytochrome gene

*Phy A* [19]. By using *Amaranthus albumin gene* [*ama 1*] expression; lysine, methionine, tyrosine, and cysteine content can be significantly increased in wheat. With the aim of increasing its antioxidant activity, maize regulatory gene named *C1*, *B-Peru* was expressed in it which were observed to be involved in anthocyanin production. To deal with the obesity issue, amylose starch content was increased by silencing the genes targeting *SBE* [*SBEIIa*].

### *9.2.1.3 BIOFORTIFIED MAIZE (ZEA MAYS)*

Prominently in developing countries, maize is considered to be another staple crop. By means of genetic engineering, maize can deal with the issue pertaining to the deficiency of vitamins, proteins, minerals, and anti-nutrient components. By expressing bacterial *crtB* and carotenogenic genes, maize endosperm can be potentially enriched with pro-vitamin A. Different researches are currently on run to supplement maize with Vitamin A and its analogs which are considered as potential antioxidant [28].

Bioavailability prospects of micro-nutrients can be further hindered by anti-nutrient entities. *Aspergillus phytase*, *Soybean ferritin*, *Aspergillus niger* phyA2 can increase the bioavailability of iron and also the silencing of ATP binding cassette transporter and drug resistance-related protein can also help in increasing its bioavailability. For example, *B VLAH 30101* variety of maize issued by Origin Agritech of China was biofortified for phytate degradation (Table 9.2).

**TABLE 9.2** Genes Altered in *Zea mays* to Increase Nutrient Content Through Biofortification

| Altered Gene | Commodity |
|---|---|
| Homogentisic acid geranyl-geranyl transferase [*HGGT*] | Tocotrienol and tocopherol |
| Dehydroascorbate reductase [*DHAR*] | Vitamin C (L-ascorbic acid) |
| *Sb401* gene | Lysine content |
| Anti-sense dsRNA targeting α-zeins | Lysine and Tryptophan |
| Cis-acting site for *Dzs10* | Amino acid |

### *9.2.1.4 BIOFORTIFIED SORGHUM (SORGHUM BICOLOR)*

Sorghum has undergone biofortification to enhance the pro-vitamin A (β-carotene) content by expressing *Homo188-A* (Table 9.3).

**TABLE 9.3** Genes Altered in *Sorghum bicolor* to Increase Nutrient Content Through Biofortification

| Altered Gene | Commodity |
|---|---|
| Lysine Protein *HT12* | Lysine |
| RNAi silencing of γ-kafirin-1, γ-kafirin-2 and γ-kafirin A1 | Grain digestibility |

## 9.2.2 BIOFORTIFIED PULSES AND LEGUMES

### 9.2.2.1 BIOFORTIFIED SOYBEAN (GLYCINE MAX)

Biofortified soybean was prepared with the aim to increase pro-vitamin A (β-carotene) and mono-unsaturated ω-9 fatty acid (oleic acid) and seed protein. *PSY* Gene was expressed in soybean to fulfill the micro-nutrient deficiency [4] (Table 9.4).

Many private companies have introduced different varieties of cultivars with increased oleic, linoleic acid, and STA. In Canada, Tapas, USA, Australia, and New Zealand G94-1, G94-19, G168 biofortified varieties of soybean rich in oleic acid are widely popular. Another variety named Vistive Gold™ (MON 87705) were introduced by Monsanto in Columbia, Canada, Indonesia, Japan, New Zealand, Philippines, Taiwan, USA, Singapore, and in many other countries as well. MON 87769 was also released by Monsanto in Australia, European Union (EU), Indonesia, South Korea, Taiwan, Vietnam, etc.

**TABLE 9.4** Genes Altered in *Glycine max* to Increase Nutrient Content Through Biofortification

| Altered Gene | Commodity |
|---|---|
| *PSY* gene [crt B, crt w, bkt 1] | Pro-vitamin A (canthaxanthin) |
| 2-methyl-6-phytyl benzoquinomethyl transferase gene [*At-VTE3*, *At-VTE-4*] | Vitamin E, delta-tocopherol |
| O-acetyl serine self-hydrolase | Cysteine content in soybean seeds |
| Maize Zain protein | Methionine and Cysteine |
| Cystathione delta-synthase | Methionine content in soybean |
| $Delta^6$ desaturase gene | γ-linolenic acid, STA (ω-3 fatty acids) |
| $Delta^{12}$ oleate desaturase [*GmFAD2-1b*] | Linolenic and palmitic acid |

#### 9.2.2.2 BIOFORTIFIED COMMON BEANS (PHASEOLUS VULGARIS)

Common bean is one of the important legumes and is highly important for human use. Generally, beans contain a high concentration of amino acids such as lysine, leucine, isoleucine, valine, and threonine. But its nutritional value is highly limited due to the small proportion of methionine and cysteine. Methionine content has been greatly elevated in common bean by targeting the expression of methionine enriched storage albumin from Brazil nut [23].

#### 9.2.2.3 BIOFORTIFIED LUPINES (LUPINUS LATIFOLIUS)

Lupin is considered another vital legume. Lupin seed protein is deficient in a sulfur-containing amino acid named methionine and cysteine. By altering the expression of the sunflower seed albumin gene, methionine content can be significantly increased.

### 9.2.3 BIOFORTIFIED VEGETABLES

#### 9.2.3.1 BIOFORTIFIED POTATO (SOLANUM TUBEROSUM)

Potato is the major source of calories and is highly used globally. Transgenic potatoes expressing *ama*1 gene encoding for Amaranth albumin has led to the elevation of protein content in its tubers and a remarkable increment in the quantity of amino acids present in it, specially methionine (Tables 9.5 and 9.6).

#### 9.2.3.2 BIOFORTIFIED CASSAVA (MANIHOTES CULENTA)

Millions of poor people worldwide are dependent on cassava because it is considered as staple food crop in major sections globally. Cassava is observed to be highly stress-tolerant crop but simultaneously lacks the presence of essential micronutrients such as Pro-Vitamin A, iron, and zinc. Biofortified cassava is enriched with above-mentioned micronutrients and can significantly eradicate the malnutrition from the community. Transgenic cassava was introduced as a part of Bio-Cassava Plus project.

With elevated expression of β-carotene in root by altering the expression of *npt II*, *crt B* and *DXS*. Carotenoid enriched Cassava varieties were also produced by facilitating the overexpression of *PSY* gene. Several field trials are also undergoing under the Bio-cassava Plus Program in African countries to provide iron, zinc, and β-carotene-rich Cassava [1].

**TABLE 9.5** Genes Altered in *Solanum tuberosum* to Increase Nutrient Content Through Biofortification

| Altered Gene | Commodity |
|---|---|
| PSY, Lycopene β-cyclase, phytoene desaturase | *PSY* gene |
| β-carotene hydroxylase gene | B-carotene |
| Lycopene β-cyclase [*StLCYb*] | Zeaxanthin |
| *Or* gene | Carotenoids |
| *GalUR* | Vitamin C (ascorbic acid) |
| Cystathionine γ-synthase (*CgSD90*) | Methionine |
| *StMGL1* | Leucine |
| Perilla CPrLeg polypeptide and Cystathione γ-synthase (*CgS*) gene | Methionine |
| Amaranth albumin (*ama* 1) | Protein content |
| Cydodextrin glycosyl transferase gene | Carbohydrate |
| CHS, chalcone isomerase (CH1), dihydroflavonol reductase | Phenolic acid, anthocyanin |

**TABLE 9.6** Different Varieties of *Solanum tuberosum* with Their Specific Enrichment of Micronutrients

| Variety Name | Company | Country | Specialty |
|---|---|---|---|
| Starch potato (AM 04-1020) | BASF | USA | Reduced amylose and increased amylopectin |
| Amflora™ (EM 92-527-1) | BASF | European Union | Reduced amylose and increased amylopectin |
| Transgenic Potato | J-R Simplot | Canada, USA | Reduced sugar content through starch degradation |

### *9.2.3.3 BIOFORTIFIED CARROT (DAUCUSCAROTA SUBSP. SATIVUS)*

Although carrot contains a high concentration of vitamins, minerals, and β-carotene but lacks the calcium content to a great extent. By expressing

Arabidopsis $H^+/Ca^{2+}$ transporter, the bioavailability of calcium can be significantly increased.

## 9.2.4 BIOFORTIFIED OILSEEDS

### 9.2.4.1 BIOFORTIFIED LINSEED (LINUM USITATISSIMUM)

Linseed edible oil is highly in demand because of its potency to be used as a nutritional supplement. Biofortified linseeds contain higher concentration of potential and stable antioxidants and were originally generated by inhibiting the expression of *CHS* gene. VLCPUFA, very long-chain unsaturated fatty acids were essential fatty acids, but their supply is very limited due to the decline in marine resources, for example, fish oil. VLCPUDFA deficiency can be overcome by its biosynthesis into oilseed crops. By seed-specific expression of cDNAs, $delta^6$ desaturated CIS fatty acids and $C_{20}$ polyunsaturated fatty acids can be overexpressed in linseed. By introducing *PS1* gene [*crtB*]. Carotenoids can be increased in flaxseed. University of Saskatchewan, Colombia, USA has released linseed rich in amino acids and are broadly popular with the name CDCTriffid Flax (FP967).

### 9.2.4.2 BIOFORTIFIED CANOLA (BRASSICA NAPUS)

Worldwide, *Cannolis* is considered an important oilseed crop and is highly popular amongst millions. With the purpose of increasing α and β-carotenes, PSY gene was altered (Table 9.7).

**TABLE 9.7** Genes Altered in *Brassica napus* to Increase Nutrient Content Through Biofortification

| Altered Gene | Commodity |
|---|---|
| PSY, phytoene desaturase, lycopene cyclase gene | β-Carotenoid |
| Idi, crt E, crt B, crt I, crt Y, crt W, crt Z | β-Carotenoid |
| RNAi silencing of lycopene E-cyclase [E-CYC] | Xanthophyll and lutein |
| Asartokinase (AK) and dihydro-dipicolonic acid synthase (DHDPS) genes | Lysine, caprylate |
| Thioesterase gene [ch FatB2] | Caprate |
| $Delta^{12}$ or $Delta^6$ desaturase gene | GLA |

#### *9.2.4.3 BIOFORTIFIED MUSTARD (BRASSICA JUNCEA)*

To attain socio-economic benefits, mustard is cultivated throughout the world. To improve the unsaturated fatty acid contents, mustard seeds were genetically engineered. Delta$^6$ FAD3 enzyme expression has been altered, ultimately leading to γ-linoleic acid production in biofortified mustards [20].

### ***9.2.5 TRANSGENIC FRUITS***

#### *9.2.5.1 BIOFORTIFIED APPLE (MALUS DOMESTICA)*

Apple is a great source of antioxidants. *SHlbene* synthase gene from grapevine (*Vitis vinifera* L.) was introduced into the apple to increase the production of resveratrol in transgenic apple, thereby enhancing its antioxidant potential.

#### *9.2.5.2 BIOFORTIFIED BANANA (MUSA ACUMINATA)*

Banana stands at 4$^{th}$ position in the list of food crops in developing countries. Banana was also subjected for bio-fortification to increase β-carotene content. PSY gene (*PSY*2a) of was introduced into banana to transform it into super banana with high concentration of β-carotene.

### ***9.2.6 BIOFORTIFIED FODDER***

#### *9.2.6.1 BIOFORTIFIED ALFALFA*

In several countries, alfalfa is considered an essential legume crop. To improve iso-flavonoid, amino acid content and its digestibility, many attempts have been made so far. By incorporating *IFS* gene transgenic alfalfa was prepared which contains increased iso-flavonoid content. However, alfalfa has a very low level of sulfur-containing amino acids such as methionine and cysteine. Recently, cystathionine γ-synthase [*AtCgS*] gene was expressed to enrich its methionine content. Furthermore, digestibility issues were dealt with by altering the expression of cytochrome P450 enzymes. Transgenic alfalfa was created with low lignin content. It has been further engineered to enhance its phytase activity and therefore ensuring its usage in livestock, poultry, and other animal feeds [27].

## 9.3 METHODS AND STRATEGIES ADAPTED BEHIND THEIR PRODUCTION

Globally different strategies have been adopted till now to produce biofortified fruits, vegetables, legumes, cereals, oilseeds, and many more to deal with micro-nutrient deficiency. In this section, we will briefly discuss about the different available techniques for preparing biofortified products and majorly it includes:

- Agronomic approach;
- Conventional breeding; and
- Genetic engineering.

### *9.3.1 BIOFORTIFICATION THROUGH AGRONOMIC APPROACH*

Production of biofortified products through agronomic methods involves the physical application of additional nutrients temporarily to uplift the nutritional content of crops which are often consumed as staple crops in one or other regions of the globe. Implication of agronomic methods is not only restricted to staple crop but also is applicable for several others as well. Organic nutrients do not face many such issues in terms of bioavailability as they can be easily absorbed by the human body and are often less excreted in comparison to the inorganic form of minerals. Toxicity symptoms are also not intensive in the case of organic nutrients. Agronomic technique relies on the addition of mineral fertilizers such as nitrogen, phosphorous, and potassium (NPK) to increase the crop yield. The use of NPK fertilizers have resolved the problem of starvation to a great extent. Low input agriculture system is not capable enough to meet the need of the growing population. Micronutrients such as zinc, copper, manganese, Mo, Co, Ni, Se, Fe, etc., are available in different concentrations in edible part of plants and is often absorbed from the soil. Soil micronutrients status can be improved by applying fertilizers to them and further can reduce their micro-nutrient deficiency in humans. Micro-nutrient present in the soil is not readily available to the crops grown and are not rapidly translocate to their edible parts. By agronomic technique these micro-nutrients are made immediately available by this practice. The present technique is highly simple and cost-effective but require special attention with regard to the source of nutrition, their application protocol, and its impact on the environment. In developed countries, it is very easy to implement the addition of nutrient fertilizers under the agronomic approach. Se, Zn fertilization in Finland, Turkey, and China, respectively, are the few examples of its success [35].

Along with fertilizers, soil micro-organisms for increasing plant growth are also essential and can significantly increase the nutrient mobility from soil to other edible parts and can therefore increase its nutritional status. *Rhizobium*, *Pseudomonas*, *Bacillus*, *Azotobacter* can be significantly preferred to enhance the bioavailability of minerals. For example, in nitrogen-limited conditions; $N_2$ fixing bacteria can play a pivotal role in enhancing crop productivity. Several crops are linked with mycorrhizal fungi, which can efficiently release side phores, organic acids, and enzymes which possess the potential to degrade the organic compounds and can further enhance the concentration of minerals in the edible plant portion. Via agronomical biofortification process, different crops have been altered to increase their nutritional status, thereby eliminating hidden hunger globally.

#### *9.3.1.1 BIOFORTIFICATION OF RICE THROUGH AGRONOMIC PRACTICES*

- To improve the iron and zinc content in rice grain, agronomical approach may play a vital role;
- Through foliar spray of iron its concentration can be significantly increased;
- To enhance bioavailability and concentration of zinc, foliar spray of zinc is often preferred, similar is the fact with selenium spray;
- Iron, zinc, and selenium addition can subsequently increase its antioxidant potential.

#### *9.3.1.2 BIOFORTIFICATION OF WHEAT THROUGH AGRONOMIC APPROACH*

- Foliar spray of zinc to wheat grains has improved its concentration and bioavailability and also reduces phytic acid popularly known as anti-nutrient factor;
- In turkey, Zn containing NPK fertilizer amount increased from 0 to 100,000 t per annum;

- In Finland, Se biofortification in wheat has remained a successful approach [24].

#### *9.3.1.3 BIOFORTIFICATION OF MAIZE THROUGH AGRONOMIC APPROACH*

- Rhizobacteria, a well-known plant growth-promoting microbe have been introduced in maize;
- Zinc and selenium concentration have increased to a great extent by applying fertilizers.

#### *9.3.1.4 BIOFORTIFICATION OF SORGHUM THROUGH AGRONOMIC APPROACH*

- Both organic and inorganic fertilizers have been applied as an additive to increase micro-nutrient concentration in sorghum;
- Arbuscular mycorrhizal fungi (AMF) and plant growth-promoting bacteria have been added into sorghum crop to increase its nutrient uptake.

#### *9.3.1.5 BIOFORTIFICATION OF CHICKPEA THROUGH AGRONOMIC APPROACH*

By using plant growth enhancer action-bacteria; zinc, calcium, iron, copper, manganese, and magnesium deficiency can be replenished. By altering *AMF* gene, iron, and zinc concentration was optimized in Chickpea to meet the complications of hidden hunger. Through the foliar spray of specific micro-nutrients, zinc, and Se could be fortified in Chickpea.

#### *9.3.1.6 BIOFORTIFICATION OF COMMON BEAN THROUGH AGRONOMIC APPROACH*

Dry grains of common beans are edible in nature. Through foliar spray of zinc fertilizers, zinc biofortification can be carried out in common beans. The application of organic acid and inorganic fertilizers can further increase the nutrient uptake in common beans [21].

Other biofortified products which can possibly be prepared by biofortification:

- Canola;
- Mustard;
- Potato;
- Sweet potato;
- Carrot;
- Lettuce;
- Tomato.

### 9.3.2 BIOFORTIFICATION THROUGH CONVENTIONAL BREEDING

Conventional breeding is considered the most promising approach of biofortification. It is a cost-effective, sustainable approach to overcome nutrient deficiency. To implement conventional breeding, genetic variation seeds to be made in the trait of interest. To improve the vitamins and mineral content, breeding programs can opt for such an efficient approach. Under the conventional breeding method, parent lines enriched with nutrients are crossed with the recipient line containing the desirable traits. Through this way, different food, vegetable, cereals, legumes, oilseeds can be prepared to contain desired nutrients and the required agronomic traits. Breeding strategies are dependent on the genetic variations currently existing in the gene pool. Whereas in certain cases, distant relative varieties can also be crossed, ultimately the desired trait will transfer into commercial cultivars. Through this approach, new beneficial traits can also be incorporated into commercial varieties via mutagenesis.

Conventional breeding is observed to be the most promising approach to improve micronutrient content. Different organizations at both the national and international levels have initiated several programs to increase their nutritional content through the conventional breeding program. In the EU, a health grain project was initiated with the collaboration of 44 partners belonging to 15 countries which aimed to invest around 10 million dollars to produce healthy and safe food ingredients to surpass the micro-nutrient deficiency in human beings. Later on, it developed into a health grain forum which involves participation from several academicians and industrialists. Above 100 publications have highlighted the introduction of bioactive components in cereals and others through the genetic variation and heritability and ultimately reducing the chances of occurrence of several acute and chronic diseases in humans [31].

Harvest Plus Program was launched by International Center for Tropical Agriculture, CGIAR, and International Food Policy Research Institute. The aim of the program is to boost the concentration of zinc, iron, and vitamin A primarily in stable crops such as rice, wheat, beans, sweet potato, cassava, etc. The prime objective of Harvest Plus is to enhance the bioavailability of essential vitamins and minerals to deal with the issue of micronutrient deficiency, especially in developing countries. For example, Bio Cassava Plus Programs aimed to increase the nutritional status of Cassava crop. Through the crop breeding technique, several crops have been targeted for biofortification so far.

### *9.3.2.1 BIOFORTIFICATION OF CEREALS THROUGH CONVENTIONAL BREEDING*

#### ***9.3.2.1.1 Rice Breeding***

Rice is the widely chosen staple crop around the world. Micro-nutrient enhancement in rice can resolve the issue of micronutrient deficiency to a great extent. With the air of nutrient enrichment, several rice varieties were screened first, and their mineral traits were adjoined with improved organic traits through the breeding method (Table 9.8).

**TABLE 9.8** Different Biofortified Varieties of Rice Produced by Breeding Methods

| Variety Name | Organization | Commodity |
|---|---|---|
| BRRIdhan62, BRRIdhan 72, BRRI dhan64 | Developed by harvest plus and released by Rice Research Institute of Bangladesh | 20–22 ppm zinc in brown rice |
| IR68144-3B-2-2-3, IR72, Zawa Bonday | India and Philippines Research Institutes | 21 ppm grain iron in brown rice |
| Jalmagna | – | Double of iron concentration as that of common rice |

### 9.3.2.1.2 Wheat Breeding

Wheat is the foremost acceptable staple crop, especially in developing countries. A vast variation in iron and zinc concentration in wheat grain in comparison to its closely linked wild species was studied. By keeping in mind, the variation issues, different varieties of wheat were released by Harvest Plus (Table 9.9) [25].

**TABLE 9.9** Different Bio-Fortified Varieties of Wheat with Enriched Micronutrients

| Variety | Commodity |
|---|---|
| BHU1, BHU3, BHU5, BHU6, BHU7, BHU18 | 4–10 ppm higher zinc content |
| NR419, 42, 421, Zincol | Increased Zn content |
| BHU1, BHU6 | Drug Resistance and higher Zn |
| WB2 | Enriched Zn and iron content |
| HI 8627 | High Pro-vitamin A |
| CN10 2217664B | Colored wheat |

### 9.3.2.1.3 Maize Breeding

Maize is a predominant cash crop widely grown to serve as animal feed and for several other industrial purposes. Extensive genetic diversity existing in maize has prepared a framework for it to undergo the breeding program. Harvest Plus have prepared several varieties of maize enriched with Vitamin A content (Table 9.10).

**TABLE 9.10** Bio-Fortified Varieties of Maize

| Variety | Commodity |
|---|---|
| GV662A, Ife Maizehyb-3, Ife maizehyb-4, Sammaz 38, Sammaz 39, CSIR-CRI Honampa | Pro-vitamin A content |
| ZS242 | Orange maize |
| ProVA bio-fortified maize | Toco-chromanols, oryzanol, phenolic compounds |
| QPM | Lysine and tryptophan |
| CML176, CML176X, CML186, HQPM4, FQH-4567, CML142×CML150, CML186×CML149, QS-7705, GH-132-28, BR-451, FOMAIAP, INIA, HQ-61, HB-Protica, NB-Nutrinta, HQINTA-993 | Carotenoids, vitamin E, phenolic, antioxidants. |

## 9.3.2.2 BIOFORTIFICATION OF LEGUMES AND PULSES THROUGH CONVENTIONAL BREEDING

### 9.3.2.2.1 Lentil Breeding

In dryland countries, lentil is used as the major pulse and is highly easy to prepare as well. Lentil was biofortified with zinc and iron by ICARDA and Harvest Plus by using the breeding process, which involved the genetic diversity carried out by using genetic diversity often stored in gene banks (Table 9.11).

**TABLE 9.11** Biofortified Varieties of Lentil Produced by Conventional Breeding Method

| Variety | Commodity |
|---|---|
| Barimasur-4, Barimasur-5, Barimasur-6, Barimasur-7, Khajurah-2, Shitil, Sisir Shekhar, Simal, L4704, Pusa Vaibhav, Aleemaya, Idlib-2, Idlib-3 | High iron and zinc content |

### 9.3.2.2.2 Cow Pea Breeding

Cowpea, which is a rich source of protein is biofortified to elevate the iron content in it through the conventional breeding method (Table 9.12).

**TABLE 9.12** Biofortified Varieties of Cow Pea Produced by Conventional Breeding Method

| Variety | Commodity |
|---|---|
| Plant Lobia-1, Plant Lobia-2, Plant Lobia-3, Plant Lobia-4 | High iron content |

### 9.3.2.2.3 Bean Breeding

Iron and zinc content could be substantially increased in common bean (*Phaseolus vulgaris*). In developing countries, iron biofortified beans are currently in progress of development by Harvest Plus (Table 9.13).

**TABLE 9.13** Biofortified Varieties of Beans Produced by Conventional Breeding Method

| Variety | Commodity |
|---|---|
| RWR 2245, RWR 2154, CAB 2, RWV 1129, RWV 3006, RWV 3317, RWV 2887 | Iron rich |
| COD MLB 001, HM 21-7, RWR 2245, PVA 1438, COD MLV 059, Nain de Kyondo, Cuarentina, Namulenga | Higher iron content |

### 9.3.2.3 BIOFORTIFICATION OF VEGETABLES THROUGH CONVENTIONAL BREEDING

#### 9.3.2.3.1 Potato Breeding

Potato tubers contain a very high concentration of antioxidants in it. Several efforts have been made for breeding of the variants containing red and purple pigment and contributing towards the higher antioxidant capacity. Harvest Plus and International Potato Center have introduced the potato varieties with increased iron and zinc level by crossing the Andean landrace potatoes with resistant tetraploid clones. Rwanda and Ethiopia have been majorly targeted to supply with biofortified potato to deal with micro-nutrient deficiency. INIA Kawsay variety was launched by National Institute for Agrarian Innovations (INIA) Potato programs in Peru and is found to contain a significant level of zinc and iron [12, 33].

#### 9.3.2.3.2 Cauliflower Breeding

Cauliflower gene pool along with *Brassica oleracea* was screened to analyze the genetic variation existing in gene content in it (Table 9.14).

**TABLE 9.14** Biofortified Varieties of Cauliflowers Produced by Conventional Breeding Method

| Variety | Commodity |
|---|---|
| Pusa β-Kesari | Pro-vitamin-A (β-carotene) |
| Purple graffiti and orange cheddar | B-Carotene and anthocyanins |

#### 9.3.2.3.3 Cassava Breeding

In Africa, the Caribbean, and Latin America, Cassava is considered a major vegetable root crop (Table 9.15).

**TABLE 9.15** Biofortified Varieties of Cassava

| Variety | Commodity |
|---|---|
| TMS 01/1368-UMUCASS 36, TMS 01/1368-UMUCASS 37, TMS 07/0220-UMUCASS 44, TMS 07/0593-UMUCASS 45, TMS 2001/1661 | High carotene, protein, iron, and zinc level |

### 9.3.3 BIOFORTIFICATION THROUGH GENETIC ENGINEERING

When enough genetic variation does not exist in specific food crops in such a scenario, the transgenic technique can be used as an alternative strategic approach in attaining the goal of biofortification. Desired genes isolated from different organisms can be introduced into the plant species by using genetic engineering. This process can either activate or elevate the expression of required genes into the target tissue. However, the mentioned technique has attained the only initial grounds and is currently under the initial stage of development. Till now, biofortified rice which contains a high level of β-carotene have been developed by transgenic approach. Rice being a major staple crop can resolve the issue of micro-nutrient deficiency to a great extent. Although the genes involved in the production of β-carotene already exists in rice but their expression is often turned off in endosperm itself. By genetic engineering, two major genes named as bacterial carotene desaturase (*CRTI*) and phytoene synthase (*PSY*) were activated, thus leading to increased production of β-carotene in biofortified rice. Very often, biofortified rice is referred to as Golden Rice, but in the beginning was observed to contain a very small concentration of β-carotene and the quantity was found to be approximately 1.6 mg/g of dry weight. Later on, with the help of further advancements second generation Golden Rice was produced by using the *PSY* gene which was originally procured from maize rather than daffodil and contains significant level of β-carotene approximately 430 mg/g of the dry weight before the storage and is considered as to be of very high nutritional value [11].

Mutagenesis is the major method of introducing genetic variations in the field of biology. With the help of chemicals and radiation technology, induction of mutagenesis can be facilitated for the production of required traits in plants. Along with mutagenesis, conventional breeding approach is also required along with it and collectively is referred mutational breeding. From the past several decades, the current approach is in practice for the production of improved varieties of crop and recently have been utilized along with biofortification to target the micro-nutrient deficiency. Biofortified wheat, maize, rice, sorghum, barley, soybean have been supplied worldwide to meet the issue of micronutrient deficiency. Phytic acid content in bio-fortified staple crops have been significantly reduced by 50–95%.

## 9.4 RATIONALE OF UNIVERSAL ACCEPTANCE OF BIOFORTIFIED FOOD

Several factors have affected the impact of biofortification. These factors can be briefly divided into two categories:

1. **Technology Efficacy:** It includes the available micronutrient contents in biofortified crops, retention ability after undergoing the processing process and ultimately their bioavailability after the consumption. Researchers across the world have verified the efficiency of biofortified staple crops to deal with micronutrient deficiency [30].
2. **Technology Coverage:** It deals with the adoption of bio-fortification concept by farmers and the consumer acceptance of these crops.

Biofortified crops can be produced by agronomic methods, conventional breeding and genetic engineering but has severely suffered strong opposition because farmers and consumers since certain biofortified crops exhibits variation in texture and color in comparison to the original crop. For example, maize, a staple crop was subjected to biofortification and were enriched with β-carotene content in it. But in Mozambique it faces lots of acceptance issues in spite of being rich in Pro-Vitamin A content. In trade experiments and trade tests, people preferred their local white maize in comparison to the orange color biofortified variety of maize. But the trend does not remain the same around the globe. In Kenya, consumers were ready to pay even extra premiums to purchase biofortified maize. In western Kenya, yellow biofortified maize was observed to be more popular than their local white maize variety, and a similar situation was also observed in the Saia district of Kenya.

Cassava, a staple crop, was subjected to biofortification to improve the β-carotene content in it. In developing countries, approximately 1 billion people are dependent upon cassava for their food, feed, and for other industrial requirements. Biofortified cassava was introduced in Brazil. The price of yellow cassava biofortified with vitamin A was observed to be 60–70% higher than the traditional local varieties. But the consumers showed a positive attitude towards even the expensive Cassava variety in Brazil. Whereas in Nigeria, the similar product has observed lots of criticism due to product color difference. Consumers were less likely willing to for Garry in Ima in Southwest, whereas in Southwest Oyo, consumers were ready to pay the premium prices to purchase yellow cassava.

In both the developing and developed countries, sweet potato was the first amongst the staple crop which was biofortified to contain higher levels of β-carotene. In Uganda, customers were ready to pay extra prices and were willing to buy the biofortified variety after making them aware about the nutritional benefits of the produce. But in Mozambique the biofortified variety could not compete with the traditional variety and faces criticism [2].

Furthermore, the transgenic approach which has been implemented behind the concept has also observed lots of criticism amongst the mass population. To enhance or uplift the nutritional status of a country, it is highly essential to gain the acceptance of farmers and consumers regarding the adaptation of biofortified varieties. At the administration level also, several challenges are already standing in queue regarding their consumption. Different countries have initiated several regulatory processes regarding their acceptance and especially for their commercialization which has restricted several organizations involved in the production of biofortified crops. These regulatory and legal complications are highly expensive, tedious, and time-consuming to deal with. For example, Bt Brinjal which was initially developed by an Indian seed company named as Mahyco did not get the clearance to get released in India itself because of certain concerns raised by Scientists, Researchers, anti-GMO activists and other whereas in Bangladesh, four biofortified varieties of Bt Brinjal have been released so far. The research efforts behind the concept are very high, but the outcome is far lower. The process involves very time-consuming steps such as identification of gene, gene modification, expression, analysis of agronomical trait and a large investment to compete these sites are also needed. For example, the study of Golden Rice took almost 8 years to complete but soon after the publication, it faced lots of negative comments due to the ethical and legal concerns over it. The government has not yet approved its mass distribution and its dissemination was held back [13].

## 9.5 IMPACT OF BIOFORTIFICATION OVER SOCIOECONOMIC DEVELOPMENT OF COUNTRY

Biofortified crops cannot be considered as ordinary crop as it not only satisfies the regular obvious needs rather helps in eliminating the hidden hunger of our society. When a customer visits the market, they may not be able to distinguish between the biofortified and non-biofortified crops, often they judge the product only by its price. Biofortified crops have an added advantage over the ordinary crops available in market:

1. Extra nutrition dimension, thereby involved in the realm of public health concerns.
2. Biofortified crops are simultaneously targeting and dealing with food security prospects.
3. Biofortified crops are quite expensive in comparison to the local availability, which has limited its beneficiary list; therefore, it can be clearly predicted that only a privileged section of the society could gain benefits out of it. People are well aware regarding their nutritional benefits but cannot afford to purchase them. Cost-effectiveness analysis needs to be carried out to draw a comparison in the outcome of biofortification along with the cost which is potentially needed for the development and for their distribution. Policymakers need to prioritize the intervention by keeping in mind the financial outcomes and is better than other currently available alternative strategies to mitigate the hidden hunger. GM biofortified crops represents themselves as robust investment plan to enhance the public health and will help to overcome the micronutrient deficiency. Major investment has been made for the biofortified crops, especially during the germplasm development. Once the germplasm is established, the profits can be reaped every year from it. Apart from the financial concerns, the acceptability of biofortified crops represents the same texture and color of locally produced varieties, and that further raised a question in front of consumers regarding its acceptability. Various other limitations are also linked with other technological interventions developed to tackle micronutrient deficiency other than bio-fortification. For example, supplementation process requires the benefices should get available health service. Very few commercial foodstuffs are generally fortified and ultimately poor people remained deprived of the nutritional benefits which can be gained from the advanced interventions. In rural areas, people cannot afford to buy enough fruits and vegetables to meet their daily nutritional requirements then how they can purchase such expensive crops. Moreover, the success of biofortification concept further depends on, how much quantity of micro-nutrient will remain

available after the consumption. It can overcome the nutrient deficiency in how many people and how much population it can really target. The similar facts can further illustrate that why the Golden Rice could not gain much attention in Philippines in comparison to India. There are several ways in which the biofortified crops has induced a significant impact on the socio-economic development of a country and has built a great challenge worldwide which needs to be cautiously dealt with [36].

## 9.6 SUMMARY

Biofortification offers a promising strategy to overcome the current situation of hidden hunger by enriching the currently available varieties of food grains, fruits, vegetables, and oilseeds. It can also be referred to as an agricultural strategy to improve the nutritional content to eliminate the issue of micronutrient deficiency. The agronomic approach, genetic engineering, and conventional breeding are the major strategic approaches to overcome the present scenario. Several initiatives at both the national and international platforms have been adopted to target the same. National agricultural boards along with International organizations such as Harvest Plus have initiated several programs to accomplish the target. Their tremendous efforts have led to the release of several varieties by altering certain genes encoding for nutritional components. Biofortified crops are not only rich with micro-nutrients but also offer higher bioavailability index. But to achieve the target of eliminating hidden hunger completely from our society, joint collaborative efforts of plant breeders and genetic engineers are needed to uplift the scale of the current attempt. There are certain ethical and cost-related issues are also associated with biofortified crops, but despite vast challenges, biofortified crops hold a bright future and can possibly emerge out as a major approach to deal with micronutrient deficiency.

## ACKNOWLEDGMENTS

The authors would like to thank all the staff members of PHRD for their tremendous support and Mr. Ajeet Chauhan for designing the superb graphics of the present chapter.

## KEYWORDS


- **Biofortification**
- **Genetic engineering**
- **Harvest plus**
- **Hidden hunger**
- **Micro-nutrients**
- **Plant breeding**


## REFERENCES


1. Aluru, M., Xu, Y., Guo, R., Wang, Z., Li, S., White, W., & Rodermel, S., (2008). Generation of transgenic maize with enhanced provitamin A content. *Journal of Experimental Botany, 59*(13), 3551–3562.
2. Austin-Phillips, S., Koegel, R. G., Straub, R. J., & Cock, M., (2001). *U.S. Patent No. 6,248,938*. Washington, DC: U.S. Patent and Trademark Office.
3. Bhutta, Z. A., Ahmed, T., Black, R. E., Cousens, S., Dewey, K., Giugliani, E., & Shekar, M., (2008). What works? Interventions for maternal and child undernutrition and survival. *The Lancet, 371*(9610), 417–440.
4. Blancquaert, D., De Steur, H., Gellynck, X., & Van, D. S. D., (2014). Present and future of folate biofortification of crop plants. *Journal of Experimental Botany, 65*(4), 895–906.
5. Bohn, L., Meyer, A. S., & Rasmussen, S. K., (2008). Phytate: Impact on environment and human nutrition. A challenge for molecular breeding. *Journal of Zhejiang University Science B, 9*(3), 165–191.
6. Bouis, H. E., (1999). Economics of enhanced micronutrient density in food staples. *Field Crops Research, 60*(1, 2), 165–173.
7. Bouis, H. E., Hotz, C., McClafferty, B., Meenakshi, J. V., & Pfeiffer, W. H., (2011). Biofortification: A new tool to reduce micronutrient malnutrition. *Food and Nutrition Bulletin, 32*(1), S31–S40.
8. Branca, F., & Ferrari, M., (2002). Impact of micronutrient deficiencies on growth: The stunting syndrome. *Annals of Nutrition and Metabolism, 46*(1), 8–17.
9. FAO, I., (2015). WFP: 2014. The State of Food Insecurity in the World 2014. *Strengthening the Enabling Environment for Food Security and Nutrition*. Rome, FAO.
10. Frossard, E., Bucher, M., Mächler, F., Mozafar, A., & Hurrell, R., (2000). Potential for increasing the content and bioavailability of Fe, Zn and Ca in plants for human nutrition. *Journal of the Science of Food and Agriculture, 80*(7), 861–879.

11. Galili, S., Lewinsohn, E., & Tadmor, Y., (2002). Genetic, molecular, and genomic approaches to improve the value of plant foods and feeds. *Critical Reviews in Plant Sciences, 21*(3), 167–204.
12. Gannon, B. M., & Tanumihardjo, S. A., (2013). *Linking Agriculture to Nutrition-The Harvest is Near and How Do We Measure Impact?* 638–725.
13. Garg, M., Sharma, N., Sharma, S., Kapoor, P., Kumar, A., Chunduri, V., & Arora, P., (2018). Biofortified crops generated by breeding, agronomy, and transgenic approaches are improving the lives of millions of people around the world. *Frontiers in Nutrition, 5*, 12.
14. Gillespie, S., Hodge, J., Yosef, S., & Pandya-Lorch, R., (2016). *Nourishing Millions: Stories of Change in Nutrition*. Intl. Food Policy Res. Inst.
15. Gonzali, S., Kiferle, C., & Perata, P., (2017). Iodine biofortification of crops: Agronomic biofortification, metabolic engineering and iodine bioavailability. *Current Opinion in Biotechnology, 44*, 16–26.
16. Gould, J., (2017). Nutrition: A world of insecurity. *Nature, 544*(7651), S6–S7.
17. Graham, R. D., & Mackill, D. J., (2003). *Biofortification a Global Challenge Program, Applications of Genomics to Rice Breeding (No. 31471 Caja (383))*. IRRI.
18. Guo, W. Z., Zhang, T. Z., Ding, Y. Z., Zhu, Y. C., Shen, X. L., & Zhu, X. F., (2005). Molecular marker-assisted selection and pyramiding of two QTLs for fiber strength in upland cotton. *Yi Chuan Xue Bao= Acta Genetica Sinica, 32*(12), 1275–1285.
19. Hefferon, K. L., (2016). Can biofortified crops help attain food security? *Current Molecular Biology Reports, 2*(4), 180–185.
20. Hoddinott, J. F., Rosegrant, M. W., & Torero, M., (2013). *Investments to Reduce Hunger and Undernutrition*, Lowell, USA.
21. Hotz, C., & McClafferty, B., (2007). From harvest to health: Challenges for developing biofortified staple foods and determining their impact on micronutrient status. *Food and Nutrition Bulletin, 28*(2), S271–S279.
22. Ismail, A. M., Heuer, S., Thomson, M. J., & Wissuwa, M., (2007). Genetic and genomic approaches to develop rice germplasm for problem soils. *Plant Molecular Biology, 65*(4), 547–570
23. Lee, S., Kim, Y. S., Jeon, U. S., Kim, Y. K., Schjoerring, J. K., & An, G., (2012). Activation of rice nicotiana mine synthase 2 (OsNAS2) enhances iron availability for biofortification. *Molecules and Cells, 33*(3), 269–275.
24. Lyons, G. H., Stangoulis, J. C., & Graham, R. D., (2004). Exploiting micronutrient interaction to optimize biofortification programs: The case for inclusion of selenium and iodine in the harvest plus program. *Nutrition Reviews, 62*(6), 247–252.
25. Meenakshi, J. V., Johnson, N. L., Manyong, V. M., DeGroote, H., Javelosa, J., Yanggen, D. R., & Meng, E., (2010). How cost-effective is biofortification in combating micronutrient malnutrition? An ex-ante assessment. *World Development, 38*(1), 64–75.
26. Muthusamy, V., Hossain, F., Thirunavukkarasu, N., Choudhary, M., Saha, S., Bhat, J. S., & Gupta, H. S., (2014). Development of β-carotene-rich maize hybrids through marker-assisted introgression of β-carotene hydroxylase allele. *PLoS One, 9*(12), e113583.
27. Nantel, G., & Tontisirin, K., (2002). Policy and sustainability issues. *The Journal of Nutrition, 132*(4), 839S–844S.
28. Newell-McGloughlin, M., (2008). Nutritionally improved agricultural crops. *Plant Physiology, 147*(3), 939–953.

29. Pérez-Escamilla, R., (2017). Food security and the 2015–2030 sustainable development goals: From human to planetary health: Perspectives and opinions. *Current Developments in Nutrition, 1*(7), e000513.
30. Rengel, Z., Batten, G. D., & Crowley, D. D., (1999). Agronomic approaches for improving the micronutrient density in edible portions of field crops. *Field Crops Research, 60*(1, 2), 27–40.
31. Römer, S., Lübeck, J., Kauder, F., Steiger, S., Adomat, C., & Sandmann, G., (2002). Genetic engineering of a zeaxanthin-rich potato by antisense inactivation and co-suppression of carotenoid epoxidation. *Metabolic Engineering, 4*(4), 263–272.
32. Saltzman, A., Birol, E., Bouis, H. E., Boy, E., De Moura, F. F., Islam, Y., & Pfeiffer, W. H., (2013). Biofortification: Progress toward a more nourishing future. *Glob. Food Sec.*, 2, 9–17.
33. Telengech, P. K., Maling'a, J. N., Nyende, A. B., Gichuki, S. T., & Wanjala, B. W., (2015). Gene expression of beta carotene genes in transgenic biofortified cassava. *3 Biotech., 5*(4), 465–472.
34. United Nations System Standing Committee on Nutrition, (2004). *5th Report on the World Nutrition Situation: Nutrition for Improved Development Outcomes* (Vol. 5). United Nations System Standing Committee on Nutrition.
35. Welch, R. M., & Graham, R. D., (2002). Breeding crops for enhanced micronutrient content. In: *Food Security in Nutrient-Stressed Environments: Exploiting Plants' Genetic Capabilities* (pp. 267–276). Springer, Dordrecht.
36. Winkler, J. T., (2011). Biofortification: Improving the nutritional quality of staple crops. *Access Not Excess* (pp. 100–112). Smith-Gordon.

# CHAPTER 10

# Global Food Traceability: Current Status and Future Prospects

ACHARYA BALKRISHNA, PARSHOTTAM RUPALA, DIVYA JYOTI, and VEDPRIYA ARYA

## ABSTRACT

Global food traceability is a prime requirement of the food industry business around the globe for the safety and quality assurance of food products. A real-time traceability mechanism is able to keep a check on the unsafe food that leads to a decrease the foodborne illness. Several traceability mechanisms adopted by different countries were reviewed and analyzed in the current manuscript. It was found that proper traceability with an approach of ICT is immediately required for the maintenance of a comprehensive system of external and internal traceability. It is also noticed that, in general, traceability is adopted for some of the agro-products and not for the others. A complete and transparent traceability mechanisms need to develop for sustainability and food safety. An overview of the current scenario and future prospects of traceability mechanism across the globe is discussed here in detail. This study recommended the development of centrally controlled traceability mechanisms from the farm to the consumer table in a transparent manner.

## 10.1 INTRODUCTION

Food safety has become a prominent issue for the residents of numerous nations. Episodes of illness in animals that could be communicated to people, for example, avian influenza, or the presence of synthetics chemical analogs above the acceptable limits in the food, can diminish both the quality and wellbeing of humans. There is an emergence to review and reintroduce the products and protect the individuals from foodborne illness. Food safety can be attained with the proper implementation of traceability mechanism for the quality assurance of the products (Figure 10.1). Traceability is a risk mitigation technique, which empowers the food business administrators or specialists to establish a real time coordination between the demand and supply. It is a foundation of any nation's food safety strategy [6]. Also, quality assurance has become a foundation of food safety strategy in the food business that began to incorporate quality and food safety management mechanisms [67]. Traceability is an effective and comprehensive idea to permit the development of a product from the crude materials to the selling stage, contemplating its total process by identifying, tracking methods, and documental entries up to the final level. The components of food traceability system are summarized in Figure 10.2. Currently, food traceability has gained its critical significance as it permits effective identification, correction, and mitigation of risk factors throughout the process, in order to convey quality-assured food items to the consumers.

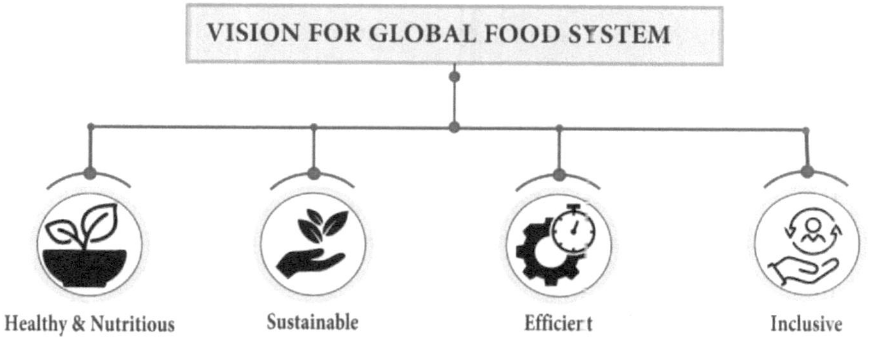


**FIGURE 10.1** Vision for global food systems.

### *10.1.1 WHAT IS FOOD TRACEABILITY?*

Traceability can be variously defined. As indicated by ISO 8402:1994 quality standards, traceability is characterized as: "the capacity to follow the set of experiences, application or area of an element by the analysis of recorded identification" [46]. In ISO 9000:2005 standards, the definition is reached out into "the capacity to follow the set of experiences, application or area of that which is in demand" [47]. ISO rules further determined that traceability may allude to the origin of sources, the dynamic history, and the dissemination and site of the item/product after its delivery conveyance. Also, the US Food and Drug Administration (FDA) proposes the accompanying definition: "the ability to recognize by means of document or electronic data about a food item and its producer from its original source to final consumer" [22]. The European Union (EU) defined traceability as "the capacity to trace a food, food-producing animal or item planned to be, or anticipated to be incorporated into a food or feed, in all progressive stages of manufacturing, processing, and dissemination" (regulation 178/2002) [26].

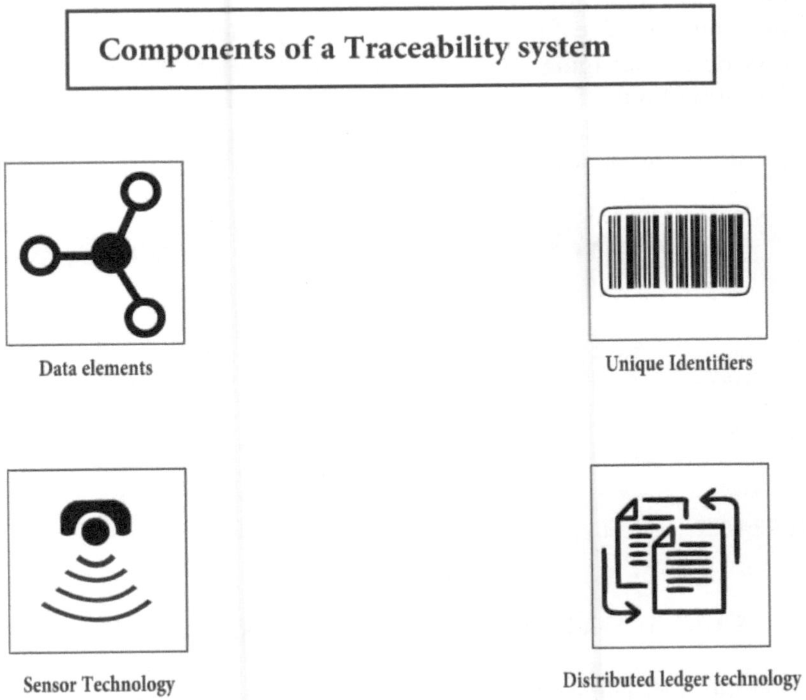


**FIGURE 10.2** Components of a traceability system.

As per the Codex Alimentarius Commission (CAC) [19], traceability or item tracing signifies "the capacity to follow the development course of a food item in different progressive stages of manufacturing, processing, and dissemination." Traceability permits the tracking of an item, following its way from crude materials until presentation for selling, up to the final consumer. Also, as indicated by ISO 22005 standard [46], traceability framework implies that all the information and activities able to maintain the required information about an item and its constituents during the entire food supply chain.

According to Bosona and Gebresenbet [12], food traceability is portrayed as a part of logistics management that catch, store, and communicate the sufficient data about a food, food-producing animal, or item/substance at different stages of the food supply chain so that the product can be verified for the quality and safety upward or downward any time [10]. For the high quality, safe, and nutritious food, the plan, and execution of entire chain of traceability from farm to final consumer has become a significant part of the comprehensive food quality assurance framework [65]. FAO [31] expressed that managing food safety and quality is a mutual obligation of all the entities in the food supply chain, including Governments, industrial partners, and customers [30]. Different approaches used for the food traceability used by various countries are summarized in Figure 10.3.

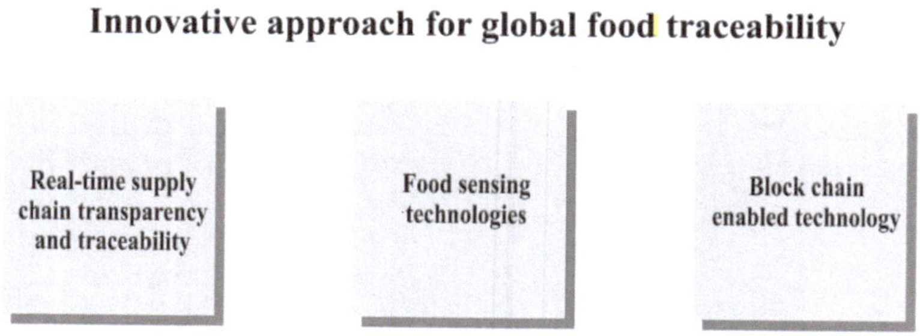


**FIGURE 10.3** Innovative approach for global food traceability.

## 10.2 SIGNIFICANCE AND NEED OF FOOD TRACEABILITY

Traceability has increased impressive significance concerning the food, especially following different food safety incidents in which traceability frameworks have been missing [34]. Alterations in the trading environment have prompted the development of the global production framework. The supply chain has evolved towards the expanded versions as per the complexity across multiple industries, and worldwide reach of food supply chains. The transparent, real time traceability framework is must for the food safety and quality assurance system. It is an approach to expand buyer certainty and to interface producers and consumers (Figure 10.4). Some of the significant features of incorporation of food traceability are discussed in subsections.

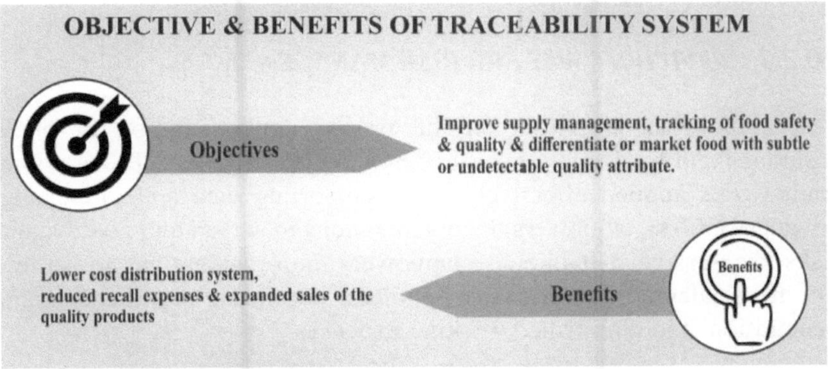


**FIGURE 10.4** Objective and benefits of traceability system.

### *10.2.1 QUALITY ASSURANCE OF PRODUCTS*

The proficiency of a traceability framework relies upon the capacity to track and follow every individual item and its distribution (logistics) unit, in a way that empowers persistent checking from the production stage until the endpoint disposal by the customer. Execution of viable traceability frameworks improves the capacity to actualize the evident wellbeing and quality consistence programs. Not just an approach to improve the food safety frameworks, traceability can likewise be viewed as a vital instrument to improve the nature of crude materials [39], to improve inventory management and competitive advantage of the products [6].

### *10.2.2 REAL-TIME DATA ANALYSIS*

The subsequent perceivability of significant data empowers agri-food organizations to considers speedy responses for crises, reviews, and withdrawals. Inclusive traceability frameworks essentially decrease the response times when a creature or a plant disease flare-up happens by giving more fast admittance to pertinent and reliable information that decides the source and area of involved items. So, the information about the animals, plant wellbeing, country's origin, and so forth at any time in the chain from the producer to the customer has become essential [7], and this can only be done by the implementation of traceability.

### 10.2.3 EFFICIENT WAY FOR RISK MANAGEMENT

Traceability can emphasis on the proper implementation of quality assurance and food safety measures by GAP, GMP, HACCP. Traceability frameworks applied effectively, with supporting data and interchanges advances (ICTs), empowers the organizations to screen and guard against risks progressively. It likewise empowers the organizations to settle on the more talented administration choices, prompting significant market penetration, and diminished working expenses.

### 10.2.4 BETTER MANAGERIAL AND ADMINISTRATIVE EFFICIENCY

- Visibility of data given by the traceability frameworks empowers the organizations to use their assets and cycles more successfully and productively and increase of their durable efficiency.
- Implementation of traceability can lessen the outdated product loss, lower stock levels, revive the identification process and supplier challenges, and raise the significance of logistics and distribution tasks.
- Transparency of a food supply chain framework is significant as all the partners of the organization have a mutual consent to access any item and process-related data without ambiguities. It also empowers them to accomplish effective reviews for any step of supply chain level when essential and supports early alerts (due to pro-active quality monitoring framework) emerges for a potent and emerging issue [9].

### 10.2.5 PRODUCT AUTHENTICITY

Improved client satisfaction assisted with marketing and improved brand value. Besides, in the case of special products from well-known sources, for example, saffron, vanilla, cloves, cacao, and different flavors and toppings that get higher values due to their extraordinary qualities, traceability can help to prevent blending with lower quality produce and contamination, ensuring item authenticity.

### 10.2.6 CLIENT SATISFACTION

From a consumer point of view, traceability accompanies with building trust, significant serenity, and increase trust in the food framework. For the cultivators, traceability is essential for a profitable quality assurance framework that can likewise aid persistent improvement and minimization of the effect of safety hazards. It additionally encourages in the fast and viable review of items and the assurance and resolution of any liabilities [70]. Traceability gives a more prominent way for the quality assurance and empowers the fast identification of issues; So, it can diminish food wastage and its deterioration. It also enables the development of a real time, strong communicating channel between the customer and the grower [86].

Conclusively, a comprehensive traceability management framework permits the follow back and forward capacities to any progressive step of the supply chain, for the quality assurance and food safety of the products [70]. This sort of comprehensive traceability system in a supply chain has been characterized in numerous terms, for example, "seed to shelf" [62], "field to plate" [70], "farm to plate" [63], "farm to fork" [74], "farm to table" [28, 72], and so on.

## 10.3 KINDS OF TRACEABILITY

Traceability can be characterized by the activity or the direction in which data is reviewed in the food chain. According to the activity in the food supply chain, three distinct parts of traceability can be recognized. Those are: back traceability or suppliers' traceability, internal traceability or process traceability, and forward traceability or client traceability [67]. Moe [60] clarified that discernibility could be found in two sorts: inward traceability that tracks inside in one of the means in the chain or chain traceability that tracks an item batch and its history through the entire, or part, of a creation chain from harvest through vehicle, stockpiling, processing, distribution, and deals [60]. Opara [70] divided traceability into six significant components on the basis of: product, genetic, process, input, disease, and pest, and measurement traceability [70].

As per the direction of movement of item in the food supply chain, traceability is of two types: backward traceability or tracing is the capacity, at each progressive step of the food supply chain, to check the source and qualitative attributes of an item or product as per one or a few given standards. Conversely, forward traceability, or tracing, is the capacity, at each progressive step of the food supply chain, to discover the locality of items or products from one or a few given standards [51].

Traceability system can be divided into two parts: logistics traceability which follows just the actual development of the item or product and treats food as a commodity and qualitative traceability that relates with the data associated with the quality assurance of food products along different progressive stages of food supply chain [27]. As indicated by Jansen-Vullers, Van Drop, and Beulens [48], traceability can be seen as active and passive on the basis of its uses. In the passive part, traceability gives the real time visibility to where the item or product at the time and its disposition. In active sense, the online following data is furthermore used to upgrade and control quality measures in and between the various progressive connections of the food supply chains with authentic records by the methods for the recorded recognizable proof [48].

Traceability can be further divided into external and internal traceability. In the external traceability, all traceable products or items are to be uniquely distinguished, and the data to be shared between all distribution channel participants. The product identification for the traceability may include:

- Unique product or item identification number;
- Batch/parcel number.

For the external traceability, traceable item or product identification numbers must be coordinated with the distribution channel members on the product labels and related papers or electronic business archives. This connects the actual products with the data prerequisites necessary for the traceability. External traceability allows following back (supplier traceability) and forward (client traceability). Internal traceability means tracking of the process at every progressive step from the raw materials to finished products within an enterprise. The process includes combining of raw materials, processing, reconfiguration, repacking, and new product development with its own Unique Product Identifier. For the traceability, a linkage is to be maintained from the original source up to the finished product. A label demonstrating the batch/Lot number of the input item should remain on the packaging until that whole traceable item is exhausted. This standard applies to the product which are a part of larger packing hierarchy.

## 10.4 FACTORS AFFECTING THE FOOD TRACEABILITY

The concept of traceability at the production level of the agri-food supply chain is connected with quality assurance schemes and not so much with logistics support schemes, such as ERP, etc. Also, customers are mainly concerned for the quality assurance and good manufacturing practices (GMP) for the products and least for the implementation of a unique traceability system. As the consumption of agri-food products mainly related to its quality and safety. Both criteria can be attained by a strong traceability framework in the food supply chain. Major factors influencing the viability of traceability are as follows.

### 10.4.1 ORGANIZATIONAL STRUCTURE OF THE SUPPLY CHAIN

The structure and organization of the supply chain and its management is very crucial for the maintenance of traceability. There are several factors that affect the food supply chain as a number of partners in the supply chain, degree of collaboration between them to maintain internal and external traceability, the ability of different entities of the food supply chain to identify the product origin and to manage traceability at various levels.

### 10.4.2 DESTINATION OF A PRODUCT

The identification of traceable lot and also the time needed for the product traceability is also an influencing factor for the traceability mechanism. In general, traceability of the products should be a quick, efficient, and accurate process. The technology link with each progressive step of the supply chain should be very specific and effective.

### 10.4.3 CREDIBILITY OF A TRACEABILITY METHOD

The traceability method should be unique and efficient and able to trace the specified product in a given time duration with authenticity. The traceability method should be customer-friendly, easy to operate, comprehensive, and in-depth. For the proper execution of the traceability mechanism, its compatibility with the already existing management system is very important.

### 10.4.4 DATA IDENTIFICATION, STANDARDIZATION, AND MANAGEMENT

The traceability mechanism largely depends on its backend support. The identification and management of data is very important for the implementation of traceability program. Data authenticity, privacy, and analysis is very important for the traceability mechanism.

### 10.4.5 LEGISLATION ON TRACEABILITY

Different countries have their own legislation and standards for the traceability mechanism. The implementation of traceability also depends on them that what kind of techniques they allow for the conceptualization of traceability.

## 10.5 ESSENTIAL STANDARDS OF GLOBAL FOOD TRACEABILITY

To bring out the uniformity and harmonization in the global food laws with consumer's rights protection and safe health, the Food and Agricultural Organization (FAO) of the United Nations has designed CODEX standards (Codex Alimentarius) to be followed by the 189 member nations of FAO (as on 1 May 2020) which has become binding on all WTO members through the implementation of 'sanitary and phytosanitary measures' (SPS agreement) [62, 89]. The idea is to build a 'One Step Forward-One Step Backward' system which provides rapid identification in food supply management system [74, 84].

Codex Alimentarius are the international food standards that are recognized worldwide for the international trade of activities related to food, food products, and food safety. It also lays down the provisions on Food's hygiene practices; use of Additives; handling the chemical residues and contamination; Labeling, Sampling, and Inspection and Rules on Import-Export certification, etc., [17]. These provisions altogether influence the composition of 'Global Food Traceability' standards. The CAC is responsible for the due implementation of all food standards legislated by FAO in conjunction with the World Health Organization (WHO) [18].

All the WTO members are expected to legislate their national food laws for internal and external purposes, e.g., Food and Drug Administration (USA), Food Sanitation Law (Japan), and Food Safety and Standards (Licensing and Registration of Food Businesses Regulations (India), etc., based on the international guidelines, standards, and recommendations as defined under CAC to maintain uniformity in food's legal and commercial compliances [87]. Additionally, the joint effort of FAO and WTO in 'Assuring Food Safety and Quality: Guidelines for Strengthening National Food Systems' should also be referred by the national authorities in drafting their food safety management codes [9]. In case of any trade dispute, the respective members can approach the WTO panel where Codex food safety text is also referred during the dispute resolution process [84]. Some of the popular Codex standards devised by FAO are summarized in Table 10.1.

### *10.5.1 GLOBAL FOOD TRACEABILITY STANDARDS*

The technological advancements have led to faster business transactions in the food industry, making the door-step delivery of even fresh and perishable food products much easier. With this one-finger tap speed, the need for traceability has also increased, but loopholes in the food supply chain management system on international trade platforms due to variance in drafted legislation of participating trade nations, have made the food inspection even more complex and difficult [36]. Hence, Global Food Traceability Regulations started gaining recognition which require Internal as well as External Traceability of Food Products. The internal compliances are the ones that need to be adhered to by the food business entities through various inspection and certification programs before

**TABLE 10.1** Some of the Popular Codex Standard Devised by FAO

| | | |
|---|---|---|
| **CXS 193-1995** (Contaminants and Toxins in Food and Feed) | **CXC 80-2020** (Food Allergen Management for Food Business Operators) | **CXC 77-2017** (Arsenic Contamination in Rice) |
| **CXC 1-1969** (Food Hygiene) | **CXS 333-2019** (Quinoa) | **CXC 78-2017** (Mycotoxins in Spices) |
| **CXS 1-1985** (labeling of Pre-packaged Foods) | **CXC 79-2019** (Reduction of 3-Monochloropropane-1,2-Diol Esters (3-MCPDEs) and Glycidyl Esters (GEs) in Refined Oils and Food Products made from it) | **CXS 329-2017** (Fish Oils) |
| **CXG 2-1985** (Nutrition Labelling) | **CXG 92-2019** (Rapid Risk Analysis on Food Contamination detection) | **CXS 325R-2017** (Unrefined Shea Butter) |
| **CXM 2** (MRLs and RMRs for Residues of Veterinary Drugs in Foods) | **CXS 330-2018** (Aubergines) | **CXS 323R-2017** (Laver Products) |
| **CXS 192-1995** (Food Additives) | **CXS 332R-2018** (Doogh) | **CXS 326-2017** (Black, White, and Green Peppers) |
| **CXG 91-2017** (Monitoring the Performance of National Food Control Systems) | **CXC 76R-2017** (Hygienic Practice for Street-Vended Foods in Asia) | **CXS 327-2017** (Cumin) |
| **CXG 88-2016** (Food Hygiene to the Control of Foodborne Parasites) | | **CXS 328-2017** (Dried Thyme) |
| **CXG 89-2016** (Exchange of information between food importing and exporting nations) | | **CXS 331-2017** (Dairy Permeate Powders) |
| | | **CXS 324R-2017** (Yacon) |
| | | **CXG 90-2017** (Pesticide residue detection in Food and Feed) |
| | | **CXC 75-2015** (Hygienic Practice for Low-Moisture Foods) |

engaging in international trade. Although, it is not mandatory to follow them but getting verified through these certifications gives accreditation to the food business entities a global appeal and accountability towards public health concern. Few examples that are recognized internationally for their assistance in internal food verification procedures are listed as follows:

1. ISO 22 000 (Food Safety Management Standards);
2. ISO 9001 (Quality Management System);
3. ISO/TS 22003:2013 (Rules for Audit and Certification of Food Management System);
4. ISO 22005:2007 (Food and Feed Traceability System);
5. British Retail Consortium (BRC);
6. Safe Quality Food (SQF);
7. International Featured Standards (IFS).

However, it is not necessary that above-stated examples are limited to only internal verification, for instance, IFS requires both internal and external traceability standards before granting the certification [20]. Furthermore, the external compliances are the ones which food business entities are expected to comply in accordance with the guidelines of international trade and food laws either in their original form or similar valid law or widely accepted and recognized trade practices at international level and shares the traceability data with the third party. E.g., Use of Identifiers for product's Package, Lot, and Batch, etc. Some of the important Global Food Traceability Standards are discussed below:

#### *10.5.1.1 GLOBAL STANDARDS 1 (GS1)*

GS1 stands for global standards 1, a non-profit international standards organization established for the purpose of bringing harmonization in product traceability by setting certain standards for business communication having around 100 countries as its members. It is basically a supply chain method which is widely used through its electronic commerce tools and identifiers such as Barcodes, global location numbers (GLN) and Global trade identification numbers, etc. GS1 is not a mandatory regulation, but its importance is intricately related with prescribed traceability schemes and standards to have uniform implementation. GS1 identification standards are largely accepted by most national and multi-national industries especially in retail sector for its high reliability of correct product information such as its original location, manufacturing company, type of product, etc. Use of GS1 standards enables significant faster and coordinated electronic commerce thus helps in boosting economy with easy traceability mechanism. Its extended application has enabled various food products, especially the fresh plants, meat, and dairy products, which could not have happened otherwise [2].

#### *10.5.1.2 GLOBAL FOOD SAFETY INITIATIVE (GFSI)*

Global Food Safety Initiative (GFSI) was introduced in May 2000, with the objective of ensure safety measurements in food supply, gain back the consumer's trust and stimulate the volume of global food trade. With increasing global integration of GFSI, as a third-party certification and verification of food management systems, it has become a genuine source of standardized procedure by majority of the countries. GFSI acts as a source of verification for both compliances (internal and external) in international food trade. However, it is not a mandatory requirement by government regulations. Popular schemes under GFSI are BRC, GLOBAL GAP, safety quality food (SQF), best aquaculture practices (BAP), Food Safety System Certification (FSSC), IFS and Canada GAP, etc. The renowned auditing companies that certify food business entities for GFSI standards gain international cognizance.

#### *10.5.1.3 HAZARD ANALYSIS AND CRITICAL CONTROL POINT (HACCP)*

International Code of Practice General Principles of Food Hygiene, proposed under Annex to CAC/RCP 1-1969 (Rev. 4-2003), set the requirements and procedure for the execution of hazard analysis and critical control point (HACCP) regulations. GFSI's guidance document also regulates the risk-based food safety requirements to be fulfilled as per the HACCP standards. The aim of HACCP standards is to remove any strong biological, chemical, or physical hazards during the food manufacturing process to ensure the safety of finished food products. These standards can be applied for internal as well as external traceability that will help in minimizing damage and enable identification of product lots and their relevant batches for further inspection purposes [40].

#### *10.5.1.4 GLOBAL GOOD AGRICULTURAL PRACTICES (GLOBAL GAPS)*

Good agricultural practices are designed by the privately-owned body 'GLOBALGAP,' which sets certain 'good' and 'safe' agricultural methods to enhance sustainability effect by farming activities. These standards are deployed by various national authorities, food industries, and non-governmental organizations, etc., to regulate the quality of food by improvising its harvesting, handling, and storage methods. In order to get certified under Global Gap, Farmers or groups of farmers need to have administrative traceability of their farming activities right from the procurement of raw materials till the final sale of matured crops. This will help farmers gain exposure at the international market and win consumer's trust [83].

#### *10.5.1.5 GOOD MANUFACTURING PRACTICES (GMPS)*

Good manufacturing practices (GMP, also known as 'cGMP' or 'current Good Manufacturing Practice') are laid down by WHO, to ensure that the quality standards are maintained during production of finished consumer goods. At first, it was applied only to pharmaceutical industry but later on it gained acceptance in all other industries including food production to stimulate international trade environment. It addresses the issues related to premises selection, basic hygiene, safe transportation, pest control and sanitation, inventory maintenance, and raw material quality standards and implemented through staff training of food business entities [52, 56].

#### *10.5.1.6 GLOBAL GOOD HYGIENE PRACTICES (GLOBAL GHPS)*

It is a consumer's basic right to have a safe qualitative meal, and in order to meet this demand, Food Business entities are expected to cater their products that comply with the food hygiene criteria set under 'Good Hygiene Practices.' These practices have gained global acceptance to maintain uniformity by removing the cultural differences without harming the sensitivity of that particular nation. For example, Codex standard CXC 76R-2017 (Hygienic Practice for Street-Vended Foods in Asia) can act as a torchbearer for Asian nations to legislated their domestic laws on hygiene practices mandatory to be followed for safe delivery of street-vended foods since it can be source of various parasitic or polluting contaminants potentially harmful to a mass population. Another text suitable for seafood safety is "good practice guidelines (GPG) on National Seafood Traceability Systems," available on the FAO website [31].

#### 10.5.1.7 MAXIMUM RESIDUE LIMITS (MRLS)

Maximum residue limit/level (MRL) is a crucial part of Good Agricultural Practices and is worth mentioning separately since pesticides' residue is the highest source of various chemical-borne diseases. It is calculated based on Acceptable Daily Intake in mg per kilogram, and for highest legally acceptable intake of pesticide is determined by the Codex Alimentarius, which is the main reference for setting MRL standards at the national level for trade purposes based on the joint recommendation of Food and Agricultural Organization (FAO) and WHO [29].

## 10.6 CURRENT STATUS OF GLOBAL FOOD TRACEABILITY

With the speeding import and export transactions of food products, especially the perishable ones, threats of microbial hazards to health have also increased, which raised concerns over the safety standards of the food production and distribution ecosystem at the global level. Technological interventions have made it easier for customers to purchase products of their choice and become fully aware of the nutritive value since health and wellness trends have shown significant surge. The current COVID-19 pandemic situation has made it even more crucial to have transparency in the food procurement cycle to ensure safety and high nutritive content of raw materials.

Countries all over the world are now more concerned and eager to comply with food safety standards than ever to reduce the potential risks of bioterrorism, lower nutrition rate, and declining quality parameters in food supply chain management. The need for traceability dates back to the cases of food contamination since middle age due to not only biological reasons but also by technological, environmental, and Product management changes [61]. Clearly, illnesses emerging out of complications in food quality, be it plant or animal, pose a major threat to human beings as business transactions are much faster now. Hence, it is crucial that we analyze the current status of traceability regulations for all countries in order to stabilize any pandemic situation at a much faster rate than its spread. The harmonization of traceability standards all over the world will result in a better understanding of each country's adopted practices, better interoperability of food business transactions and will lead to the removal of delays and trade restrictions by developing mutual trust among nations globally.

Ten fastest growing economies of the world (Brazil, Canada, Public Republic of China, France, Germany, Italy, India, Japan, United Kingdom and United States of America (USA)) as enlisted on the basis of their nominal GDP value in Dollars by 'The Global Economy' [41]. These economies have much faster import-export business transactions, and the rest of the 172 countries contribute only 1/5$^{th}$ portion to the world economy [78]. For a comparative study, agricultural food products, including all perishable, non-perishable, processed, and packaged foods and any such eatable item intended for human consumption are taken into consideration for the investigation and comparison of above mentioned 10 countries' traceability rules and regulations [66]. Ranking of countries on each parameter is tagged as 'Adequate,' 'Moderate' and 'Regressed' as per higher to lower compliance, respectively. Most of the selected nations have set up their own traceability mechanisms for the food and agricultural products as per their national laws and regulations. For example, Brazil has formalized a mandatory implementation of 'Brazilian System of Identification and Certification of Origin for Bovine and Buffalo' (SIS-BOV) to ensure Livestock's easy traceability and identification [11, 64]. Canada has the traceability policies regulated by Health of Animals Regulations and enforced by the CFIA (2013a). Consumer Packaging and Labeling Act and Regulations, enables the traceability of processed food products, Food Safety Enhancement Programs (FSEP) aid for traceability meat products and other Specified acts are available for specific commodities; However, not all food products have regulatory procedures in Canada yet [43]. People's Republic of China has accelerated its food safety measure through strict implementation of Food Safety Law, Articles 36–41 of which requires all the food producers, processors, packers, and retailers to carry out testing and record-keeping activities for all the input-output transactions and to keep them in database for at least 2-years duration [33]. For identification for certain animals like Pig, Cattle, and Sheep, China's Ministry of Agriculture has mandated under Decree No. 67 (2006) to tag them by ear using a two-dimensional barcode (Ben-Hai) and other such methods [92].

In India, as per the report of Confederation of Indian Industry, Food, and Agricultural Centre of Excellence in India and GS_ India, states that Until the recent Food Safety and Standards (Food Recall Procedure) Regulations (2017), food safety in India was governed by Section 28 of the Food Safety and Standards Act (2006). Under the Food Safety Standards Act (2006), the responsibility of the food safety lies absolutely on the business owners. Few exceptions to this are the restaurants, caterers, and takeaway joints, unless they own multi-outlet food business franchises having integrated manufacturing and distribution network. This Regulation is applicable to food or food products that are detected or prima facie observed unsafe and/or as may be specified by FSSAI every so often However, there is still a lot of need for improvements, especially in traceability animals and crops and their by-products [35].

The Japanese Ministry of Agriculture, Fishery, and Forestry (JMAFF, 2008) have constituted 'The Japanese Handbook for Introduction of Food Traceability Systems' as a comprehensive guidance on food safety standards and made it mandatory to be followed domestically especially for 'Beef traceability' [44, 55, 80]. Apart from this, rice being their main dietary ingredient, Japan has implemented a Rice Traceability Regulation 1 July, 2011 (USDA/ERS, 2014) under the Rice Traceability Act [80].

Regulation (EC) No. 178/2002 of the European Parliament and of the Council of 28 January 2002, relying on 'one step forward-one step backward' approach mandates that business operators shall be:

i. Responsible to identify the product suppliers on 'to and from' basis;
ii. Required to have all the necessary traceability systems and procedures for auditing by competent authority whenever needed [73].

In order to comply with Article 18, it is not explicitly mentioned but is necessary to have information on 'Name and Addresses of Suppliers, Customers, Date of purchase/sale and Volume/Quality of product' ready to be produced for inspection if needed; Failure in doing so would constitute an offense. Furthermore, EU legislation also provides traceability requirement on Seafood products under Article 58 of EC 1224/2009 [21]. Since its exit from EU on 31 January 2020, the United Kingdom will remain in transition period until the end of 2020 and all the EU Laws and regulations including Food and feed will continue to apply on the United Kingdom during this phase. However, after the negotiation phase is over, we might see some significant changes in updated rules and regulations adopted by UK in the future [23].

The United States is much progressive in terms of taking cognizance of potential threat to safety and security of its nation and hence legislated the Bioterrorism Act of 2002 (BT Act), which enables the food product tracing for FDA-regulated products by requiring the proper documentation of food distribution chain and root cause investigation in case of illness outbreak to find out the source contamination [71]. Also in 2011, the U.S. Department of Agriculture (USDA) under the Food Safety Modernization Act (FSMA), has mandated the 'Animal Disease Traceability requirements' for livestock that get transported across its state boundaries [37].

Traceability regulations and policies in these 10 countries are legislated and managed by their respective government departments that deal with food safety, agriculture, and health sectors, and the ministers of these portfolios. Brazil has the federal department of 'Ministry of Agriculture, Livestock, and Food Supply' (Ministério da Agricultura, Pecuária e Abastecimento, abbreviated as MAPA); Canada has 'Agriculture and Agri-Food Canada' and 'Canadian Cattle Identification Agency'; People's Republic of China has 'China Food and Drug Administration (CFDA)' and 'Administration of Quality Supervision, Inspection, and Quarantine (AQSIQ)'; EU has the agency of 'European Food Safety Authority (EFSA); India has 'Food Safety and Standards Authority of India (FSSAI)' under its Ministry of Health and Family Welfare; Japan has 'Ministry of Health, Labor, and Welfare' for its food safety and USA has 'FDA,' USDA, and USDA's Animal and Plant Health Inspection Service (APHIS) are responsible for drafting their national and international food traceability standards as per their need, convenience, and trade obligations [3, 5, 12, 16, 25, 32, 58, 59, 81, 82, 84].

Most of the selected countries are relying on the electronic databases to monitor the majority of their business transactions. Where some countries are going a step further by micro-managing for their animal traceability data such as Canadian Livestock Tracking System in Canada, Germany's Herkunftssicherungs-und Informations System fur Tiere (HIT), France's National Livestock Database (BDNI), and the UK's cattle tracing system (CTS) while some like India and China still in struggle to do so [13, 45, 85]. However, during pandemic situation, the Electronic National Agricultural Market platform of India has seen a significant growth in business transactions and hopefully it can prove to be as a reliable source of food grain tracking. No data could be found on Brazil's electronic database of their CTS. Furthermore, the EU has the rapid alert system for food and feed (RASFF) for reporting and raising alarms in hazardous situations related to food and feed products which entails the quick exchange of information among EU Member States and helps in rapid identification of root cause ensuring minimal damage [24]. Although in USA, the United States Department of Agriculture (USDA) had formulated four goals to achieve animal traceability via establishing federal animal health events repository (AHER), promoting the use of electronic identification tags, and designing an animal tracking database from their birth to slaughter for better traceability [6]. But no significant pragmatic results have been found so far on this topic. In sum, none of the countries have a systematic database for plant-based food products so far (Table 10.2). All selected nations being the member of World Trade Organization (WTO) are bound by International Trade Law and hence bound to follow the food safety measures and policies as per the agreements on the implementation of 'SPS measures' (SPS agreement) and 'technical barriers to trade' (TBT agreement) [90]. In case of any conflict, WTO encourages the concerned countries to find mutually agreed solutions. Other judicial solutions like Arbitration, Mediation, and Conciliation, etc., are available for international conflict resolutions and all WTO members are responsible for providing transparency and necessary information during resolution procedures [89–91].

## 10.7 CHALLENGES FOR THE IMPLEMENTATION OF FOOD TRACEABILITY

The developing significance of the food safety and quality in business implement all the partners in the food supply chain to implement traceability from farm to fork; however, there are several challenges to execute traceability mechanisms.

### *10.7.1 COST OF TRACEABILITY FRAMEWORKS*

First, the costs associated with establishing the traceability frameworks is a critical challenge specifically for the micro or small enterprises of developing countries. However, the advantages picked from the traceability far exceed the expenses of traceability. Many developing countries lag in creating and actualizing the food safety and quality standards and

**TABLE 10.2** Status of Global Food Traceability based on Different Standards*

| Nation | Based on GS1 Standards [2, 9, 19, 21, 22, 24, 36, 41, 88] | Based on Availability of GFSI Auditing Companies [42–44, 59] | Based on Their National Traceability Regulations | Based on Traceability Regulations for Imported Products | Self-Governing Traceability Policies | Based on Products Traceability Regulation | Based on Their Use of Labelling and Identifiers | Based on Their Electronic Food Tracking Systems |
|---|---|---|---|---|---|---|---|---|
| Brazil | Adequate | Adequate (SGS) | Moderate | Moderate | Moderate | Moderate (Livestock only) | Moderate | Regressed |
| Canada | Adequate | Adequate (NSF, SGS) | Moderate | Adequate | Adequate | Moderate (cattle, swine regulation) | Adequate | Moderate |
| China | Moderate | Moderate (NSF, SGS, AQSIQ) | Regressed | Regressed | Regressed | Regressed | Moderate | Regressed |
| France | Adequate | Adequate (NSF, SGS, COFRAC) | Adequate | Adequate | Adequate | Adequate | Adequate | Adequate |
| Germany | Adequate | Adequate (NSF, SGS) | Adequate | Adequate | Adequate | Adequate | Adequate | Adequate |
| Italy | Adequate | Adequate (NSF, SGS, ACCREDIA) | Adequate | Adequate | Adequate | Adequate | Adequate | Adequate |
| India | Moderate | Moderate (NSF, SGS, FSSAI) | Regressed | Regressed | Regressed | Moderate (Buffalo Meat, Vegetables) | Moderate | Regressed |
| Japan | Adequate | Moderate (NSF, SGS, JFSM) | Moderate | Adequate | Moderate | Moderate (domestic beef, rice) | Adequate | Adequate |
| United Kingdom | Adequate | Adequate (NSF, SGS, UKAS) | Adequate | Adequate | Adequate | Adequate | Adequate | Adequate |
| United States | Adequate | Adequate (NSF, SGS, ANSI) | Adequate | Moderate | Adequate | Adequate | Adequate | Regressed |

*Ranking of countries on each parameter is tagged as 'Adequate,' 'Moderate' and 'Regressed' as per higher to lower compliance.

thus limit the exports of food products from their countries. Digital databases for traceability are more expensive to implement in hardware and software, including skilled manpower [49, 50].

### 10.7.2 DATA SHARING AND MANAGEMENT

Another challenge is a problem with the exchange of data in a standardized format between different connections links in the supply chain. Data exchange should be done with a precise, confidential, secured, and electronic way [60]. Traceability frameworks are critically responsible on the effectiveness of data storage and its proper management. Also, the processing of data into specific formats is difficult and challenging according to different enterprises.

### 10.7.3 TRENDS OF CUSTOMER PREFERENCES

As per the innovative technological interventions, customer also shifts their expectations and preferences. As in business, product need to be revised or upgraded from time to time, traceability mechanisms should also be feasible to timely updated.

### 10.7.4 GLOBAL STANDARDS AND INTERVENTIONS

Overlapping and conflicting demands from different global regulators around the world is also a major challenge for food traceability. Different regulatory standards on various components also differ at the global level. Food adulteration and market potential for economic benefit is also a global challenge for food traceability.

### 10.7.5 PRODUCT AND ENTERPRISES WISE TRACEABILITY

Different products and enterprises need different kinds of traceability mechanisms. Like in the Agri-food industry, farming, fishery, food manufacturing, processing, retailing, transportation, food service, and distribution needs different ways of traceability at each progressive step.

## 10.8 RECENT TECHNOLOGICAL INTERVENTIONS FOR FOOD TRACEABILITY IN INDIA: A CASE STUDY

India's foodgrains production is assessed as 291.95 million tons (2019–2020) crop year (July-June), beating the previous year produce (i.e., 290 million tons). This expansion is due to the increment within the quantity of farms and enhancement in the development strategies [75]. However, at the present, India does not have any mandatory traceability framework [76], in recent years, the Indian government has begun to figure with private accomplices, various state and central governments, which include FSSAI, APEDA, GS1 India, NABARD (National Bank for Agriculture and Rural Development), FPO (fruit products order), ITC's eChaupal, and Reliance, etc., for the development of traceability framework inside the Indian food industry and food supply chain [68].

### *10.8.1 FOOD TRACEABILITY TECHNIQUES UTILIZED IN INDIA*

In India, existing item identification advanced technologies are alphanumerical codes, Hologram, Barcode, radio frequency identification (RFID) tags, and the geographical indication (GI) tag. Also, innovative food traceability techniques, such as Bio tracing, Nanosensor, global positioning system (GPS) and geographic information system (GIS) are implemented in various fields by the Indian Government. As it is extremely clear that for the real-time traceability, and efficient harmonization between different parameters of the food supply chain is required. Additionally, new and proficient traceability frameworks are adequately equipped for controlling the human mistakes and making more mindfulness about the standard affirmation of the food items [38]. In India, the food business, including traceability is developing fast among the clients and business leads. As Cargill announced that they are getting to build an efficient and real-time traceability framework for the palm oil in India in few years [15]. Also, they successfully run the program for the food safety and sustainability under the scheme of 'Surakshit Khadya Abhiyan' [14]. In continuation, the Tea Board of India introduced, Trustea, and rain forest alliance (RFA) certifications, which are compulsory for all tea manufacturers so as to determine the transparency and traceability in both the National and International Market [52]. Recently, for the food safety and quality purpose, McDonald restaurants implemented a potatoes traceability framework which keep a track on the 40 different suppliers around the country [54]. Numerous farms produce like banana, mango, grapes, onion, soybean, potato, and poultry products are ready to exponentially increase the farmer's economy and work to develop the Indian farming sector, whereas few of them are certified by the APEDA [7]. Major food processing companies [77] are using different advanced techniques as the barcode and 2-D quick response (QR) code technology so as to develop a promising and competent product solution and also lead to consumer satisfaction. The traceability mechanism and quality assurance system are summarized in Figure 10.5.

Recently, APEDA adopted the GS1 norms, while the overwhelming majority of the apparent and valuable applications are accomplished through the utilization of GS1's item identifiers in barcoding for Grapenet, Anarnet, and Tracenet. Also, APEDA, an agro trade promotional organization of the Indian Government, already providing the traceability systems to enhance the import mechanism in several countries about the agro-produce of India [42]. Modern traceability frameworks that are currently getting used across the Indian food industry are summarized in Table 10.3.

An ICT-based traceability mechanism is proposed that is able to maintain both internal and external traceability within and outside the enterprises (Figure 10.6). This mechanism governs the complete transparent and effective mechanism for overall traceability from the production stage to the consumer. However, this is yet to be standardized for different agro-products. Recently, one such kind to initiative is done by Bharwa Solutions Pvt Ltd, India by piloting a traceability mechanism for agro products named "Annadata."

## 10.9 SUMMARY

Food sustainability and quality assurance of the products are receiving significant attention from policymakers, researchers, and nutritionists. Execution of the traceability framework in the food supply chain is vital to scale back the 'foodborne diseases,' 'Low quality of Food materials,' and 'Unsafe food.' Traceability will emerge as a new concept and effective index of quality and a real-time basis for food business in the current scenario. Moreover, situations like the COVID-19 pandemic also motivated all of us to move forward for the quality assured product lineage with proper traceability mechanisms. The traceability framework needs

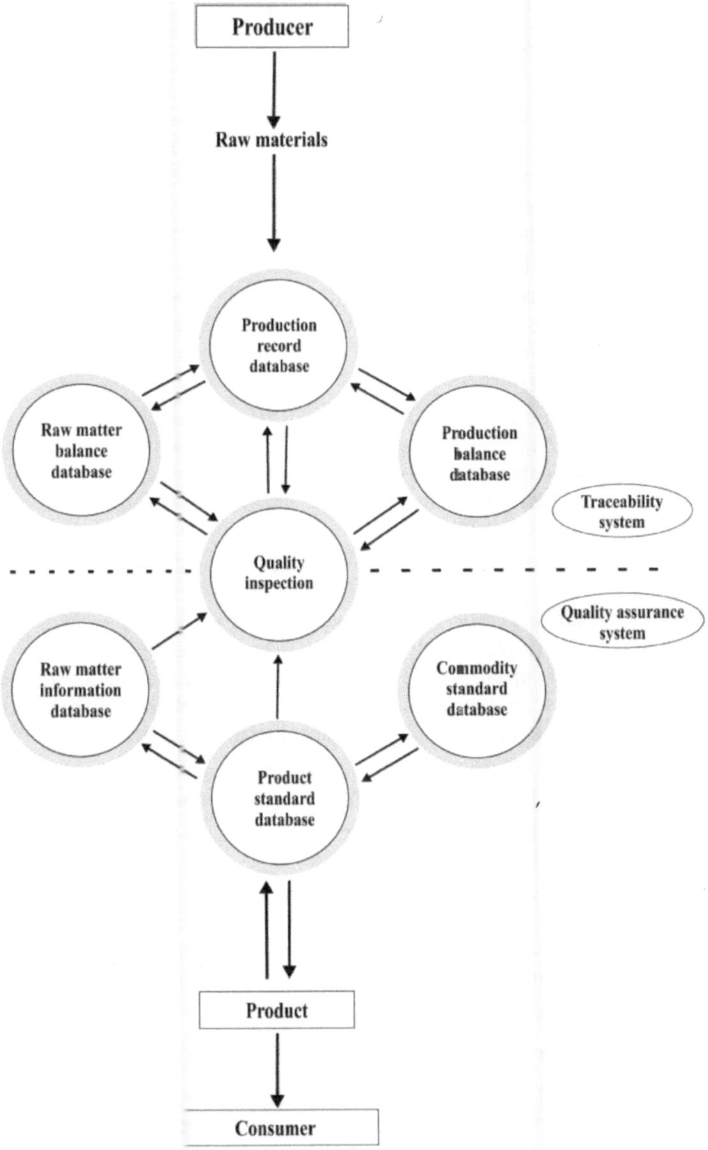


**TRACEABILITY & QUALITY ASSURANCE SYSTEM**

**FIGURE 10.5** Traceability and quality assurance system.

**TABLE 10.3** Food Traceability Techniques Used in India

| Techniques Used for Food Traceability | Features |
|---|---|
| Alphanumerical codes | A blend of the alphabetic and numeric characters of varied sizes, which is usually found on the item/products label [1]. |
| Hologram | A body that diffracts light into an image, while it alludes to both the encoded material and therefore, the resulting picture Barger and White [8–11]. |
| Barcode | Barcode is an openly machine-readable information/data which is printed over the products/objects, whereas by means of electronics barcode readers are often easily encode, store, and review data [93–95]. |
| Radiofrequency identification (RFID) | Radiofrequency identification (RFID) utilized radiofrequency technique for the merchandise identification. It helps to stop the fabrication in the dissemination of ration, which is fixed remittance of provisions or foods like oil, sugar, etc., from the ration shops with ration cards [4, 5]. |
| Document-based (paper/electronic documents) | Indian software companies which are being guided toward using the traceability framework within the sort of Enterprise Resource Planning (ERP) systems, which is employed for storing data and inventory control, warehouse management, accounting, and asset management. ERP frameworks can read the standardized information/data from the barcodes and RFIDs, including global trade item numbers (GTIN) and global location numbers (GLN) [49, 50]. |
| Nanotechnology | India is being progressed in the field of nanotechnology, but nevertheless very difficult to estimate the particular situation due to the unavailability of data and reports from leading Indian food companies and laboratories [69]. |
| Information and communication technology (ICT) | In India, ICTs provides a simple solution to the farmer, trader, suppliers, and even manufacturers for the fast, reliable, efficient service and real-time data/information in term of the standard and quantitative agricultural products marketing [53]. |

cautious planning for guaranteeing the real-time perceivability all through the food supply chain, which includes consistency with all the set norms. Presenting the framework through the food chain and setting up powerful associations among all the sub-systems requires both a steady methodology of the food traceability framework executed by every food administrator and a typical arrangement with respect to food traceability among all food business administrators.

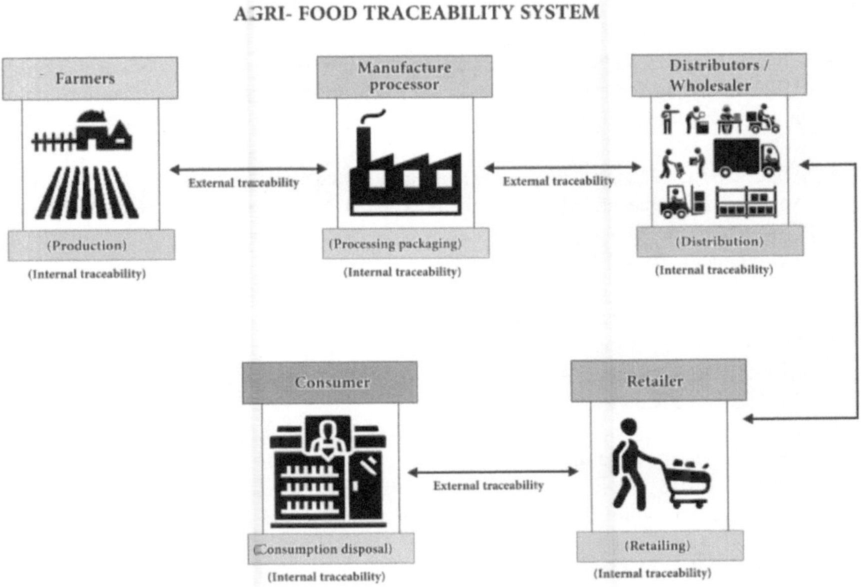


**FIGURE 10.6** Demonstration of agri-food traceability system.

## ACKNOWLEDGMENTS

The authors are thankful to Mr. Sanjay Aggarwal (Secretary, MAgri, and FW) for his invaluable support and guidance. We express our sincere thanks to Mr. Ajeet Chauhan for designing the graphics of the present chapter.

## KEYWORDS

- **Barcodes**
- **Cattle Tracing System**
- **Food traceability**
- **Global standards**
- **Hologram**
- **ICT based Technology**

## REFERENCES


1. Abad, E., Palacio, F., Nuin, M., De Zarate, A. G., Juarros, A., Gomez, J. M., et al., (2009). RFID smart tag for traceability and cold chain monitoring of foods: Demonstration in an intercontinental fresh fish logistic chain. *Journal of Food Engineering, 93*(4), 394–399.
2. About GS1 (2021). https://www.gs1.org/about (accessed on 4 May 2021).
3. Agarwal, M., Sharma, M., & Singh, B., (2014). Smart ration card using RFID and GSM technique. In: *Confluence the Next Generation Information Technology Summit (Confluence), 2014 5th International Conference* (pp. 485–489).
4. *Agriculture and Agri-Food Canada*. (2021). https://www.agr.gc.ca/eng/agriculture-and-agri-food-canada/?id=1395690825741 (accessed on 12 May 2021).
5. Alfaro, J. A., & Rábade, L. A., (2009). Traceability as a strategic tool to improve inventory management: A case study in the food industry. *International Journal of Production Economics, 118*(1), 104–110.
6. *Animal Disease Traceability: USDA APHIS*. (2021). https://www.aphis.usda.gov/traceability/ (accessed on 12 May 2021).
7. APEDA, (2013). *Agriculture and Processed Food Products Export Development Authority*. Traceability. http://apeda.gov.in/apedawebsite/index.html (accessed on 12 May 2021).
8. Administration of Quality Supervision, Inspection and Quarantine of the People's Republic of China (AQSIQ) Association. (2021). https://www.aqsiq.net/ (accessed on 12 May 2021).
9. *Assuring Food Safety and Quality: Guidelines for Strengthening National Food Systems*. ISSN: 0254-4725. http://www.fao.org/3/y8705e/y8705e00.htm (accessed on 12 May 2021).
10. Barger, M. S., & White, W. B., (2000). *The Daguerreotype: 19th-Century Technology and Modern Science*. JHU Press.
11. Beulens, A. J. M., Broens, D. F., Folstar, P., & Hofstede, G. J., (2005). Food safety and transparency in food chains and networks. *Food Control, 16*(6), 481–486.
12. Bosona, T., & Gebresenbet, G., (2013). Food traceability as an integral part of logistics management in food and agricultural supply chain. *Food Control, 33*, 32–48.
13. *Brazilian System of Identification and Origin Certification*. https://www.cpap.embrapa.br/agencia/congressovirtual/pdf/ingles/04en05_1.pdf (accessed on 12 May 2021).
14. Canadian Cattle Identification Agency. https://www.canadaid.ca/ (accessed on 12 May 2021).
15. *Canadian Livestock Tracking System*. https://www.clts.canadaid.ca/CLTS/secure/user/home.do (accessed on 12 May 2021).
16. Cargill, India, (2015). *Cargill Launches Surakshit Khadya Abhiyan, a Nationwide Initiative on Food Safety Awareness in India, New Delhi*. https://www.cargill.co.in/en/news/NA31873256.jsp (accessed on 12 May 2021).
17. Cargill, (2014). *Tracking Progress Toward Sustainable Palm Oil*. http://www.cargill.com/corporate-responsibility/sustainable-palm-oil/traceability/index.jsp (accessed on 12 May 2021).
18. China Food and Drug Administration (CFDA). (2003). https://www.allfordrugs.com/sfda-china/ (accessed on 12 May 2021).

19. *Codex Alimentarius Commission*. (2013). http://www.fao.org/fao-who-codexalimentarius/committees/cac/about/en (accessed on 12 May 2021).
20. Codex Alimentarius Commission, (2007). Joint FAO/WHO food standards program, and World Health Organization. Codex aliment Arius Commission: Procedural manual. *Food and Agriculture Org*.
21. *Codex Alimentarius Standards*. http://www.fao.org/fao-who-codexalimentarius/home/en/ (accessed on 12 May 2021).
22. *Comparing Global Food Safety Initiative (GFSI) Recognized Standards*, (2011). https://www.sgs.com/~/media/Global/Documents/White%20Papers/sgs-global-food-safety-initiative-whitepaper-en-11.ashx (accessed on 12 May 2021).
23. Council Regulation (EC) No. 1224/2009, (2009). https://eur-lex.europa.eu/legal-content/EN/TXT/?uri=CELEX%3A02009R1224-20190814 (accessed on 12 May 2021).
24. *Dispute Settlement*. https://www.wto.org/english/tratop_e/dispu_e/dispu_e.htm (accessed on 12 May 2021).
25. *Dispute Settlement Without Recourse to Panels and the Appellate Body*. https://www.wto.org/english/tratop_e/dispu_e/disp_settlement_cbt_e/c8s1p2_e.htm (accessed on 12 May 2021).
26. Ene, C., (2019). The relevance of traceability in the food chain. *Economics of Agriculture, 60*(2), 287–297.
27. *EU Legislation and UK Law*. https://www.legislation.gov.uk/eu-legislation-and-uk-law#:~:text=As%20a%20member%20of%20the,action%20required%20by%20the%20UK (accessed on 12 May 2021).
28. EU (2002). *Regulation (EC) No. 178/2002 of the European Parliament and of the Council*.
29. *European Food Safety Authority*. https://www.efsa.europa.eu/ (accessed on 12 May 2021).
30. FAO and WHO, (2003). *Assuring Food Safety and Quality: Guideline for Strengthening National Food Control System*. Joint FAO/WHO Publication. http://www.fao.org/3/y8705e/y8705e00.htm (accessed on 12 May 2021).
31. FAO, (2003). *FAO's Strategy for a Food Chain Approach to Food Safety and Quality: A Framework Document for the Development of Future Strategic Direction*. http://www.fao.org/DOCREP/MEETING/006/y8350e.HTM (accessed on 12 May 2021).
32. Folinas, D., Manikas, I., & Manos, B., (2006). Traceability data management for food chains. *British Food Journal, 108*(8), 622–633.
33. *Food Safety and Standards of India*. (2020). https://www.fssai.gov.in/ (accessed on 12 May 2021).
34. *Food Safety Law of the People's Republic of China*, (2015). https://www.hfgip.com/sites/default/files/law/food_safety_-_16.02.2016.pdf (accessed on 12 May 2021).
35. Food Traceability in India. (2020). *A Study Report by CII, FACE & GS1 India*. http://face-cii.in/sites/default/files/final_report-version_2.pdf (accessed on 12 May 2021).
36. *Food Traceability*. (2020). https://globalfoodsafetyresource.com/food-traceability/ (accessed on 12 May 2021).
37. FSA, (2002). *Traceability in the Food Chain a Preliminary Study*. UK: Food Standard Agency. https://www.adiveter.com/ftp_public/articulo361.pdf (accessed on 12 May 2021).

38. *Full Text of Food Safety Management Act (FSMA)*. (2020). https://www.fda.gov/food/food-safety-modernization-act-fsma/full-text-food-safety-modernization-act-fsma (accessed on 12 May 2021).
39. Furness, A., Osman, K. A., & Lees, M., (2003). Developing traceability systems across the supply chain. *Food Authenticity and Traceability*, 473–495.
40. Galvão, J. A., Margeirsson, S., Garate, C., Viðarsson, J. R., & Oetterer, M., (2010). Traceability system in cod fishing. *Food Control, 21*, 1360–1366.
41. *General Principles of Food Hygiene, CAC/RCP 1-1969*. https://www.loex.de/files/downloads/lebensmittel/Codex%20Alimentarius%20(EN).pdf (accessed on 12 May 2021).
42. *Global Agricultural Practices and Global Food Handling Practices*. (2015). https://www.ams.usda.gov/services/auditing/gap-ghp (accessed on 12 May 2021).
43. *Good Hygiene Practices and HACCP*. http://www.fao.org/food/food-safety-quality/capacity-development/haccp/en/ (accessed on 12 May 2021).
44. *Good Manufacturing Practice*. (2020). https://www.who.int/news-room/q-a-detail/medicines-good-manufacturing-processes#:~:text=What%20is%20GMP%3F-,Good%20manufacturing%20practice%20(GMP)%20is%20a%20system%20for%20ensuring%20that,through%20testing%20the%20final%20product (accessed on 12 May 2021).
45. Gross Domestic Product, Billions of U.S. Dollars, (2019). *Country Rankings*. https://www.theglobaleconomy.com/rankings/GDP_current_USD/ (accessed on 12 May 2021).
46. GS1 India, (2012). *The Global Language of Business: India*. http://gs1india.org/ (accessed on 12 May 2021).
47. *Guide to the Consumer Packaging and Labelling Act and Regulations*. https://www.competitionbureau.gc.ca/eic/site/cb-bc.nsf/eng/01248.html (accessed on 12 May 2021).
48. *Handbook for Introduction of Food Traceability Systems (Guidelines for Food Traceability)*. https://www.maff.go.jp/j/syouan/seisaku/trace/pdf/handbook_en.pdf (accessed on 12 May 2021).
49. Traceability and Information System for Animals. (2019). *Herkunftssicherungs-Und Informations System Fur Tiere*. https://www.hi-tier.de/ (accessed on 12 May 2021).
50. ISO 8402, (1994). http://www.scribd.com/doc/40047151/ISO-8402-1994-ISO-Definitions (accessed on 12 May 2021).
51. ISO 9000, (2005). Retrieved from: http://www.pqm-online.com/assets/files/standards/iso_9000-2005.pdf (accessed on 12 May 2021).
52. Jansen-Vullers, M., Van, D. C. A., & Beulens, A. J. M., (2003). Managing traceability information in manufacture. *International Journal of Information Management, 23*(5), 395–413.
53. Karippacheril, T. G., Rios, L. D., & Srivastava, L., (2011). Global markets, global challenges: Improving food safety and traceability while empowering smallholders through ICT. In: *ICT in Agriculture, The World Bank's E-Sourcebook Report No. 64605* (pp. 285–310).
54. Karippacheril, T., Rios, L., & Srivastava, L., (2012). *Module 12: Improving Food Safety and Traceability*. http://www.ictinagriculture.org/sourcebook/module-12-improving-food-safety-and-traceability (accessed on 12 May 2021).
55. Kelepouris, T., Pramatari, K., & Doukidis, G., (2007). RFID-enabled traceability in the food supply chain. *Industrial Management and Data Systems, 107*(2), 183–200.

56. Kumar, V. S., (2015). *Increasing Exports Critical to Boost S Indian Tea Prices.* The Hindu: Business Line. http://www.thehindubusinessline.com/economy/agri-business/increasing-exports-critical-to-boost-s-indian-teaprices/article7550818.ece (accessed on 12 May 2021).
57. Lashgarara, F., Mohammadi, R., & Najafabadi, M. O., (2011). Identifying appropriate information and communication technology (ICT) in improving marketing of agricultural products in Garmsar City, Iran. *African Journal of Biotechnology, 10*(55), 11537–11540.
58. *MAPA Brazil.* https://extranet.who.int/sph/donor/mapa (accessed on 12 May 2021).
59. *Maximum Residue Limits.* http://www.fao.org/fao-who-codexalimentarius/codex-texts/maximum-residue- limits/en/#:~:text=A%20maximum%20residue%20limit%20(MRL,accordance%20with%20Good%20Agricultural%20Practice (accessed on 12 May 2021).
60. McDonald, (2015). *Farm to Fork: Potato.* http://www.mcdonaldsindia.com/farmtofork.html (accessed on 12 May 2021).
61. *Meat Traceability in Japan.* https://www.card.iastate.edu/iowa_ag_review/fall_03/article2.aspx (accessed on 12 May 2021).
62. *Members of Codex Alimentarius Commission.* (2020). http://www.fao.org/fao-who-codexalimentarius/about-codex/members/en/ (accessed on 12 May 2021).
63. *Ministry of Health.* Labor and Welfare, Japan. (2020). https://www.mhlw.go.jp/english/ (accessed on 12 May 2021).
64. Moe, T., (1998). Perspectives on traceability in food manufacture. *Trends in Food Science and Technology, 9*(5), 211–214.
65. Montet, D., & Dey, G., (2018). *History of Food Traceability.* https://www.taylorfrancis.com/ chapters/edit/10.1201/9781351228435-1/history-food-traceability-didier-montet-gargi-dey (accessed on 12 May 2021).
66. Morris, C., & Young, C., (2000). "Seed to Shelf," "teat to table," "barley to beer," and "womb to tomb": Discourses of food quality and quality assurance schemes in the UK. *Journal of Rural Studies, 16*(1), 103–115.
67. Mousavi, A., Sarhadi, M., Lenk, A., & Fawcett, S., (2002). Tracking and traceability in the meat processing industry: A solution. *British Food Journal, 104*(1), 7–19.
68. *National Livestock Database (BDNI).* (2020). http://dahd.nic.in/about-us/divisions/national-livestock-mission (accessed on 12 May 2021).
69. *OECD Environmental Indicators: Development, Measures and Uses.* http://www.oecd.org/environment/indicators-modelling-outlooks/24993546.pdf (accessed on 12 May 2021).
70. Opara, L. U., (2003). *Traceability in Agriculture and Food Supply Chain: A Review of Basic Concepts, Technological Implications, and Future Prospects.* https://agris.fao.org/agris-search/search.do?recordID=FI2016100260do? recordID=FI2016100260 (accessed on 12 May 2021).
71. Perez-Aloe, R., Valverde, J. M., Lara, A., Carrillo, J. M., Roa, I., & Gonzalez, J., (2007). Application of RFID tags for the overall traceability of products in cheese industries. In: *1st Annual RFID Eurasia, Istanbul*, 1–5.
72. Pinto, D., Castro, I., & Vicente, A., (2006). The use of TIC's as a managing tool for traceability in the food industry. *Food Research International, 39*(7), 772–781.

73. Pradhan, N., Singh, S., Ojha, N., Shrivastava, A., Barla, A., Rai, V., et al., (2015). Facets of nanotechnology as seen in food processing, packaging, and preservation industry. *BioMed Research International*.
74. *Principles for Traceability/Product Tracing as a Tool Within a Food Inspection and CertificationSystem, CAC/GL60-2006*. http://www.fao.org/fao-who-codexalimentarius/sh-proxy/en/?lnk=1&url=https%253A%252F%252Fworkspace.fao.org%252Fsites%252Fcodex%252FStandards%252FCXG%2B60-2006%252FCXG_060e.pdf (accessed on 12 May 2021).
75. *Public Health Security and Bioterrorism Preparedness and Response Act*, (2002). https://www.govinfo.gov/content/pkg/PLAW-107publ188/pdf/PLAW-107publ188.pdf (accessed on 12 May 2021).
76. *RASFF-Food and Feed Safety Alerts*. https://ec.europa.eu/food/safety/rasff_en (accessed on 12 May 2021).
77. Raspor, P., (2008). Total food chain safety: How good practices can contribute? *Trends in Food Science and Technology, 19*(8), 405–412.
78. *Regulation (EC) No 178/2002 of the European Parliament and of the Council* (2002). https://eur-lex.europa.eu/legal-content/EN/TXT/?uri=CELEX%3A02002R0178-20190726 (accessed on 12 May 2021).
79. Ruiz-Garcia, L., Steinberger, G., & Rothmund, M., (2010). A model and prototype implementation for tracking and tracing agricultural batch products along the food chain. *Food Control, 21*(2), 112–121.
80. Saxena, M., & Gandhi, C., (2014). *Indian Horticulture Database*. http://nhb.gov.in/area-pro/NHB_Database_2015.pdf (accessed on 12 May 2021).
81. Schroeder, T. C., & Tonsor, G. T., (2012). International cattle ID and traceability: Competitive implications for the US. *Food Policy, 37*(1), 31–40.
82. Shah, A., (2011). *List of Top Food Processing Companies in India-Equipment and Industry Growing at a Rapid Clip*. http://www.greenworldinvestor.com/2011/07/06/list-of-top-food-processing-companiesin-india-equipment-and-industry-growing-at-a-rapid-clip/ (accessed on 12 May 2021).
83. *The 10 Fastest-Growing Economies in 2019*. https://groww.in/blog/the-10-fastest-growing-economies/ (accessed on 12 May 2021).
84. *The Importance of Food Product Traceability*, (2020). https://www.siroccoconsulting.com/the-importance-of-food-product-traceability/ (accessed on 12 May 2021).
85. *Traceability and Food Labeling of Rice in Japan*, (2017). https://ap.fftc.org.tw/article/1181 (accessed on 12 May 2021).
86. *U.S. Department of Agriculture*, (2021). https://www.usda.gov/ (accessed on 12 May 2021).
87. *U.S. Food and Drug Administration*, (2016). https://www.fda.gov/home (accessed on 12 May 2021).
88. *Understanding Codex*, (2016). http://www.fao.org/3/a-i5667e.pdf (accessed on 12 May 2021).
89. *Understanding the Codex Alimentarius*, (2003). http://www.fao.org/3/y7867e/y7867e08.htm (accessed on 12 May 2021).
90. *Understanding the WTO Agreement on Sanitary and Phytosanitary Measures*. https://www.wto.org/english/tratop_e/sps_e/spsund_e.htm#:~:text=The%20TBT%20

(Technical%20Barriers%20to,defined%20by%20the%20SPS%20Agreement (accessed on 12 May 2021).
91. *USDA's Animal and Plant Health Inspection Service (APHIS),* (2021). https://www.aphis.usda.gov/aphis/home/ (accessed on 12 May 2021).
92. *Use Cattle Tracing System (CTS) Online*. https://www.gov.uk/cattle-tracing-online (accessed on 12 May 2021).
93. Wilson, T. P., & Clarke, W. R., (1998). Insights from industry food safety and traceability in the agricultural supply chain: Using the Internet to deliver traceability. *Supply Chain Management*, *3*(3), 127–133.
94. Xiong, B., Luo, Q., Yang, L., & Pan, J., (2011). Traceability system of pig-raising process and quality safety on 3G. In: *International Conference on Computer and Computing Technologies in Agriculture* (pp. 115–123). Springer, Berlin, Heidelberg.
95. Zare, M. Y., (2010). Coupling RFID with supply chain to enhance productivity. *Business Strategy Series, 11*(2), 107–123.

# CHAPTER 11

# Virtual Farmers' Market: A Single-Step Solution to Sustainable Agriculture and Food Security

ACHARYA BALKRISHNA, KAILASH CHOUDHARY, ASHISH DHYANI, and VEDPRIYA ARYA

## ABSTRACT

Virtual farmers' market (VFM) is a technological intervention associated with the E-Commerce platform for the buying and selling of farmers' agri-produce. Rural smallholder farmers faced numerous obstacles, such as they struggle to reach markets for better prices of their crops. In fact, the current pandemic situation accelerated the critical situation of agri-business to a greater extent. VFM provides a transparent, trustworthy, and open platform for farmers and consumers and encourages them for fair prices and deals. Nowadays, this approach has been utilized all across the globe by different countries. This new approach is also beneficial for the consumer because they are getting agricultural goods from some distant farmers with a fair price. VFM is growing as a sustainable agri-business model and is widely adopted for food security purposes. The current study summarizes the need, importance, and status of VFM systems around the world, with special mention of challenges and benefits associated with this system.

## 11.1 AGRICULTURE MARKETING: AN INTRODUCTION

Agriculture plays a key role in the development of the nation. In this world, food is the most essential thing, and man cannot survive without food. "Agriculture" and "marketing" are the two main components of agriculture marketing. Agriculture marketing is based on two concepts;

1. Producing agricultural products with the use of natural factors for human welfare. It completely depends upon natural processing; and
2. Marketing refers to the promotion of goods and services to provide a target to the customers that were done by the different business organizations. The customers who are targeted can be attracted to the company by creating strong customer value for them [42].

Farmers' market is defined as a physical marketplace projected to sell the farm products such as animals and vegetables, raw food products, prepared foods, and beverages directly to the target customers. It can occur indoors or outdoors and is usually made up of booths, tables, or stands and reflects the local culture and economy in most countries. Farmers' markets may vary in size (some stalls range from large to many city blocks). Due to their nature, they are less rigorously regulated than retail produce shops and differ from public markets that typically operate in permanent structures, are open year-round, and in a variety of non-farmer/non-producers offer vendors packaged. Foods and non-food products (Figure 11.1) [4].

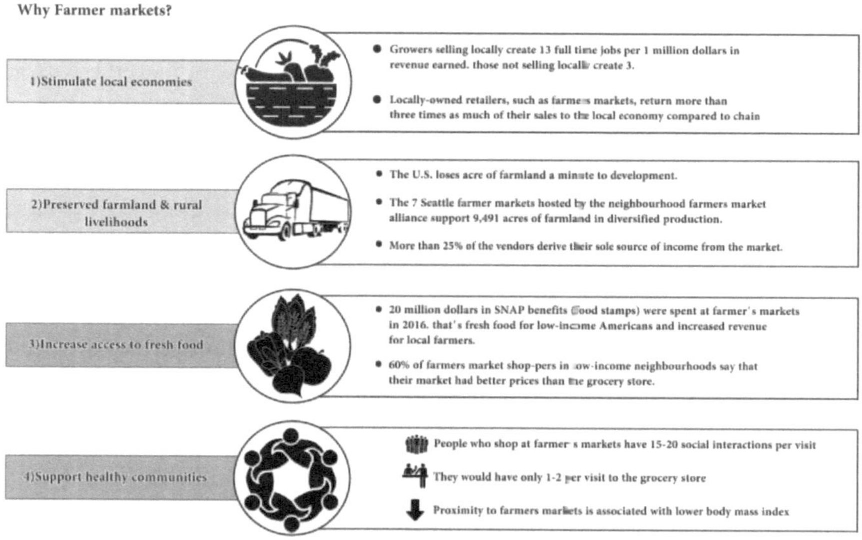


**FIGURE 11.1** What is the need for a farmers' market?

These markets operate on the idea of a direct relationship between local producers and their customers and exclude middlemen. This can help to increase the profit of the farmer by direct sales and reduce transportation costs, handling, preservation, and storage. Producers can also make a good profit by selling their agricultural produce to food processing firms [31]. Along with this, the consumer also benefited from the farmers' market, such as easy use of fresh, healthy, seasonal, organic foods, meats, eggs, poultry, cheese, etc. [20]. In the agricultural sector, the implementation of this innovative technology is considered to increase farm production and accelerate the economy. Agribusiness moved e-commerce or e-marketing, or internet marketing beyond traditional marketing to manufacturers and consumers, and this online platform is known as the virtual farmer market.

## 11.2 FUNCTIONARIES OF AGRICULTURAL MARKET

Agricultural marketing is classified into three market stages based on the participation of functionaries. They are:

1. **Primary Market:** The farmer pre-harvest worker, itinerary business, and transport agents.
2. **Secondary Market:** In addition to primary market functionaries, agents for finance and agents for processing are also included in the secondary market functionaries.
3. **Terminal or Export Market:** Agents of shipping and commercial analysts are also included in the primary and secondary market executives as well as terminal market functionaries.

The main functions of marketing involved in agricultural marketing are:

1. **Concentration:** The initial work to be done in agricultural marketing is to collect agricultural products which are ready to sell economically at a central main place to buy.
2. **Grading of Agro Produce:** The process of separating huge quantities of products into different categories based on variety, quality, and size is called grading. It helps to set standards for those products.
3. **Processing:** In the processing phase, agricultural commodities or raw materials are being converted into products that are consumable by the customer. For example, rice is obtained by processing paddy.

4. **Warehousing:** The process of storage of agricultural goods starting from production to final consumption. This is mainly to protect the goods from spoilage or contamination.
5. **Packaging:** Packing the products is an essential function that helps in handling the goods easily, which can prevent the goods from getting spoiled and attract customers.
6. **Distribution:** The last work done in the marketing of goods is the distribution of goods from warehouse to market for final consumption [42].

## 11.3 VIRTUAL FARMERS' MARKETS

The concept of an online virtual farmers' market (VFM) for locally produced foods appears to be a very recent innovation. Based on our internet research in the last 10 to 15 years, and mostly in urban locations with large populations. At its core, an online virtual farmers market allows many farmers to sell produce to multiple customers using the internet, using a centralized distribution point (s) in the physical world, so that the pick-up and pick-Both of them can be arranged. Many farmers, who all arrive at the same centralized location, form a critical mass, which allows through the maintenance of tariffs in the market, refrigeration at the distribution site, and other storage equipment. Despite the recent development of the concept, there are many examples of online markets at which consumers can shop at their convenience and pick up later or have their food items already at their doorstep [34].

In Brooklyn, California, Los Angeles, Louisiana, New Orleans, New York, and San Francisco, The Good Eggs Virtual Market has been satisfying customers for many years and growing continuously [48]. Wortham [48] studied that companies like Urban Organic and Next-Door Organics in New York as well as companies such as Quinciple, which specialize in artisan food distribution. It appears that most of the distribution points of these examples exist in established urban areas, suggesting that densely populated areas provide the critical mass needed for a viable market. That said, the local food marketplace (established in Eugene, Oregon in 2009) provides online connections to farms, food hubs, and marketplaces; however, as exemplified on their website, their platforms are workable in both urban and rural areas. White [45] notes that there is an aptitude for both consumers and farmers when buying and selling through an online virtual farmer market. Customers do not have to come to the farmers' market in the morning to get the best produce. Additionally, customers are not required to accept random items as they would on the part of Agriculture Supported Agriculture (CSA). For farmers, instead of estimating the amount of produce they can sell in the in-person market, they can only harvest the crop that was ordered to destroy any wasted product. Farmers have the facility to know which crop they should produce and in what quantity. They also have the added advantage of selling the remaining produce at the traditional farmers' market or any other place. But for any of it Transpire, software that allows digital connections between producers and customers.

Perhaps the earliest example of this type of software and its associated market is around 2002, when Eric Wagner of Athens, Georgia, launched http://locallygrown.net/. A University of Tennessee Extension study outlined the primary features of Locallygrown.net and provided a limited case study of its use [17]. According to the profile, if customers are allowed to pay online, and customers are allowed to pay online, all customers pay in cash and roughly 7% (a combination of software costs, security fees, and transaction fees). The service charge for the markets is 3% for joining markets. According to a 2011 *Mother Earth News* article, Wagner stated that his Athens market had added weekly sales of US $8,000 to US $12,000 depending on the season [45]. It also described how he copied the platform to create a template that any entrepreneurial farmer or farm market manager could use.

On the basis of Wagoner's website, 300 online markets are currently running nationally, and 140 are under development [26]. Over time, other online platforms have emerged. In 2009, Farmigo created a software system that would allow online ordering to any local farm. They have evolved to serve more than 300 farms in more than 20 states [48]. In Africa, the Virtual Farmers Market (VFM) app helps farmers advertise and sell surplus crops [47]. The VFM app was launched in May 2017. The developers targeted 2,500 Zambian farmers who intend to connect them with 70 national and international buyers and become sustainable after three to 5 years. The application allows farmers to negotiate fair prices and deal in a transparent manner. Additionally, farmers also get bargaining power and high-profit potential through real-time pricing information provided by the app. A possible criticism of the idea of online VFMs is the ability of the so-called "digital divide" among more affluent and less affluent customers.

The former possess smartphones and are accustomed to purchasing goods online, but the less affluent customers are often less financially and culturally able to purchase goods online. Given the recent evolution of farmers' markets, research on this conflict is limited. Freedman et al. [13] indicated that traditionally, markets can create targeted marketing, especially in targeted languages, and provide tours to local residents to increase customer diversity. The most digital nature of a market using this model may restrict the ability of well-intentioned market managers to reach out to these potential customers. However, Skizim et al. [35] analysis of social marketing for farmers' markets in a low-income region of Louisiana suggests that Internet and social media access is not a significant barrier for low-income individuals to access information about farmers' markets.

### 11.3.1 IMPORTANCE OF AGRICULTURAL MARKETING

The importance of agricultural marketing is as follows:

- Breaking the cruel circle of poverty;
- Utilizing agricultural resources in an optimum manner;
- Enhances the living standard;
- basis for opportunities of employment;
- basis of developing industries;
- Creation of utilization;
- basis of trade with foreign countries;
- Source of national revenue.

### 11.3.2 NEED OF VIRTUAL FARMERS' MARKET (VFM)

Purchase for Progress (P4P) is the major program of WFP which connects small farmers to markets. VFM builds on this concept. Through WFP's reputation as a reliable buyer of quality crops, VFM's virtual market infrastructure will help bring farmers, buyers, and other stakeholders together. The VFM platform supports small farmers organized by lead farmers (VFM Ambassadors) to estimate the production and selling prices of their community, and advertise this information on the online market. Allowing buyers to access this information and communicate directly with the farmers gives them the opportunity to increase their business by accessing larger volumes of quality produce. As smallholder farmers become more visible to new buyers, VFM increases competition between buyers for farmers' produce, thus helping farmers get better prices and more favorable marketing options Figure 11.2.

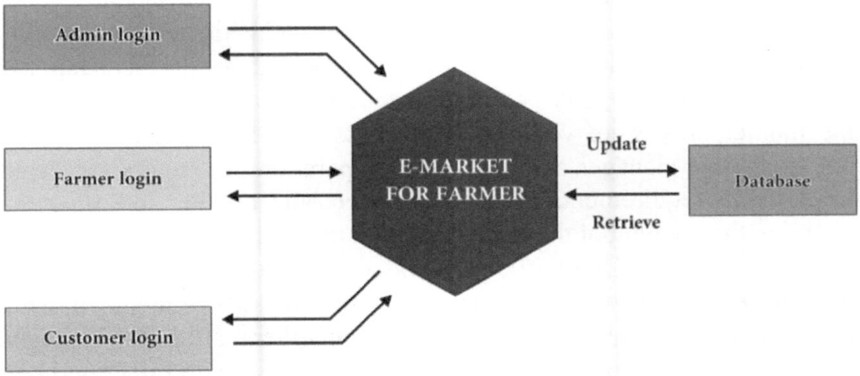


**FIGURE 11.2** How e-market for farmers' work.

### 11.3.3 SUSTAINABLE BUSINESS MODEL

VFM essentially follows the same 'three-way handshake' modality used by successful online businesses like Airbnb, Uber, and Alibaba. Considering the fast-growing revenues of these and similar companies, this model also has the potential to revolutionize WFP's work in providing market access and fair business opportunities for smallholder farmers. VFM would become sustainable through a 5% transaction fee paid by buyers on each purchase. At the moment, smallholder farmers are not paying anything to use the marketplace, thereby increasing client acquisition and the speed with which the system will have the critical mass of data and users to spur its growth.

### 11.3.4 PROBLEMS FACED BY VIRTUAL FARMERS' MARKET (VFM)

Participants were asked to list the major obstacles they faced as farmers' market managers. Some indicated that their markets ran smoothly and had no major problems to report. However, three participants mentioned parking to be a constraint, either because it was insufficient or because it drew objections from local merchants. Another three respondents stated that they needed more farmers for their operations. Other problems mentioned were getting farmers to pay the fees and fill out application forms, lack of volunteers, lack of funding to promote consistently, tardiness of farmers, farmers' rivalry, and drawing customers to downtown areas [16].

When asked to share some experiences they had with their municipality, vendors, or other entities, two respondents indicated that the city supported the program entirely, and one participant added that good relations with the municipality were vital for without its cooperation markets would not exist. One manager mentioned the preparation of vendors for Health Department requirements as an experience, and another indicated that planning the opening of the market (getting permits, licenses, etc.) was the greatest challenge. One participant was concerned with the competition with privately-owned produce markets on the outskirts of the town, while one complained about the Boro/Chamber members' attitudes. Managers and coordinators believe that the New Jersey Agricultural Experiment Station/Rutgers Cooperative Extension could best assist them in the following ways:

- By contacting farmers who are willing to participate in their markets;
- By designing the best layout for a new market;
- By giving ideas to make markets better and more profitable;
- By supporting the New Jersey Farmers' Market Council;
- By continuing to advertise farmers' markets;
- By organizing workshops or seminars where all market managers could get together to share information;
- By providing information on growing seasons and products 8) By setting up a stand in the market to educate the public;
- By communicating new ideas and suggestions to those participating in the Farmers' Market Council.

### *11.3.5 GLOBAL STATUS OF VIRTUAL FARMERS' MARKET (VFM)*

The invention of the internet has a crucial role in the development and transformation of the world economy. In the Agriculture field, implementation of this innovative technique is considered for the enhancement in farm output and accelerate economy. Agribusiness accepted e-commerce or e-marketing, or internet marketing for mounting the manufacturers and consumers beyond conventional marketing. The efficiency of e-commerce in the agriculture sector has been studied in various studies [18]. According to Zapata et al. [50], farmers experienced that there is both numbers of customers and sales increased after utilizing electronic market maker system in market policy [18]. Gumirakiza and Vanzee [17] observed that both male and female express their interest in the online purchase of locally grown fresh food [18], and several consumers are still in favor of supermarkets for the shopping of food products [46]. The virtual market platform has various advantages over the traditional agriculture market (Table 11.1).

**TABLE 11.1** Comparative Assessment between Traditional Market v/s E-Market Agriculture

| E-Market | Traditional Market | Marketing Mix Elements [24] |
|---|---|---|
| • Very economical<br>• Fast | • Very expensive<br>• Time taking | **Produce:** Variety, quality, design, feature, brand name, packaging, sizes, services, warranties. |
| • Very useful for product promotion<br>• Low cost | • Very expensive | **Rate:** List, price, discounts, allowances, payment period, and credit terms |
| • Less manpower requires | • Excess manpower requires | **Promotion:** Advertising, personal selling, sales promotion, public relations, and direct marketing |
| • Open 24 hours round the year | • Open with a defined time period | **Location:** Channels, coverage, assortments, locations, inventory, transportations, and logistics |
| • Paying professional E-market company is very economical | • Paying renowned advertising and marketing is very costly | – |

In 2010, Bozkurt conducted comparative research on the shopping habit between the physical store and online grocery shoppers and revealed that the consumer still depends on shopping for their fresh foodstuff from large grocery stores and online shoppers had higher education. Gumirakiza and Vanzee, in 2017, also observed that 44% of customers consider grocery stores as the most preferred market for fresh food while only 33% of shoppers are depending on online trading [18, 19]. At present, in COVID-19 pandemic situation, online virtual farmers market play a role in its prevention by providing physical distancing and consumer safety. The consumer chooses orders online and thus avoided interaction with farmer [34]. Different countries have their own virtual farmer markets. Table 11.2 represent virtual farmer market in the world (Figure 11.3).

### *11.3.5.1 AFRICA*

In Africa, the mobile market is the fast-growing marketplace which is quite bigger than Asia and around half of Africa's 1 billion population is friendly with mobile phones. They use their smart mobile phone, technology, and applications for banking and agri-business purposes. However, smallholder farmers from rural locations face a number of problems, mainly as they struggle to reach near markets to collect better prices for their crops and other foodstuffs such as preserved food and eggs. To solve these issues World Food Program (WFP) has started an online-based virtual farmer market app on a digital platform, which provides information and communications uprising for new business openings and upgraded income options for smallholder farmers [47].

This Virtual farming market system is built under the P4P flagship program of WFP's that connect smallholder farmers to markets mandate. The virtual farming market platform supports smallholder growers organized by lead farmers (VFM Ambassadors) who calculate their community's production, selling price and upload or advertise this valuable data on the online market, which allow buyers to use particular information and communicate directly with the producers provides an opportunity to improve their business. This market platform increases competition among the buyers for farmers' produce and thus supports farmers to get good prices and additional satisfactory marketing options. WFP also focused on the new market access policies and initiatives for providing unbiased business opportunities to smallholder farmers. Virtual farmer market becomes sustainable by imposing a 5% transaction fee on buyers in each purchase. MAANO seems like the most popular virtual farmer market app [47].

Maano supports smallholder farmers to access their target markets via mobile and offers a transparent, open, and trustworthy place to negotiate product prices and deals. It facilitates the smooth sale of farm products by using an online gateway as a payment method. In the initial phase, good progress of this app is recorded, and 70 consumers from both the national and international levels express huge interest in procuring target crops.

**TABLE 11.2** List of Virtual Farmer Market in the World

| SL. No. | Country | Virtual Platform | Technology | Rating | References |
|---|---|---|---|---|---|
| 1. | Australia | Farmhouse | Online | NA | [12] |
| 2. | Canada | Green circle food hub | Online | | [43] |
| | | Steve and Dans Online Market | Online | | [36] |
| 3. | China | Alibaba 'ET Agricultural Brain' | Online | | [14] |
| 4. | Europe | WeFarm | Online | | [40] |
| 5. | India | Electronic-National Agricultural Markets (e-NAM) | Mobile | 3.6 Star | [7, 15] |
| | | E-Choupals | Mobile | 3.8 Star | [3, 15] |
| | | SmartCrop | Online | 3 Star | [44] |
| | | Dhaan Mandi | Mobile | 4 Star | |
| | | Agribuzz-AgriApp | Mobile | 4 Star | |
| | | Digital Mandi | Mobile | 3 Star | |
| | | Gramseva: Kisan (Mandi Prices) | Mobile | 3 Star | |
| | | Mandi trades | Mobile | 4 Star | |
| | | Agri Market | Mobile | 2 Star | |
| | | Krishidirect | Mobile | 4.3 | [15] |
| 6. | London | WeFarm | Online | NA | [40] |
| 7. | Mexico | Chipotle Virtual Farmers Market | Online | | [6] |
| 8. | Russia | Agro. Club | Mobile | | [2] |
| 9. | United Arab Emirates | Agriota | Online | | [1] |

**TABLE 11.2** *(Continued)*

| SL. No. | Country | Virtual Platform | Technology | Rating | References |
|---|---|---|---|---|---|
| 10. | United Kingdom | Harvest Bundle | | | [21] |
| | United States of America | Indigo Agriculture | Online | | [23] |
| | | Thrive Market | Online | | [38] |
| | | LocallyGrown.net | Online | | [26] |
| | | Farm Fresh Ex | Online | | [10] |
| | | Our harvest | Online | | [32] |
| | | Farm Fresh to You | Online | | [11] |
| 11. | Zambia | MAANO | Mobile | 4.8 Star | [15, 47] |

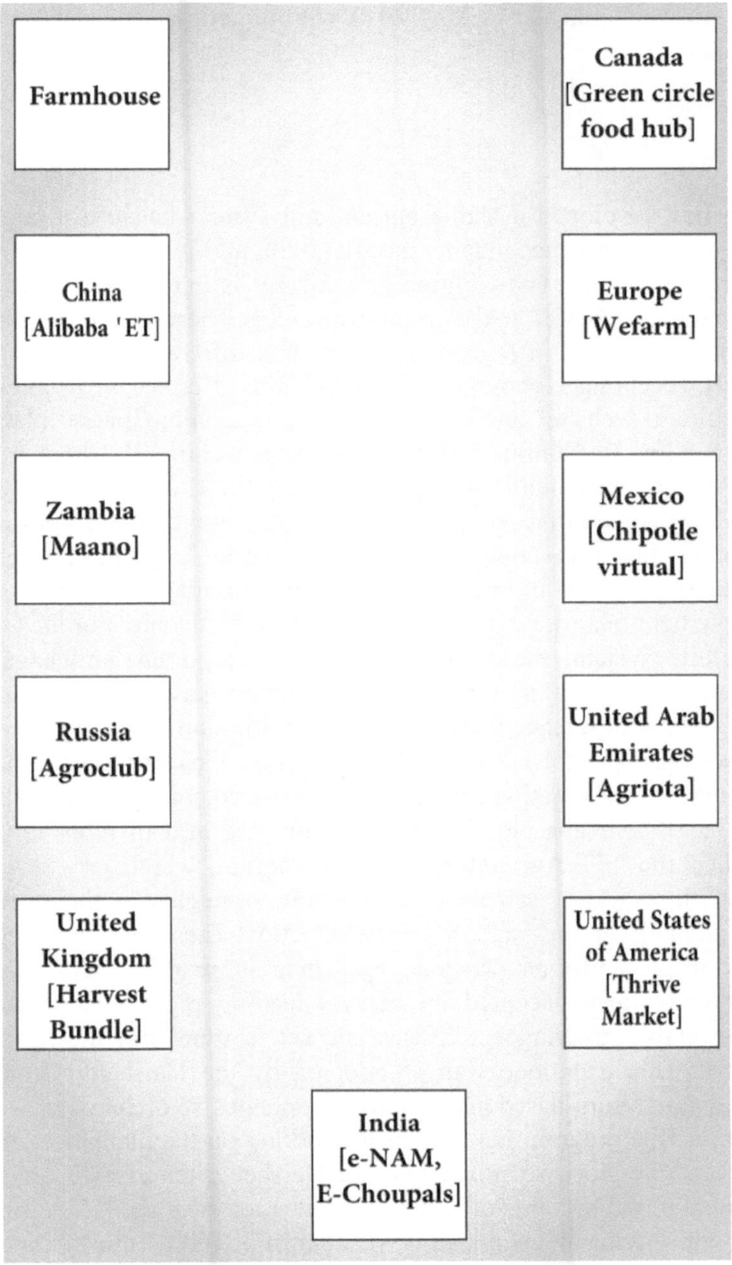


**FIGURE 11.3** Classification of socio-economic issues for the success of e-markets.

This app was launched in May 2017 with targeting 2,500 farmers from Zambia [47].

### 11.3.5.2 CHINA

China has developed a stable agricultural system because a large part of its population is dependent on agriculture, and it can be better for the stability of society. The Chinese government is highly concerned about their rural economies and agriculture sector. There online marking of agricultural stuff is in a growing phase, and different organizations are actively accepting e-commerce. Several parts of china set up their own "agricultural website" and have created various online market platforms, e.g., "Nanjing BaiYunting," "Fuzhou peak," etc. Currently, there are more than 10,000 agricultural websites available in China. These websites provide valuable information regarding agricultural demand, aquaculture, economy, and investment facts, fruits, livestock, poultry, trees, and vegetables. It plays an important role in the circulation, promotion, and improvement of agricultural products and farmer income. For the Chinese agriculture system, these websites are very important and have great influence in the China agricultural information network (www.agri.gov.cn), agricultural science and technology information network (www.caas.net), seed group co., LTD. (www.chinaseeds.com.cn), and the North China seed industry information center (www.northseed.com.cn), etc., [9].

In 2018 'Alibaba cloud, a cloud computing arm of Alibaba Group' launched the "ET Agricultural Brain" program, which provides information that allows farmers to apply new strategies in their effective planting methods and increase crop production. These are the agricultural e-commerce platforms provided by China in marking. This program has already been accepted by several leading pig farming and fruit and vegetable growing enterprises and set to transform the agronomic industry. This technique was 1st adopted by the Haisheng Group, the leading agriculture-based industry which operates 55 orchards up to 4,000 hectares. This program has also been working on the plantation of fruits and vegetables for two producers. For Haisheng, the growth limitations of around 4,000 hectares of apple are converged with the ET. Agriculture Brain with saving an estimated RMB 20 million (€2.64 mln) in operating costs annually. Alibaba Cloud supports Chinese farmers to sell fruits and vegetables more effectively, in order to increase their income [14].

### 11.3.5.3 EUROPE

At present, several European countries (Sweden to Spain) are gaining popularity in terms of the online supermarket for connecting their valuable consumers with local farm suppliers and supermarket chains and providing a safe and convenient grocery experience [25]. For example, Crisp is a mobile-based app from Amsterdam, Netherlands, offering a wide variety of local seasonal products to consumers with addressing awareness on the food and its ecological impact. This app provides next-day delivery to all over the Netherlands and concentrates on the food quality collected from more than 200 local farms. It collected €3 million seed capital in 2018, which recorded up to €5 million in summer 2019.

WeFarm is a virtual marketplace and networking site designed for small-scale farmers, rises worth $13 million funding and plan to use this money to uplift farmers income in Europe. At present, there are approximately 1.9 million registered users in WeFarm and provide a suitable marketplace to support farmers for their trade with local suppliers of goods [40].

### 11.3.5.4 INDIA

In India, the virtual farmers market is in its growing stage, which provides a digital platform to improve the livelihood of farmer communities and also support the country's GDP. Indian Government has launched an internet-based system (AGMDRKNET) that provides actual data related to the farmer market across the country [30]. Online marketing platforms reduce the appearance of middlemen who are involved in the smooth functioning of agriculture trading. These marketing platforms are free of cost and can be used from anywhere with local or understandable language to farmers. In these platforms, any transaction criteria are quite easy and recorded in a transparent manner. The food quality in supply chains is improved and produces within the stipulated time and supports changing lifestyle of farmers [8]. There are several mobiles as well as online-based virtual farmer market are located in India; some are as follows:

1. **SmartCrop:** It is an online portal that involves the trading of crops all over the country. In this system, farmers post their products with their price and attract buyers and save their precious time as well as money. In SmartCrop, the farmer can adjust their crop

price according to the market rate and reduce the interference of middlemen or agents by providing private chat options [44].

2. **Agribuzz-AgriApp:** It is a free mobile app, and it is a transparent platform that connects the farming community and supports them in retailing, purchasing, exchanging agriculture merchandise and amenities locally without middlemen uploading their product from their mobile. In this app, farm products are classified into more than 12 groups and 110 sub-groups. It provides a separate Agribuzz-Chat chat option to seller or buyer where they can chat with already prepared text templates, can share their details such as name, contact number shared via e-mail, and then negotiate a deal on call [44].
3. **Digital Mandi India:** It is also a mobile application in which all data is data is synchronized from the Indian government portal Agmarknet.nic.in-Powered by National Informatics Centre. This app helps their user to examine the latest Indian agricultural commodities with mandi prices across all states and cities. Users can also browse through various commodity and states category. It has a simple flow to reach the selected commodities mandi price, making it user-friendly [44].
4. **Gramseva: Kisan (Mandi Prices):** This mobile application is geared towards farmers and provides market data for farm produce. It provides actual market prices from the government's website, data.gov.in. With the graphical representation of commodity price trends can help in the decision-making. In the end, users can share final prices to their producers via mail and short message service (SMS) [44].
5. **Mandi Trades:** It is an agricultural goods trading platform, which connects farmers with their buyers and enables direct transactions between buyers and growers. In this app, buyer can get mandi prices, rate alert, and food product in demand and product prices and farm data from the Government of India. It has a Geo-tagging system of farm product that enables map-based identification of procurement and direct transactions between farmers and traders [44].
6. **Kisan Suvidha:** It is an omnibus mobile application developed to support farmers by providing related information to them quickly. Farmer can get information related to weather of current as well as

next 5 days, dealers, market prices, agro advisories, plant protection, IPM Practices, etc., by login in this app. This app provides the comparative price of the nearest area and the maximum price in the state as well as the country level and allows farmers to get the best price in a possible manner [44].

7. **Annadata:** It is a marketplace platform to connect farmers and consumers with ICT-based technological intervention. This product is established by Bharwa Solutions Pvt. Ltd., India. This marketplace platform is used to connect buyers and sellers at a single place by the information technology (IT) led interventions. In particular, this product provides a multiplatform integrated facility agro-business sector by bridging the gap between the buyer and seller. In India, a lot of farmers are gaining benefits from this platform with real-time doubling of their income.

#### *11.3.5.5 UNITED KINGDOM*

After the establishment of 1st farmers' market in the United Kingdom in 1997, around 550 new markets are developed across the countries. Several factors such as increasing education as well as awareness of consumers due to articles, books, cookery, gardening programs and magazine creates great focus on food safety and quality. Customers were concerned about the farming practices through which food is produced, processed in a health food. At present, virtual farming market provides an online platform which combines current 3D world with traditional online retail website to improve trading and user experience which is created by the digital agencies [22].

Virtual farming market in the United Kingdom allows virtual research on the traditional Farmers' market with specialist food and drink producers in 3D manner without going outside. £4.4 billion worth was recorded from the online grocery market in 2009, which was expected to reach around £6.9 billion in 2014. According to Marcus Carter, the virtual farming market provides an opportunity to their consumers to purchase the desired foodstuff directly from the producer and also help to connect remote location across the countries by using internet and build a friendly relationship between buyers and producers. In 2010, 1st virtual farming market started in London [37].

## 11.4 TECHNOLOGY INTERVENTION

To generate positive influence of technology in agriculture area required proper implementation of a specific technology with other factors that transform the basic conditions for identifying the opportunities for increasing social, economic, and environmental growth. However, the benefits of digital technology in agriculture are still not proven, which is due to several challenges like difficulty using software, data usage concerns, disparate, propriety data formats, uncertain return on investment, poor network infrastructure, and limited capital of emerging economy. As per the Trendov et al. [39], several novel technologies such as big data, internet of things (IoT) and artificial intelligence (AI) plays a vital role by incorporating information of different sections throughout the entire value chain, by means of operational use of inputs that fulfill consumers requirements and in reduction of environmental and climate risks in production. Different types of digital platform used in the area of agriculture and food business Figure 11.4. Several factors or technologies that facilitate Virtual farmer market are:

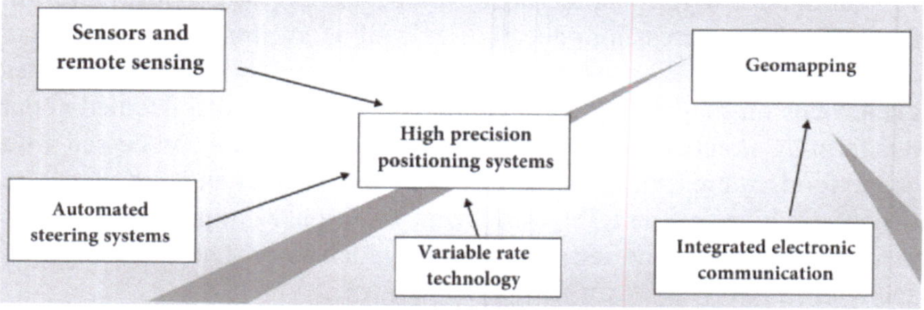


**FIGURE 11.4** Key precision agriculture technologies.

### *11.4.1 PRECISION AGRICULTURE (PA)*

In the agricultural sector, precision agriculture (PA) is one of the known IoT applications. PA employs technologies such as Radio-frequency identification (RFID), global navigation satellite system (GNSS), variable rate technology (VRT) and drones that connected with sensors to communicate farm environments, measure spatial variability, plan irrigation and harvesting, and thus eliminate human intervention to a large extent. PA sensors data may be shared with the different stakeholders either by the local server or the cloud, which retrieved via smartphones. These data help in the increment trading between producers and consumers. RFID tags enhance barcodes and enable them to find out actual information about particular products [27].

### *11.4.2 INFORMATION AND COMMUNICATIONS TECHNOLOGY (ICT)*

Information and Communications technology (ICT) includes all the techniques used for telecommunications, broadcast media, intelligent building management systems, audiovisual processing, and transmission systems, and network-based control and monitoring functions, etc. This technology is seeming to an important tool in agricultural sectors in developing countries where resources are very poor, small-scale farmers who face high transaction costs and poor access to information which limit their market participation. Effective use of ICT devices such as mobile phones is considered as the ideal device that reduces irregularities about information between buyers and producers [5].

Mobile or smartphones are the most widely used digital technology in the world. At present, most of the population is depends on these devices for their routine works such as online shopping, etc. In continuation with this, a number of applications are widely available for agricultural trading which offers information, deliver transactional facilities, provide advisory services for decision-making and complete analysis of stocktaking exercise [33]. The following benefits of these apps are:

- Provide direct access to market information that offers higher product prices, weather, and pest and diseases related information;
- Provide precise guidance for farming practices and business;
- Provide good connectivity with market and distribution networks such as it links producers, suppliers, and buyers, value chains in a more transparent way by excluding intermediaries;
- This app also offers funding, insurance opportunities, and alternative payment methods by which farmers can increase their crop yields with expanding market [33].

### 11.4.3 ARTIFICIAL INTELLIGENCE (AI)

Artificial intelligence (AI) is the simulation of human intelligence processes by machines, especially computer systems. These processes include learning (the acquisition of information and rules for using the information), reasoning (using rules to reach approximate or definite conclusions), and self-correction. It is a constellation of technologies that enable machines to act with higher levels of intelligence and emulate the human capabilities of sense, comprehend, and act. In the agriculture field, this technology can support increment in crop yield by offering actual advisory, crop price prediction, and early detection of pest attacks, precision farming, and others [41].

### 11.4.4 VIRTUAL REALITY

Virtual reality is a computer-based technology to create a simulated environment. In the agriculture sector, this technique is utilized for scientific research, teaching, agriculture resources, planning, production, circulation of products, and designing of agricultural machinery manufacturing and online marketing. It offers a 3D picture with complete detail of the desired product to its buyer and increases the buyer's interest in trading [41, 49].

## 11.5 CHALLENGES AND BENEFITS

There are a number of challenges and benefits are associated with operating online farmer market platforms. Potential benefits in the selling of farm products using online farmers market include (Figure 11.5) [28, 29]:

- Producers or farmers can only harvest and deliver products that have been sold previously. There is no need for much time, market staff, and product marketing materials, and infrastructure for trading. Online marking reduces the wastage of products and simultaneously increases farmers' market increases marketing efficiencies.
- Farmers can enhance their farm revenue by increasing their consumer base to customer sections, which value the accessibility

of modified online ordering and effective product pick-up locations. Consumers actively search the market platform, read product descriptions, and purchase suitable products which are packaged, labeled, and delivered by the farmers and send to a pre-specified pick-up location.

**FIGURE 11.5** Represents different countries name and logo of e-marketing.

- Consumer simply finds and purchases a wide range of locally grown products, which increases competition between farmers.
- Online farmers markets can also emphasize more on offering consumers with a novel experience that facilitates businesses between farmers and consumers. Several farmers can value virtual markets that require less farm-to-buyer interaction to conduct trade.
- Online farmers' markets are less likely to be exposed to bad weather as compared to traditional farmers' markets. In online trading, customer orders their prepackaged, labeled, and sorted product for suitable pick-up location on delivery day.
- Producers and managers of the online platform have instant access to daily and old sales and important accounting data information for analyzes. On the other hand, online farmers' markets can have the same organizational issues as a traditional market.

- At the initial level, a virtual farmer market platform may require a significant time investment upfront.
- As in the traditional market, online markets also require a committed team of market managers and volunteers for proper functioning and a permanent location for product sorting and pick-up.
- Consumer are not able to sense the smell and touch their desired product before purchase in online farmers markets. Thus, producers face the challenge to create adequate explanations of products and building trust with consumers.
- A well-organized online farmers markets alleviate these issues by providing proper product list, detail, and farm profiles.

## 11.6 SUMMARY

Virtual Farmer market is the need of enormously growing population today. As the increase in population led to several challenges like food, hunger, livelihood, etc. Food security and sustainability of the agriculture is very important sector to meet the requirements of the citizens. In the current pandemic situation, farmers are hardly suffered by the less or no consumption or sale of their products due to lockdown across the globes. Minimum support prices of the crops are always a big concern for the farmers. At present, virtual platform is a nice technological-based intervention, used for the upliftment of the agribusiness. It also paves the way for the farmers as well as for the consumers to engage in a fair deal of the agri-produce. A lot of countries today are shifting for the virtual system of agri-marketing. For this, a number of ICT-based platforms are launched by different countries as per their need. It is the most authentic and transparent way for sustainable agriculture and food security.

## ACKNOWLEDGMENTS

The authors are highly thankful for the vision and mission provided by the Swami Ramdev Ji for improving the status of farmers for sustainable agri-resources. The ICT-based data and technological information provided by Mr. Kavindra Singh and Dr. Rishi Kumar are also gratefully acknowledged. We are also thankful to Mr. Ajeet Chauhan for the graphical artwork.

## KEYWORDS


- **Agri-business**
- **Agro-produce**
- **Global navigation satellite system**
- **Information and communications technology**
- **Technological interventions**
- **Virtual farmer market**


## REFERENCES


1. Agrigal Global, (2019). *Agriota e-Marketplace: Here is all you Need to Know About Agri-Commodity Trading and Sourcing Platform*. https://agrigateglobal.com/asia/agriota-e-marketplace-heres-all-you-need-to-know-about-agri-commodity-trading-and-sourcing-platform/ (accessed on 12 April 2021).
2. Agro.Club, (2020). *Agro.Club-Connecting the Ag Industry to Buy, Sell, Earn*. https://agro.club/about (accessed on 12 April 2021).
3. Bowonder, B., Gupta, V., & Singh, A., (2003). *Developing a Rural Market E-Hub the Case Study of E-Choupal Experience of ITC*. Planning Commission of India. https://niti.gov.in/planningcommission.gov.in/docs/reports/sereport/ser/stdy_ict/4_e-choupal%20.pdf (accessed on 12 April 2021).
4. CBC/Radio-Canada, (2020). *People are Being Duped': CBC Exposes Homegrown Lies at Farmers' Markets*. https://www.cbc.ca/news/business/farmers-markets-lies-marketplace-1.4306231 (accessed on 12 April 2021).
5. Chikuni, T., & Kilima, F. T., (2019). Smallholder farmers' market participation and mobile phone-based market information services in Lilongwe, Malawi. *The Electronic Journal of Information Systems in Developing Countries, 85*(6), e12097.
6. Chipotle Farmers Market, (2020). *Chipotle Farmers Market-Chipotle Mexican Grill*. https://farmersmarket.chipotle.com (accessed on 12 April 2021).
7. *E-NAM Portal this Month is a Landmark Achievement in E-NAM History*, (2019). Retrieved from: https://pib.gov.in/Pressreleaseshare.aspx?PRID=1561146#:~:text=The%20start%20of%20online%20inter,achievement%20in%20e%2DNAM%20history&text=The%20e%2DNAM%20platform%20has,state%20trade%20in%20January%202019.&text=60%2C000%20crore%20has%20been%20recorded%20on%20e%2DNAM%20platform (accessed on 12 May 2021).
8. Espolov, T., Espolov, A., Kalykova, B., Umbetaliyev, N., Uspanova, M., & Suleimenov, Z., (2020). Asia agricultural market: Methodology for complete use of economic resources through supply chain optimization. *International Journal of Supply Chain Management, 9*(3), pages 408.

9. FAOSTAT, (2014). (http://www.fao.org/faostat/en/#data/QV).www.fao.org). *China Internet Network Development Statistic Report*. http://www.cnnic.net/hlwfzyj/hlwxzbg/ (accessed on 12 April 2021).
10. Farm Fresh Ex., (2020). https://farmfreshex.com/about/ (accessed on 12 April 2021).
11. Farm Fresh to You, (2016). *Farm Fresh to You-Home Page*. https://www.farmfreshtoyou.com (accessed on 12 April 2021).
12. Farmhouse, (2020). *Farmhouse-How it Works*. https://www.farmhousedirect.com.au/p-47-how-it-works.html (accessed on 12 April 2021).
13. Freedman, D. A., Vaudrin, N., Schneider, C., Trapl, E., Ohri-Vachaspati, P., Taggart, M., & Flocke, S., (2016). Systematic review of factors influencing farmers' market use overall and among low-income populations. *Journal of the Academy of Nutrition and Dietetics, 116*(7), 1136–1155.
14. Freshplaza.com. (2020). *Alibaba Cloud Launches ET Agricultural Brain*. https://www.freshplaza.com/article/2196039/alibaba-cloud-launches-et-agricultural-brain/ (accessed on 12 April 2021).
15. Google, (2020). *Krishi Direct*: *Online Grocery, Fruits and Vegetables*. https://play.google.com/store/apps/details?id=com.app.krishidirect&hl=en (accessed on 12 April 2021).
16. Govindasamy, R., Zurbriggen, M., Italia, J., Adelaja, A. O., Nitzsche, P., & Van, V. R., (1998). *Farmers Markets: Managers' Characteristics and Factors Affecting Market Organization*. https://ideas.repec.org/p/ags/rutdps/36723.html (accessed on 12 April 2021).
17. Grigsby, C., & Bruch, L. M., (2016). *E-Commerce for Direct Farm Marketing: An Overview of Locallygrown.net and Case Studies of Online Markets in Tennessee*. University of Tennessee Extension. https://extension.tennessee.edu/publications/Documents/PB1857.pdf (accessed on 12 April 2021).
18. Gumirakiza, J. D., & Schroering, M. E., (2019). Do online shoppers attend farmers' markets? *Journal of Agricultural Science, 11*(10).
19. Gumirakiza, J. D., & Vanzee, S. M., (2017). Most preferred market venues for locally grown fresh produce among online shoppers. *Journal of Agricultural Science, 9*(10), 26–35.
20. Halweil, B., & Prugh, T., (2002). *Home Grown: The Case for Local Food in a Global Market*. Washington, DC: World watch Institute. https://library.uniteddiversity.coop/Food/Home_Grown-The_Case_For_Local%20Food_In_A_Global_Market.pdf (accessed on 12 May 2021).
21. *Harvest Bundle*, (2020). https://www.harvestbundle.co.uk/ (accessed on 12 April 2021).
22. Holloway, L., & Kneafsey, M., (2000). Reading the space of the farmers' market: A preliminary investigation from the UK. *Sociologia Ruralis*, *40*(3), 285.
23. Indigo Ag, Inc. (2020). *Indigo Transport Provides Exclusive Access to Loads from Indigo Marketplace, a Grain Transaction Platform with Thousands of Growers and Buyers*. https://www.indigoag.com/for-carriers/indigo-transport (accessed on 12 April 2021).
24. Kaur, P., Pathak, A., & Kaur, K., (2015). E-marketing a global perspective. *Journal of Engineering Research and Applications*, *5*(2&5), 116–124.
25. Kholod, D., (2020). *10 Online Supermarket Startups Zooming Through 2020*. https://www.eu-startups.com/2020/05/10-online-supermarket-startups-zooming-through-2020/ (accessed on 12 April 2021).

26. LocallyGrown.Net. (2020). *LocallyGrown.Net- Small Farms Making a Difference.* https://locallygrown.net/markets/list (accessed on 12 April 2021).
27. Maksimović, M., Vujovic, V., & Omanovic-Miklicanin, E., (2015). A low-cost internet of things solution for traceability and monitoring food safety during transportation. *International Conference on Information and Communication Technologies in Agriculture, Food and Environment*, (pp. 583–593).
28. Matson, J., (2011). *Virtual Food Hubs Tap into Local Food Markets.* USDA rural cooperatives. https://www.rd.usda.gov/files/CoopMag-may11.pdf (accessed on 12 April 2021).
29. Matson, J., Thayer, J. & Shaw, J., (2015). *Running a Food Hub: A Business Operations Guide* (Vol. 2). USDA rural development, cooperatives program. Washington, DC Service Report, 77. https://www.rd.usda.gov/files/SR_77_Running_A_Food_Hub_Vol_2.pdf (accessed on 12 April 2021).
30. Naik, S. T., (2011). *Emergence, Growth, and Impact of Agricultural Commodity Marketing Center: A Study of Gadhinglaj.* (Published Doctoral Thesis). Shivaji University, Maharashtra, India. https://shodhganga.inflibnet.ac.in/bitstream/10603/111579/8/08%20chapter%202.pdf (accessed on 12 April 2021).
31. NPR, (2020). *Local Food May Feel Good, but it Does not Pay.* https://www.npr.org/2013/03/18/174655719/local-food-may-feel-good-but-it-doesnt-pay (accessed on 12 April 2021).
32. OurHarvest, (2020). *Where Farmers Market Meets Online Grocery.* https://ourharvest.com (accessed on 12 April 2021).
33. Qiang, Z. C., Kuek, C. S., Dymond, A., & Esselaar, S., (2012). *Mobile Applications for Agriculture and Rural Development.* Washington DC: World Bank.; https://openknowledge.worldbank.org/handle/10986/21892 (accessed on 12 April 2021).
34. Raison, B., & Jones, J., (2020). Virtual Farmers Markets. *Journal of Agriculture, Food Systems, and Community Development, 9*(4), 1–12.
35. Skizim, M., Sothern, M., Blaha, O., Tseng, T. S., Griffiths, L., Joseph, J., & Nuss, H., (2017). Social marketing for a farmers' market in an underserved community: A needs assessment. *Journal of Public Health Research, 6*(3), 164–168.
36. *Steve and Dans Online Market*, (2020). https://steveanddansonlinemarket.ca/ (accessed on 12 April 2021).
37. Thomas, J., (2010). *World's First' Virtual Farmers Market Launches.* https://www.campaignlive.co.uk/article/worlds-first-virtual-farmers-market-launches/976212#:~:text=LONDON-Theworld'sfirstinteractive,oftheonlin (accessed on 12 April 2021).
38. Thrive Market, (2020). *Thrive Market-Healthy Living Made Easy.* https://thrivemarket.com/web/membership/welcome (accessed on 12 April 2021).
39. Trendov, N. M., Varas, S., & Zeng, M., (2009). *Digital Technologies in Agriculture and Rural Areas-Status Report.* Rome License. http://www.fao.org/3/ca4985en/ca4985en.pdf (accessed on 12 April 2021).
40. Verizon Media, (2020). *WeFarm Rakes in $13M to Grow its Marketplace and Network for Independent Farmers.* https://techcrunch.com/2019/10/29/wefarm-rakes-in-13m-to-grow-its-marketplace-and-network-for-independent-farmers/ (accessed on 12 April 2021).

41. Verma, J., (2020). *E-Technology in The Aid of Farmers Insights-Technology and Agricultural.* https://www.jatinverma.org/e-technology-in-the-aid-of-farmers (accessed on 12 April 2021).
42. Vignesh, C. C., & Kalaikumaran, T., (2019). E-marketing for farmers. *International Journal of Scientific Research and Review, 7*(3), 70–76.
43. Virtual Farmers Market, (2020). *Green Circle Food Hub.* https://virtualfarmersmarket.ca/pages/about-us (accessed on 12 April 2021).
44. Vodafone Idea CSR, (2020). *Social App Hub.* https://www.socialapphub.com/app/mandi-trades (accessed on 12 April 2021).
45. White, M., (2011). *Virtual Farmers Markets-The Future.* Mother earth news. https://www.motherearthnews.com/homesteading-and-livestock/virtual-farmers-markets-the-future (accessed on 12 April 2021).
46. Wolf, M. M., Spittler, A., & Ahern, J., (2005). A profile of farmers' market consumers and the perceived advantages of produce sold at farmers' markets. *Journal of Food Distribution Research, 36*(1), 192–201.
47. World Food Program, (2020). *Maano-Virtual Farmers Market.* https://innovation.wfp.org/project/virtual-farmers-market (accessed on 12 April 2021).
48. Wortham, J., (2013). *Good Eggs, a Virtual Farmers' Market, Delivers Real Food.* The New York Times. https://www.nytimes.com/2013/11/14/technology/personaltech/goodeggs-a-virtual-farmers-market-delivers-real-food.html (accessed on 12 April 2021).
49. Yu, F., Zhang, J. F., Zhao, Y., Zhao, J. C., Tan, C., & Luan, R. P., (2009). The research and application of virtual reality (VR) technology in agriculture science. In: *International Conference on Computer and Computing Technologies in Agriculture* (pp. 546–550).
50. Zapata, S. D., Isengildina-Massa, O., Carpio, C. E., & Lamie, R. D., (2016). Does e-commerce help farmers' markets? Measuring the impact of market maker. *Journal of Food Distribution Research, 47*(856-2016-58222), 1–18.

# Index

## B

## C

## N



## O



## P

## V



## W



## X



## Y



## Z